OBJECTIVE BOTANY

PLANT TAXONOMY, ANATOMY, ECONOMIC BOTANY, EMBRYOLOGY, PLANT BREEDING, PLANT PATHOLOGY

OBJECTIVE BOTANY

Plant Taxonomy, Anatomy, Economic Botany, Embryology, Plant Breeding, Plant Pathology

M. S. Shekhawat

Assistant Professor of Botany
K.M. Centre for Post Graduate Studies
Pondicherry

MJP PUBLISHERS

Chennai New Delhi Tirunelveli

ISBN 978-81-8094-169-6

Printed and bound in India

MJP PUBLISHERS
New No. 5, Muthu Kalathy Street
Triplicane
Chennai 600 005

Publisher : J.C. Pillai

To

My Nanosa
Th. Himmat Singh Rathore

Preface

The present book has been written to meet the specific needs of students appearing for the competitive examinations, like CSIR-NET, GATE, TIFR, SLETs, NCCS, IITs, IISc., ICMR, Union and State Civil Services, common entrance test for M.Sc./Ph.D. in Botany/Biotechnology and other entrance tests like, Pre Medical Tests conducted by various organizations/universities in the field of Plant Biology (Botany).

The book is intended to provide wide knowledge on the subject since the field has grown enormously. It is expected that the book would also be a useful guide for the undergraduate and postgraduate students of Botany.

I would like to acknowledge the assistance of my colleagues in the possible task of keeping me up-to-date in the subject. I am grateful to Prof. N.S. Shekhawat, Biotechnology Centre, JNV University, Jodhpur, for his motivation and generous support. I am thankful to my wife Mrs. Saroj Shekhawat for her help throughout the progress of the book, particularly in the preparation of the figures. I express warm appreciation to my parents and my son Anantpal for their encouragement and patience during the course of writing this book.

I would also like to express my appreciation for the generous support provided by the editorial team of MJP Publishers for their cooperation in completion of this work with perfection.

Constructive criticism and suggestions for improvement of this book are welcome.

M. S. Shekhawat

CONTENTS

1

PLANT TAXONOMY

MULTIPLE CHOICE QUESTIONS

1. According to Oswald Tippo, the Angiosperms are classified under the phylum
   (a) Spermatophyta (b) Embryophyta (c) Tracheophyta (d) Angiospermae
2. In the classification of plants, the presence of the following is a primitive character
   (a) Zygomorphic flower (b) Epigynous flower
   (c) Unisexual flower (d) Bisexual flower
3. The Division of Anthophyta refers to
   (a) Dicots (b) Monocots (c) Gymnosperms (d) All Angiosperms
4. The most important characteristic difference in between Angiosperms and Gymnosperms taxonomically is
   (a) Seed enclosed in fruit (b) Triploid endosperm
   (c) Presence of vessels in xylem (d) None of these
5. Diadelphous condition is found in
   (a) Rosaceae (b) Papilionaceae (c) Cucurbitaceae (d) Leguminosae
6. Liliaceae and Amaryllidaceae resemble each other in which of the following characters?
   (a) Placentation is axile
   (b) The perianth lobes are usually 3+3 and may be free or united
   (c) Both have herbaceous plants that perennate by means of bulbs and rhizomes
   (d) All of these characters
7. Famous publication 'Species Plantarum' was written by
   (a) John Ray (b) Charles Darwin (c) Joseph Hooker (d) C. Linnaeus
8. In Fabaceae the floral structures which enclose the essential organs are
   (a) Posterior petals (b) Lateral petals (c) Anterior petals (d) Sepals

9. The characteristic fruit of the family Asteraceae is
   (a) Cremocarp (b) Siliqua (c) Cypsela (d) Silicula
10. Hutchinson's system of classification is a type of
    (a) Natural (b) Artificial (c) Phylogenetic (d) None of these
11. Which of the following is a phylogenetic system of classification
    (a) Hutchinson (b) Linnaeus
    (c) Lindley (d) Bentham and Hooker
12. Linnaeus system of plant classification is
    (a) Phylogenetic (b) Natural (c) Artificial (d) None of these
13. The Angiosperms are also known as
    (a) Rosophyta (b) Malvophyta (c) Magnoliophyta (d) Gramophyta
14. Androgynophore is absent in members of family
    (a) Fumariaceae (b) Cruciferae
    (c) None of these (d) Both Fumariaceae and Cruciferae
15. Which of the following statements about *Euphorbia* is incorrect?
    (a) Flowers are usually pedicillate and zygomorphic
    (b) Flowers are inconspicuous and very much reduced
    (c) Flowers are arranged in distinct and specialized inflorescence
    (d) Flowers are unisexual
16. Who was first to distinguish between flowering and non-flowering seed plants
    (a) C. Linnaeus (b) De Candolle (c) Robert Brown (d) John Ray
17. Floral features are chiefly used in Angiosperms identification because
    (a) Flowers are nice to work with
    (b) Reproductive parts are more stable and conservative than vegetative parts
    (c) Flowers can be safely pressed
    (d) None of these
18. The classification formed by C. Linnaeus is basically based on
    (a) Anatomy (b) Embryology (c) Morphology (d) None of these
19. Obliquely zygomorphic nature of the flower in Solanaceae is due to one of the following whorls
    (a) Gynoecium (b) Androecium (c) Calyx (d) Corolla
20. Earliest maximum written work in plants was done by
    (a) Bessey (b) Linnaeus (c) Eichler (d) Theophrastus
21. The embryo in sunflower has
    (a) One cotyledon (b) Two cotyledons

(c) Many cotyledons (d) No cotyledons

22. A plant with 4 sepals, 4 petals, 6 stamens (two outer short and four inner long) and one-celled ovary later becoming two-celled belongs to the family

    (a) Ranunculaceae (b) Solanaceae (c) Cruciferae (d) None of these

23. The lemma in family Poaceae are characterized by the presence of

    (a) Awn (b) Glume (c) Palea (d) Spike

24. Which of the following about Engler and Prantl's system of plant classification is correct?

    (a) They placed Monocotyledons before dicotyldedons

    (b) Apetalous and catkin bearing dicotyledons are primitive to those which bear petals and simple unisexual flowers

    (c) Considered orchids to be more evolved than members of Poaceae

    (d) All of these

25. Free petals are present in which one of the following family

    (a) Compositae (b) Solanaceae (c) Annonaceae (d) Cruciferae

26. Artificial system of classification was proposed by

    (a) Engler and Prantl (b) Theophrastus

    (c) De Candolle (d) Bentham and Hooker

27. *Trigonella* belongs to the following family

    (a) Brassicaceae (b) Caesalpiniaceae (c) Asteraceae (d) Fabaceae

28. The book 'Families of Flowering Plants' was written by

    (a) Bentham and Hooker (b) Hutchinson

    (c) Engler and Prantl (d) C. Linnaeus

29. In which of the following plants, reticulate venation, unisexual flowers and stipular tendrils are found

    (a) *Allium* (b) *Smilax* (c) *Raphanus* (d) None of these

30. In family Cruciferae, the two carpels in each flower are placed

    (a) Obliquely (b) Transversely (c) Anteroposteriorly (d) Separately

31. Vexillary aestivation is observed in the members of family

    (a) Poaceae (b) Fabaceae (c) Cucurbitaceae (d) Rosaceae

32. The inflorescence in case of *Yucca* belongs to Liliaceae is

    (a) Umbel (b) Catkin

    (c) Panicle (d) Compound umbel

33. Which of the families is characterized by the presence of a perianth

    (a) Solanaceae (b) Liliaceae (c) Compositae (d) None of these

34. Name the type of inflorescence in *Ficus*

    (a) Verticillaster (b) Capitulum (c) Hypanthodium (d) Cyathium

35. The system of classification proposed by C. Linnaeus was artificial because

    (a) It took into account the physiological facts along with the morphological characters
    (b) It was based on evolutionary relationship of plants
    (c) It was based on similarities and differences in floral and other morphological characters only
    (d) None of these

36. Silicula fruit is a characteristic of

    (a) Beet root (b) *Raphanus* (c) Candytuft (d) Cabbage

37. The condition of androecium in Cucurbitaceae is

    (a) Syngenesious (b) Diadelphous (c) Synandrous (d) Free stamens

38. *Allium* species differ from rest of the Liliaceae in

    (a) Presence of a centric leaves and perianth
    (b) Presence of bulbs and umbel type of inflorescence
    (c) Absence of umbel type of inflorescence and bulb
    (d) Absence of umbel type of inflorescence

39. $P_{3+3} A_{3+3} G_{(3)}$ is the floral formula of

    (a) Gramineae (b) Liliaceae (c) Solanaceae (d) Cruciferae

40. In Engler and Prantl's system of classification

    (a) Dicots precede monocots (b) Monocots precede dicots
    (c) Gymnosperms are in between docots and monocots (d) None of these

41. Bicarpellary ovary with parietal placentation and tetradynamous condition of stamens are characteristic features of family

    (a) Cruciferae (b) Liliaceae (c) Leguminoseae (d) None of these

42. Which of the following statements about Linnaeus is incorrect?

    (a) He was first to use binomial nomenclature
    (b) His system of plant classification was artificial
    (c) He recognized 24 classes
    (d) His system of plant classification was complex and therefore, could not survived for a longer time

43. Accrescent calyx found in the members of the family

    (a) Malvaceae (b) Leguminosae (c) Solanaceae (d) None of these

44. The type of placentation in the family Cruciferae is
 (a) Axile (b) Basal (c) Marginal (d) Parietal
45. Magnoliales and Ranales are considered most primitive orders of the Dicotyledons according to which system of classification?
 (a) Hutchinson's system of classification
 (b) Engler and Prantl's system of classification
 (c) Bentham and Hooker's system of classification
 (d) Bessey's system of classification
46. Angiosperms were classified into Archichlamydeae and Metachlamydeae by
 (a) De Candolle (b) Cronquist (c) Hutchinson (d) Angler and Prantl
47. Family Cruciferae is characterized by
 (a) Didynamous stamens (b) Cypsela fruit
 (c) Marginal placentation (d) Bicarpellary ovary with parietal placentation
48. Oblique septum and swollen placenta are characteristic features of the gynoecium of
 (a) *Althea rosea* (b) *Dalbergia sissoo*
 (c) *Capsicum* sp. (d) None of these
49. The inflorescence in rice is
 (a) Panicle (b) Verticillaster (c) Catkin (d) Racemose
50. Fruit type in family Compositae
 (a) Cypsela (b) Achene (c) Siliqua (d) Lomentum
51. The most important point of Bentham and Hooker's classification is
 (a) Based on evolution (b) Natural system of classification
 (c) Classification of taxa based on actual examination (d) None of these
52. Bicarpellary, syncarpous gynoecium, tetradynamous stamens and siliqua fruits are the characteristic features of family
 (a) Malvaceae (b) Cruciferae (c) Compositae (d) Liliaceae
53. The carpels of *Solanum* flower are obliquely placed because
 (a) Posterior and anterior carpels move to the left
 (b) Posterior carpels turn to right and anterior to the left
 (c) Posterior carpels turn to left and anterior to the right
 (d) None of these
54. How many families have been recognized by Bentham and Hooker in their plant classification?
 (a) 100 (b) 300 (c) 411 (d) 200

55. Name the part of floret which is not included in Poaceae florets
(a) Lemma (b) Palea (c) Lodicules (d) Scutellum

56. In family Ranunculaceae, the stamens are
(a) Spirocyclic, monothecous, extrorse (b) Spirocyclic, dithecous, introrse
(c) Spirocyclic, dithecous, extrorse (d) Spirocyclic, monothecous, introrse

57. The inflorescence in sugarcane is
(a) Spike (b) Panicle (c) Catkin (d) Hypanthodium

58. Epipetalous stamens with free filaments but fused anthers are found in
(a) Asteraceae (b) Convolvulaceae (c) Asclepiadaceae (d) None of these

59. For declaration of new species what characters should be used
(a) Characters of endosperm (b) Floral characters
(c) Physiological characters (d) Anatomical characters

60. Which system of classification is non-phylogenetic
(a) Pre-Darwinian systems (b) Post-Darwinian systems
(c) Neo-Darwinian systems (d) None of these

61. Spirocyclic numerous free stamens and apocarpous carpels are found in the family
(a) Cruciferae (b) Ranunculaceae (c) Labiatae (d) None of these

62. Which of the following statement is correct about the ray-florets of *Helianthus*?
(a) These are generally pistilate or neuter
(b) These are arranged along in the rim of the receptacles
(c) These are zygomorphic with ligulate corollas
(d) All of these

63. Members of family Asteraceae can be easily recognized by the characters
(a) Head inflorescence, syngenesious anthers, superior gynoecium
(b) Fruit cypsela, superior gynoecium
(c) Syngenesious anthers, inferior gynoecium, head inflorescence
(d) Fruit cypsela, inflorescence umbel

64. Which vegetative character is used in the classification of plants?
(a) Shape of leaves (b) Habit of plants
(c) Type of stem and root (d) Venation of leaves

65. The family in which marginal placentation and articular condition is seen is
(a) Cruciferae (b) Compositae (c) Leguminosae (d) Solanaceae

66. Climbing with the help of tendrils is met within some members of the family
(a) Leguminosae and Liliaceae (b) Leguminosae and Solanaceae

(c) Liliaceae and Malvaceae (d) Solanaceae and Malvaceae

67. Name of the family which is based on inflorescence is

(a) Leguminosae (b) Malvaceae (c) Euphorbiaceae (d) Asteraceae

68. Who was first time use the characteristics of vascular tissues in the classification of plants

(a) Engler and Prantl (b) De Candolle (c) Hutchinson (d) None of these

69. Which of the following is the second largest family of the Indian flora?

(a) Rosaceae (b) Polygonaceae (c) Brassicaceae (d) Orchidaceae

70. Which of the following is considered as more evolved?

(a) Dicot plants (b) Monocot plants (c) Both (a) and (b) (d) None of these

71. Mimosaceae can be distinguished from Caesalpiniaceae on the basis of

(a) Cohesion of anthers (b) Adhesion of anthers
(c) Aestivation of calyx and corolla (d) Characters of carpels

72. In which of the following families, the calyx modified into a pappus

(a) Asteraceae (b) Solanaceae (c) Cucurbitaceae (d) None of these

73. Name the plant in which stipules are modified into tendrils and belongs to the family Liliaceae

(a) *Lilium* (b) *Smilax* (c) *Yucca* (d) *Gloriosa*

74. Nectaries are present on the leaves of

(a) *Cannabis* (b) *Morus* (c) *Ricinus* (d) None of these

75. Which of the following is found in the family Malvaceae?

(a) Fruit is a capsule (b) Flower is unisexual
(c) Stamens are tetradynamous (d) Gynoecium is monocarpellary

76. A ligulate corolla found in the family Compositae is also known as

(a) Two lipped (b) Masked (c) Wheel speed (d) Strap-shaped

77. The family Compositae can be distinguished from the family Solanaceae in which the following characters?

(a) Epipetalous stamens (b) Capitulum
(c) Gamopetalous corolla (d) Bicarpellary gynoecium

78. Bentham and Hooker classified dicots into

(a) Polypetalae, gamopetalae and glumiflorae
(b) Polypetalae, gamopetalae and monochlamydae
(c) Archichlamydae, monochlamydae and glumiflorae
(d) Archichlamydae, monochlamydae and polypetalae

79. Tetramerous flowers are observed in which of the gamopetalous family

(a) Fabaceae (b) Brassicaceae (c) Rutaceae (d) Rubiaceae

80. The family which shows monothecous anthers and monadelphous stamens is
(a) Liliaceae (b) Cucurbitaceae (c) Malvaceae (d) None of these

81. The inflorescence in sunflower is
(a) Spikelet (b) Capitulum (c) Spike (d) Racemose

82. The disc in the family Rutaceae, found below the ovary, represents
(a) Vestigial whorl of calyx (b) Vestigial whorl of carpels
(c) Vestigial whorl of stamens (d) Vestigial whorl of corolla

83. *Avicennia* is a
(a) Hydrophyte (b) Mangrove (c) Epiphyte (d) Mesophyte

84. The condition in which the stamens are in two whorls, the outer whorl is opposite to the petals, is known as
(a) Obdiplostemonous (b) Diplostemonous
(c) Epipetalous (d) None of these

85. In Bentham and Hooker's classification, the two major groups of plant kingdom are
(a) Cryptogams and phanerogams (b) Unicellular and multicellular plants
(c) Gymnosperms and Angiosperms (d) Monocotyledons and dicotyledonous plants

86. Which family has maximum number of genera
(a) Cruciferae (b) Cucurbitaceae (c) Asteraceae (d) None of these

87. Polyadelphous androecium is found in which of the following family
(a) Liliaceae (b) Papilionaceae (c) Compositae (d) Rutaceae

88. The number of carpels in all the members of family Malvaceae is always-
(a) Equal to the number of stigmas (b) Double the number of stigma
(c) Equal to or half the number of styles (d) Half the number of locules

89. Polycarpellary, syncarpous and unilocular gynoecium is present in family
(a) Papaveraceae (b) Brassicaceae (c) Liliaceae (d) Ranunculaceae

90. Which of the following is not a characteristic of family Lamiaceae?
(a) Flowers are zygomorphic and unisexual (b) Ovary is superior with axile placentation
(c) Leaves are opposite and decussate (d) Inflorescence is verticillaster

91. The largest variation from polyandry to synandry exists in the members of the family
(a) Leguminosae (b) Cucurbitaceae (c) Liliaceae (d) None of these

92. Rice flower is peculiar in family Poaceae because it has
(a) Three lodicules (b) Six stamens
(c) Three empty glumes (d) Bifurcated awns

93. Three subfamilies of family Leguminosae are distinguished by character of
    (a) Sepals (b) Fruits
    (c) Corolla (d) Corolla and androecium
94. *Lantana* is an example of a
    (a) Twiner (b) Prickly shrub (c) Big tree (d) Herb
95. Pulses are a group of plants belongs to the family
    (a) Asteraceae (b) Cruciferae (c) Papilionaceae (d) None of these
96. Which of the following is an incorrect statement?
    (a) In Liliaceae, some plants have underground parts
    (b) Many of the compound leaf bearing trees found in desert are members of family Leguminosae
    (c) Fiber yielding plants are found in Malvaceae
    (d) Trees with large flowers and seeds are found in Compositae
97. The type of inflorescence in family Poaceae is
    (a) Verticillaster (b) Spike of spikelet (c) Capitulum (d) None of these
98. Indian telegraph plant (*Desmodium gyrans*) belongs to
    (a) Fabaceae (b) Leguminosae (c) Liliaceae (d) None of these
99. The main character in all the subfamilies of family Leguminosae is
    (a) Marginal placentation (b) Free stamens
    (c) Legume fruit (d) Non-albuminous seeds
100. Members of Apocynaceae differ from Asclepiadaceae in having
    (a) Corona (b) Latex
    (c) Bicarpellary ovary (d) Translators
101. Bicarpellary ovary with parietal placentation is found in
    (a) Cruciferae (b) Solanaceae (c) Convolvulaceae (d) Lamiaceae
102. Parallel venation occurs in
    (a) Grasses (b) Mosses (c) Cycads (d) Ferns
103. The type of fruit in Lamiaceae is
    (a) Achene (b) Capsule (c) Folicle (d) Nutlets
104. Abutilon of the family Malvaceae does not show the following prominent trait-
    (a) Presence of epicalyx (b) Alternate leaves
    (c) Monadelpous stamens (d) Axile placentation
105. Nectarines are present on the leaves of
    (a) *Cannabis* (b) *Morus* (c) *Ricinus* (d) None of these

106. Sheathed leaf bases and ligule are characteristic features of
(a) *Cycas* (b) Grass (c) Fern (d) Banana

107. Alternate leaves are common character in the family
(a) Malvaceae (b) Lamiaceae (c) Rubiaceae (d) Solanaceae

108. The family which shows monothecous anthers and monadelphous stamens is
(a) Liliaceae (b) Poaceae (c) Malvaceae (d) Ranunculaceae

109. In Poaceae, the perianth is represented by small scaly lodicules which are generally
(a) Two (b) Three (c) Four (d) Five

110. The stamens in *Coleus* are
(a) Monadelphous (b) Diadelphous (c) Polyadelphous (d) Epipetalous

111. Which of the following order is not a member of Division Corolliferae?
(a) Dioscoreales (b) Zingiberales (c) Palmales (d) Iridales

112. Verticillaster type of inflorescence is observed in
(a) *Ocimum* (b) *Salvia* (c) *Lamium* (d) All of these

113. Butterfly like flowers are found in family
(a) Mimosaceae (b) Papilionaceae (c) Caesalpiniaceae (d) None of these

114. Which of the following is the characteristic feature of family Malvaceae
(a) Polyandrous condition (b) Monoadelphous androecium
(c) Apocarpous gynoecium (d) Polyadelphous androecium

115. Name the end of the translator which sticks to the body of the insect is
(a) Pollinium (b) Retinaculum (c) Corpusculum (d) Caudicle

116. The families Liliaceae, Malvaceae, Leguminosae, Solanaceae can be distinguished on the basis of
(a) Stamens and placentation (b) Placentation and union of floral parts
(c) Placentation and number of floral parts (d) Placentation only

117. Which one of the statement is correct while evaluating the relative degree of advancement of flowering plants by Takhtajan?
(a) Solitary flowers are more advanced than cymose inflorescence
(b) Woody plants are considered most primitive than herbaceous plants
(c) Primitive flowers have more or less indefinite and variable number of distinct parts arranged spirally
(d) All of these

118. When one specimen or other element is designated by the author as the nomenclatural type, it is known as
(a) Holotype (b) Prototype (c) Neotype (d) Syntype

119. The flowers of Liliaceae are
    (a) Dicyclic (b) Tricyclic (c) Tetracyclic (d) Pentacyclic
120. Plants are always herb in family
    (a) Malvaceae (b) Leguminosae (c) Cruciferae (d) Cucurbitaceae
121. In family Asclepiadaceae the stamens are modified and known as
    (a) Androphore (b) Translator (c) Syngenesious (d) Sinuous
122. When a specific epithet exactly repeats the generic name it is known as
    (a) Tautonyms (b) Taxa (c) Principle (d) Priority
123. The classification of plants as runners, climbers and parasites is
    (a) Natural (b) Artificial (c) Phylogenetic (d) Anatomical
124. The correct sequence of families according to Bentham and Hooker's system is
    (a) Cucurbitaceae, Malvaceae, Poaceae, Asteraceae
    (b) Malvaceae, Cucurbitaceae, Asteraceae, Poaceae
    (c) Poaceae, Asteraceae, Malvaceae, Curubitaceae
    (d) Poaceae, Malvaceae, Asteraceae, Cucurbitaceae
125. Indefinite stamens are characteristic of the family
    (a) Poaceae (b) Cruciferae (c) Malvaceae (d) None of these
126. Which one of the following shows apocarpous conditions?
    (a) Cotton (b) Datura (c) Mustard (d) Ranunculus
127. When stamens are connate by their anthers and their filaments remain free, this condition is called as
    (a) Epipetalous (b) Syngenesious (c) Synandrous (d) Diadelphous
128. Which of the following lines represent the conditions observed in the family Asteraceae
    (a) Superior ovary, single basal ovule (b) Superior ovary, axile placentation
    (c) Inferior ovary, axile placentation (d) Inferior ovary, single basal ovule
129. The umbel inflorescence is the characteristic of the family
    (a) Clusiaceae (b) Fabaceae (c) Asteraceae (d) Apiaceae
130. Herbarium is
    (a) Collection of dried plants (b) Collection of animals
    (c) Collection of living plants (d) All of these
131. Family Solanaceae and Asteraceae can be differentiated by which of the following characteristics?
    (a) Bicarpellary ovary (b) Epipetalous stamen
    (c) Placentation (d) Cymose inflorescence

132. Androgynophore is seen in the members of family
    (a) Capparidaceae (b) Papaveraceae (c) Arecaceae (d) Brassicaceae
133. Basically connate corolla is found in
    (a) Brassicaceae (b) Malvaceae (c) Solanaceae (d) Papilionaceae
134. In *Brassica* flowers are
    (a) Pentamerous (b) Perigynous
    (c) Ebracteate (d) Slightly zygomorphic
135. Cloves of commerce is obtained from member of the family
    (a) Rosaceae (b) Meliaceae (c) Anacardiaceae (d) Myrtaceae
136. The characteristic fruit of family Cucurbitaceae is
    (a) Caryopsis (b) Siliqua (c) Drupe (d) Pepo
137. Out of four most widely accepted systems of classifications one remains less phylogenetic but more natural, find out that one
    (a) Bessey's (b) Engler and Prantl
    (c) Bentham and Hooker (d) Hutchinson
138. Bicapellary ovary with parietal placentation is observed in
    (a) Convolvulaceae (b) Cucurbitaceae (c) Lamiaceae (d) Brassicaceae
139. In mustard, flowers are
    (a) Perigynous (b) Pentamerous
    (c) Ebracteate (d) Slightly zygomorphic
140. Monocarpellary, superior ovary with marginal placentation is found in
    (a) Mimosaceae (b) Papilionaceae (c) Caesalpiniaceae (d) All of these
141. The condition of androecium in family Cucurbitaceae is
    (a) Corolla (b) Gynoecium (c) Calyx (d) Androecium
142. Name the family in which members contain numerous stamens
    (a) Rutaceae (b) Brassicaceae (c) Nymphaeaceae (d) Papilionaceae
143. Pulses are obtained from the members of family
    (a) Poaceae (b) Mimosaceae (c) Rosaceae (d) Papilionaceae
144. The cremocarp type of fruits are observed in the family
    (a) Rubiaceae (b) Poaceae (c) Asteraceae (d) Apiaceae
145. The type of fruit in family Cucurbitaceae is
    (a) Samara (b) Hesperidium (c) Pepo (d) None of these
146. Name the type of inflorescence in *Leucas aspera*
    (a) Raceme (b) Thyrsus (c) Panicle (d) Verticillaster

147. Which statement among these is wrong about family Asteraceae
(a) Syngenesious anthers (b) Hypogynous flowers
(c) Head inflorescence (d) Fruit a cypsela

148. Malvaceae are related to Sterculiaceae and Tiliaceae from which they differ in the presence of
(a) Axile placentation (b) Dithecous anthers
(c) Monothecous anthers (d) 2-many carpellary gynoecium

149. Cyathium type of inflorescence is the characteristic feature of genus
(a) *Ocimum* (b) *Tagetes* (c) *Ficus* (d) *Euphorbia*

150. The members of Brassicaceae have tetradynamous condition, it means
(a) Outer 2 stamens long and inner 2 stamens short
(b) Outer 2 stamens short and inner 4 stamens long
(c) Outer 4 stamens long and inner 2 stamens short
(d) Outer 4 stamens short and inner 2 stamens long

151. *Sida*, a member of family Malvaceae, does not show this prominent character
(a) Axile placentation (b) Monadelphous stamens
(c) Alternate stipulate leaves (d) Epicalyx

152. Select the largest dicotyledon family from the list
(a) Rubiaceae (b) Asteraceae (c) Cucurbitaceae (d) None of these

153. The word taxonomy first time used by
(a) C. Linnaeus (b) J.D. Hooker (c) John Ray (d) A.P. de Candolle

154. Alternative name of the family Guttiferae is
(a) Brassicaceae (b) Arecaceae (c) Apiaceae (d) Clusiaceae

155. The condition of stamens in the *Gossypium* flower is
(a) Monadelphous (b) Free (c) Diadelphous (d) Polyadelphous

156. Which of the following is a false statement regarding Compositae family
(a) Head inflorescence (b) Hypogynous flowers
(c) Fruit cypsela (d) Syngenesious flowers

157. Obdiplostamenous condition of androecium is commonly present in the members of family
(a) Rubiaceae (b) Mimosaceae (c) Rutaceae (d) Rosaceae

158. A specimen or other element selected to serve as nomenclature type as all of the material on which the description of the new species was based is missing, is known as
(a) Isotype (b) Lectotype (c) Holotype (d) Neotype

159. Tetradynamous condition of androecium is the main character of which family
(a) Brassicaceae (b) Rubiaceae (c) Papilionacae (d) Caesalpiniaceae

160. Name a family in which the ovary is unilocular in the beginning, but it becomes bilocular by the formation of false septum
(a) Brassicaceae (b) Poaceae (c) Papaveraceae (d) Caryophyllaceae

161. Tetramerous flowers are seen in the gamopetalous family of
(a) Rutaceae (b) Rubiaceae (c) Cucurbitaceae (d) None of these

162. The presence of androphore, gynophores and gynandrophore are the characteristic features of family
(a) Rosaceae (b) Cucurbitaceae
(c) Capparidaceae (d) Caryophyllaceae

163. Polypetalous, irregular corolla is observed in family
(a) Rosaceae (b) Malvaceae (c) Cucurbitaceae (d) Papilionaceae

164. When a style arises from the base of the ovary is known as
(a) Terminal (b) Linear (c) Plumose (d) Gynobasic

165. Monocarpellary condition is one of the striking character of family
(a) Rubiaceae (b) Rutaceae (c) Papilionaceae (d) Cucurbitaceae

166. Marginal placentation is observed in which of the family
(a) Fabaceae (b) Brassicaceae (c) Rutaceae (d) Nymphaeaceae

167. Which statement is correct to the family Myrtaceae
(a) Inferior ovary (b) Hypogynous flower
(c) Zygomorphic flowers (d) Stamens limited

168. Distribution of alkaloids proved usefulness in the taxonomy of the
(a) Ranunculaceae (b) Nymphaeacaeae (c) Fabaceae (d) None of thee

169. The type of fruit in *Mangifer indica* is
(a) Berry (b) Pepo (c) Drupe (d) Capsule

170. *Rubus* a member of family Rosaceae produces
(a) An etaerio of drupes (b) An etaerio of follicles
(c) An etaerio of achenes (d) An etaerio of berries

171. The type of placentation in Caryophyllaceae is
(a) Axile (b) Parietal (c) Free-central (d) Laminar

172. According to the ICBN, publication of new names and description are effective when
(a) The printed matter is distributed to the general public or the at least ten botanical institutions
(b) A nomenclature type is indicated for a new taxon of the rank of family or below published on or after 1 January, 1958
(c) It is accompanied by a Latin description or diagnosis
(d) All of these

173. Which family has gynobasic style arise from the base of ovary?

(a) Solanaceae (b) Asteraceae (c) Ranunculaceae (d) Lamiaceae

174. Hutchinson's system of classification is based on

(a) de Candolle system (b) Eichler's system
(c) Bessey's system (d) None of these

175. Hutchinson in his book, recognized how many families of flowering plants?

(a) 380 (b) 411 (c) 200 (d) 166

176. Synandrous and sinuous anthers are common in one of the following families

(a) Rosaceae (b) Malvaceae (c) Cucurbitaceae (d) Papilionaceae

177. One of the character is not a correct statement about family Anacardiaceae

(a) Well developed intra-staminal disc (b) Presence of lateral style
(c) Hesperidium type of fruit (d) Resin passages in the parenchyma

178. Floral features are chiefly used by the researchers in angiosperms identification because

(a) Flowers can be studied easily
(b) Flowers have various colors
(c) Reproductive parts are more stable and conservative than vegetative parts
(d) Flowers can be safely pressed

179. Binomial nomenclature proposed by C. Linnaeus consists of

(a) Genus and family (b) Species and order
(c) Order and family (d) Genus and species

180. A group of plants with similar characteristics of any rank is termed as

(a) Order (b) Taxon (c) Genus (d) Species

181. Cypsela fruit developed from

(a) Superior syncarpous ovary (b) superior metacarpous ovary
(c) Both (a) and (b) (d) None of these

182. Largest herbarium in our country is located in

(a) Calcutta (b) Lucknow (c) Dehradun (d) Delhi

183. Diadelphous condition is present in

(a) *Citrus* (b) *Bombax* (c) *Brassica* (d) *Pisum*

184. Inferior ovary is present in

(a) Compositae (b) Malvaceae (c) Mimosoideae (d) Solanacae

185. National Botanical Garden (now called National Botanical Research Institute) is situated at

(a) Simla (b) Delhi (c) Lucknow (d) Calcutta

186. Floral formula fails to indicate
   (a) Epiphylly and epipetaly (b) Aestivation and placentation
   (c) Position of ovary on the thalamus (d) Cohesion of stamens and carpels
187. Tepals are found in family
   (a) Compsitae (b) Papilionaceae (c) Liliaceae (d) Poaceae
188. Epipetalous stamens with free filaments but fused anthers are found in
   (a) Solanaceae (b) Asteraceae (c) Asclepiadaceae (d) Convolvulaceae
189. Basal placentation is found in
   (a) Leguminosae (b) Compositae (c) Malvaceae (d) Cruciferaceae
190. It is not true that
   (a) Underground parts of some plants of Liliaceae are useful as food for man
   (b) Fibre yielding plants are found in Malvaceae
   (c) A tree with large flowers and oil seeds is found in Compositae
   (d) Many trees found in deserts having compound leaves are the members of Leguminosae
191. Artificial system of nomenclature was first given by
   (a) Bentham and Hooker (b) Engler and Prantl
   (c) C. Linnaeus (d) None of these
192. Which of the following is the correct sequence of taxa?
   (a) Division, class, order, family (b) Division, order, class, family
   (c) Class, order, division, family (d) Family, class, order, division
193. System of classification proposed by Bentham and Hooker is
   (a) Artificial (b) Natural (c) Numerical (d) Phylogenetic
194. Medicinal plant *Asparagus* belongs to the family
   (a) Fabaceae (b) Solanaceae (c) Liliaceae (d) None of these
195. What is the position of Gymnosperms in the classification of Bentham and Hooker?
   (a) Takes precedence over dicots
   (b) Placed between dicots and monocots
   (c) Placed after monocots
   (d) Not found as division in the taxonomic classification of angiosperms
196. Classification system proposed by C. Linnaeus was artificial because
   (a) It was based on evolutionary trends
   (b) It took into account similarities and differences in floral and other morphological characters
   (c) It was based on physiological characters only
   (d) None of the above

197. Obliquely zygomorphic nature of the flower in Solanaceae is due to

(a) Corolla (b) Androecium (c) Calyx (d) Gynoecium

198. The number of carpels in family Asteraceae is

(a) One (b) Two (c) Three (d) Many

199. A group of genera which show general structural resemblance to one another mainly in their floral organs is referred to as

(a) Order (b) Tribe (c) Family (d) Class

SMALL ANSWER QUESTIONS

1. Define systematics.

   Ans. It is a branch of life sciences which deals with the kind and diversity of all organisms and the existing relationships among themselves.

2. Which is the oldest botanical work in existence as a book?

   Ans. Historia Plantarum.

3. What is the main feature of family Asclepiadaceae that distinguishes it from the Apocynaceae.

   Ans. Presence of gynostegium.

4. Who coined the term new systematics?

   Ans. Sir J. Huxley.

5. Name a family of Polypetalae in which free-central plancentation is observed.

   Ans. Caryophyllaceae.

6. What is allopatric species?

   Ans. the species

7. Who did demonstrate conclusively that Gymnosperms are distinct group separate from Angiosperms?

   Ans. Robert Brown.

8. Name the taxonomist who placed Gymnosperms between Dicots and Monocots in the classification of plants.

   Ans. Bentham and Hooker.

9. The endosperm developed after fertilization in Angiosperms is-

   Ans. Triploid.

10. Lady's slipper orchid belongs to which genus?

    Ans. *Cypripedium.*

11. What is taxonomy?

    Ans. the study of all organisms, their identification nomenclature and classification is called taxonomy.

12. In 1883, who for the first time classified seed plants as Angiosperms and Gymnosperms?

    Ans. Eichler.

13. Define sympatric species.

    Ans. Species having same or overlapping areas of geographic distribution is called sympatric species.

14. Flora of British India was compiled by-

    Ans. J.D. Hooker.

15. Out of Bentham and Hooker, who did not agree with Darwin's theory of evolution?

    Ans. Bentham only.

16. Name the type of classification proposed by Tippo in 1942.

    Ans. Phylogenetic type of classification.

17. The nature of androecium in pea flower is-

    Ans. Diadelphous.

18. What is the new definition of species?

    Ans. It is a biological unit that has evolved from a series of often identical ancestors, by which the smaller taxa of plants may be circumscribed by a combination of general, morphological and ecological criteria.

19. What is phonetics?

    Ans. It is also called numerical taxonomy. All the characters of the organisms are given equal importance and the maximum number of similarities of characters of the species is calculated by statistical methods.

20. Define Angiosperms.

    Ans. A group of flowering plants whose seeds is enclosed within carpel or carpels and the upper part of which differentiated in stigma and style.

21. The best known artificial system of classification was proposed by-

    Ans. C. Linnaeus in 1735.

22. What are the names of sub-families of family Fabaceae?

    Ans. Papilionaceae, Caesalpiniaceae and Mimoseae.

23. What is keystone species?

    Ans. It is a species which determine the biotic structure of the entire community by predation or by any control mechanism.

24. Who gave the term Angiospermae?

    Ans. O. Tippo in 1942.

25. Name the crucifer in which the stem is reduced greatly.

    Ans. *Raphanus*.

26. What is the time of origin of Angiosperms?

    Ans. Cretaceous age.

27. Who was Theophrastus?

    Ans. He was Greek philosopher, also called as father of Botany. He has classified plants into four groups on the basis of their habit (herbs, under-shrubs, shrubs and trees).

28. Historia Plantarum was written by-

    Ans. Theophrastus

29. Write a note on Historia Plantarum.

    Ans. It is also referred as Enquiry into Plants. This is the oldest botanical work in existence. Theophrastus described about 480 plant species.

30. Name the type of corolla found in sub-family Papilionaceae.

    Ans. Papilionaceous.

31. Name the most important feature to distinguish Asclepiadaceae from Apocynaceae.

    Ans. Presence of gynostegium.

32. Write the time of origin of dicots.

    Ans. Mesozoic era.

33. What is morphospecies?

    Ans. species erected on the basis of morphological characters only.

34. Who was Andrea Caesalpino?

    Ans. He was an Italian botanist, and the first to give a methodical classification of plants based on definite morphological criteria. He divided plants in woody and herbaceous habit.

35. Materia Medica was written by-

    Ans. Dioscorides.

36. What is cladogram in taxonomy?

    Ans. A diagram indicating evolutionary relationships of a group is called as cladogram.

37. The book 'English Herbal' was written by-

    Ans. William Turner.

38. Who divided herb in Imperfectae and Perfectae?

    Ans. John Ray.

39. The third sterile stigma in the family Orchidaceae is known as-

    Ans. Rostellum

40. What is the botanical name of congress weed?

    Ans. *Parthenium hysterophorus*.

41. Who gave major stress on position and structure of ovule in taxonomy in 1915?

    Ans. Charles Erwin Bessey.

42. Name the types of systems of classifications.

    Ans. There are three major types of systems of classification, 1. Artificial, 2. Natural, and 3. Phylogenetic.

43. Name the botanist who first time classified seed plants into monocot and dicot.

    Ans. John Ray.

44. Who wrote Historia Plantarum Universalis?

    Ans. J. Bauhin.

45. What is biospecies?

    Ans. Species erected on the basis of reproductive and genetic isolation in a natural population is called biospecies.

46. What is cline?

    Ans. Gradual changes in population over a geographical area due to change in environmental gradient is called cline.

47. Who first time proved sexuality in plants?

    Ans. Camerarius.

48. 'An Introduction to Botany' was written by-

    Ans. G. Bauhin.

49. What is artificial system of classification?

    Ans. It is based only on one or a few characters.

50. What is type of fruit observed in the members of family Apiaceae?

    Ans. Cremocarp.

51. Who first time considered the use of stamens as an important character in taxonomy?

    Ans. C. Linnaeus.

52. Name the books published by C. Linneaus.

    Ans. Genera Plantarum in 1737 and Species Plantarum in 1753.

53. *Cuscuta* is a total stem parasite belongs to which family?

    Ans. Convolvulaceae.

54. Who is known as father of botany?

    Ans. Theophrastus.

55. How many classes were given by C. Linnaeus?

    Ans. 24 classes.

56. Who divided plant kingdom in two groups, Herbae and Arbores?

    Ans. John Ray.

57. The phenomenon of double fertilization and triple fusion is the characteristic of-

    Ans. Angiosperms.

58. Define cladistics?

    Ans. The genetics and evolutionary relationships among the organisms for the purpose of classification into different taxa is called cladistics. It is also called as phylogenetic classification.

59. C. Linnaeus system of classification also known as-

    Ans. Sexual system because he gave more importance on reproductive parts of flower.

60. The book 'Elements de Botanique' was written by-

    Ans. J.P. de Tournefort.

61. Name the plant in which under-developed inflorescence is edible and belongs to family Brassicaceae.

    Ans. *Brassica oleracea* var. *botrytis*.

62. Who proposed first time the 'the binomial system of classification'.

    Ans. C. Linnaeus.

63. Name the family in which Oenothera type of embryo sac is most common.

    Ans. Onagraceae.

64. Who are the important taxonomists whose systems are included in natural system of classification?

    Ans. Jussieu, de Candolle and Bentham and Hooker.

65. Who gave the concept of continuity of plant kingdom?

    Ans. C. Darwin.

66. Who has given major emphasis on number of cotyledons, presence and absence of petals and position of petals and stamens in his system of classification?

    Ans. A. L. de Jussieu.

67. What is B taxonomy?

    Ans. It deals with collection and identification on the basis of morphological evidences from genetics, cytology, anatomy, physiology etc.

68. Whose system of classification is also known as numerical system of classification?

    Ans. C. Linnaeus.

69. Modified calyx into bristles as in many Composites is called-

    Ans. Pappus.

70. What are ovules?

    Ans. These are minute bodies found inside the ovary, which after fertilization develop into seed.

71. What is the type of classifications proposed by Hallier (1912), Bessey (1915) AND Hutchinson (1926)?

    Ans. Phylogenetic.

72. Who proved experimentally that pollen is necessary for fertilization and seed formation in the life of plants?

    Ans. Camerarius.

73. The location of world's largest herbarium is-

    Ans. At Royal Botanical Garden, Kew.

74. What is phylogenetic system of classification?

    Ans. In this system the taxonomic affinity based on evolutionary as well as genetic relationship amongst organisms are taken into account.

75. Define alpha taxonomy.

    Ans. It deals with collection and identification of organisms on the basis of gross morphology.

76. Who divided monocots in Calyciflorae, Corolliflorae and Glumiflorae?

    Ans. Hutchinson.

77. The biggest Indian herbarium is located at-

    Ans. Botanical Survey of India, Colcutta, it holds more than 13 lakh specimens.

78. Define numerical taxonomy.

    Ans. The application of simple mathematical principles or techniques or methods in taxonomical studies of plants is called numerical taxonomy.

79. What is orthotropous type of ovule?

    Ans. It is a straight ovule with the micropyle opposite to the chalaza.

80. What are the special characters of Calyciflorae?

    Ans. Perianth is biseriate, being distinguishable into calyx and corolla.

81. Name the taxonomists whose classification is considered as phylogenetic.

    Ans. Engler and Prantl, Eichler, Hutchinson etc.

82. Write a note on Robert Brown.

    Ans. He was the first to demonstrate that Gymnosperms are a group discrete from the Angiosperms.

83. What is trinomial nomenclature?

    Ans. Some of the organisms are named by three words, e.g. *Puccinia graminis tritici* for genus, species and variety.

84. Who gave the concept that the evolution may be progressive or retrogressive?

    Ans. Hutchinson.

85. Define phylogeny.

    Ans. It is the evolutionary history of a taxon, and attempts to account for its origin and development.

86. How many types of International Codes of Nomenclature are in existence?

    Ans. Five types.

87. What is Omega taxonomy?

    Ans. It is based on all microscopic observations and biochemical evidences.

88. Who proposed five kingdom system of classification?

    Ans. Whittaker in 1969.

89. Define taxon.

Ans. Any taxonomic category like a phylum, class, order, family, genus or a species can be referred to as a taxon.

90. Who introduced the term taxonomy?

Ans. de Candolle.

91. What is herbarium?

Ans. It is collection of plants, which have been dried, pressed, mounted on herbarium sheets, identified and classified according to some approved system of classification.

92. What is meaning of revision in taxonomy?

Ans. Naming a species and understanding evolutionary relationship amongst a group of species is called revision of the group.

93. The type of fruits in family Graminae is-

Ans. Usually caryopsis in which pericarp completely united with the seed coat, rarely a nut or berry as in *Bambusa.*

94. What are correlated characters?

Ans. Species belonging to genus shows a number of similar characters. Such groups of common features are termed correlated characters.

95. Who was the first to use the characteristics of vascular bundles in the classification of plants?

Ans. de Candolle.

96. Define Linnaean hierarchy.

Ans. In taxonomy there are different levels of categories in which the organisms are grouped in hierarchial order, as it was first given by Linnaeus it is called as Linnaean hierarchy.

97. How many obligate hierarcheal categories in taxonomy?

Ans. Seven (Kingdom- Phylum – Class- Order-Family tribe – Genus- Species).

98. Define holotype.

Ans. It is the specimen or other element used by the author as the nomenclature type.

99. What is isotype?

Ans. It is the duplicate of holotype, like several branches of a trees are collected at the same time; one specimen may be chosen as a type and rest as isotype.

100. Define paratype.

Ans. It is a specimen cited with the original description other than holotype or isotype.

101. What type of flower is present in family Poaceae?

Ans. Sessile, bracteate (the bracts are represented by lemma and palea), hermaphrodite or unisexual, zygomorphic, hypogynous and irregular.

102. What is Neotype specimen?

Ans. It is a specimen or other element selected to serve as nomenclature type as long as all of the material on which the names of the taxon was based is missing.

103. What is lectotype specimen?

Ans. It is a specimen selected from original material to serve as nomenclature type when there is no holotype.

104. Define syntype specimen.

Ans. It is any of the two or more specimens cited by author when there is no holotype.

105. What are the special characters of order Magnoliales?

Ans. Hypogynous, bisexual, becoming unisexual with reduction of perianth, apocarpous, petals rarely absent, stamens numerous, usually free, endosperm copious, not ruminate, embryo minute, stipules present or absent.

106. What are the most important characteristic features of order Graminales?

Ans. Flowers subtended by two bracts (lemma and palea); perianth usually represented by lodicules; fruit a caryopsis, rarely a nut, berry, or utricle; stems mostly hollow between the nodes.

TRUE OR FALSE TYPE QUESTIONS

1. The stamens are usually in polyadelphous condition in family Papilionaceae.

   Ans. False

2. An aestivation with the segments of calyx or corolla are so placed that their edges touch each other but not overlapped is called valvate.

   Ans. True

3. Orchidaceae are characterized by having tricarpellary, syncarpous and inferior ovary.

   Ans. True

4. In umbel type of inflorescence aerial branches are of equal length and arise from a common point.

   Ans. True

5. Members of Poaceae are a great source of pulses.

   Ans. False

6. Gamopetalous flowers are never found in the family Cucurbitaceae.

   Ans. False

7. A leaf like out-growth at the base of the petiole is called stipule.

   Ans. True

8. Pepo is the characteristic feature of Cucurbitaceae.

   Ans. True

9. Spathe is a large showy bract usually covering the whole inflorescence.

   Ans. True

10. Alternate, exstipulate and mostly pinnate leaves are generally present in Meliaceae.

    Ans. False

11. Polypetaly is more primitive than sympetaly.

    Ans. True

12. In Rubiaceae, the ovary is usually inferior and bilocular with axile placentation.

    Ans. True

13. Spadix is a type of inflorescence with succulent axis possessing male and female flowers.

    Ans. True

14. Evolution involves necessarily all organs at a time.

    Ans. False

15. In the genus *Raphanus* of family Brassicaceae, the fruit is a lomentum.

    Ans. True

16. Grasses are insect pollinated.

    Ans. False

17. Spike is a racemose elongated inflorescence bearing sessile flowers.

    Ans. True

18. Apocynaceae family is closely related to Asclepiadaceae from which it differs in the absence of gynostegium.

    Ans. True

19. *Hibiscus cannabinus* provides a type of hemp, which can be used as a substitute of jute.

    Ans. True

20. The members of family Liliaceae have epigynous flowers.

    Ans. False

21. Rudimentary stamens which do not bear fertile pollen are called staminate condition.

    Ans. False

22. The presence of swollen placenta and oblique septum is the characteristic feature of Solanaceae.

    Ans. True

23. Solanaceae belongs to the cohort Polemoniales.

    Ans. True

24. Septal nectarines occur in Liliaceae.

    Ans. True

25. A fruit which open in two or more distinct portions each with its separate wall is called schizocarp.

    Ans. True

26. *Coffea arabica* is the botanical name of coffee belongs to family Rubiaceae.

    Ans. True

27. In *Lepidium sativum* (Cruciferae), the gynoecium is tricapellary.

    Ans. True

28. The seeds of Euphorbiaceae are often carunculated.

    Ans. True

29. The members of Amaranthaceae have sepaloid perianth parts.

    Ans. True

30. An imbricate aestivation, with 5 petals out of which two are exterior, two interior and the fifth has one margin exterior and one interior called quincuncial condition.

    Ans. True

31. The basic unit of biological classification is a species.

    Ans. True

32. The members of Meliaceae differ from Rutaceae in the absence of glands on the leaves and the filaments being united into a stamina tube.

    Ans. True

33. Raceme is an inflorescence with stalked flowers, in which the main axis continues to grow and the lowest flowers are the oldest and open first.

    Ans. True

34. The type of fruits in family Cruciferae is lomentum.

    Ans. False

35. Taxonomic group belonging to any one of the taxonomic categories is known as taxon.

    Ans. True

36. Perianth is the floral envelopes including both calyx and corolla no differentiation into calyx and corolla.

    Ans. True

37. In family Cucurbitaceae, the number of female flowers is more in number as compared to male flowers.

    Ans. False

38. In members of Caryophyllaceae, cincinnus type of inflorescence is found.

    Ans. True

39. Gymnosperms have been placed in between Dicots and Monocots in the classification of plants by Bentham and Hooker.

    Ans. True

40. A compound leaf in which the leaflets are arranged on the two sides of the rachis is called pinnate leaf.

    Ans. True

41. Liliaceae belongs to series Coronariae.

    Ans. True

42. Members of family Nymphaeaceae are perennial aquatic herbs with a creeping or erect sympodial rhizome.

    Ans. True

43. The anthers are extrorse in family Papaveraceae.

    Ans. True

44. The Capparidaceae resemble to Brassicaceae in the presence of bicarpellary ovary with parietal placentation and siliquose fruit.

    Ans. True

45. The flower buds of *Bauhinia variegata* are used as vegetables.

    Ans. True

46. The alternative name of Umbelliferae is Arecaceae.

    Ans. False

47. The classification of flowering plants proposed by J. Hutchinson was a natural classification.

    Ans. False

48. Bentham and Hooker recognized 200 families of flowering plants.

    Ans. True

49. In members of Annonaceae, numerous free stamens occur arranged spirally on a large convex receptacle.

    Ans. True

50. Adnation of petioles and peduncles with the stem is a characteristic feature of the family Cucurbitaceae.

    Ans. False

51. The ICBN is divided into three parts- Principles, Rules and Recommendations.

    Ans. True

52. Unisexual flowers are observed in Cucurbitacae.

    Ans. True

53. Tetradynamous androecium is seen in family Lamiaceae.

    Ans. False

54. Gynobasic style is the characteristic feature of family Lamiaceae.

    Ans. True

55. Genus *Urenea* belongs to family Malvaceae, and the calyx in this genus form a tube like structure.

    Ans. True

56. Didynamous is observed in the members of Cucurbitaceae.

    Ans. True

57. The collection or depository of dried plant specimen is called herbarium.

    Ans. True

58. Neotype is the one specimen or other element designed by the author as the nomenclature type.

    Ans. False

59. *Coronopus* contains only two stamens.

    Ans. True

60. The Solanaceae differ from Scrophulariaceae in the presence of bicollateral vascular bundles, usually zygomorphic corolla and four or five stamens.

    Ans. False

61. Corymb is form of racemose inflorescence in which the several pedicils arising at different levels reach more or less the same at the top.

    Ans. True

62. Didynamous stamens are seen in members of Lamiaceae.

    Ans. True

63. Translators present in *Calotropis* are spoon shaped.

    Ans. True

64. *Solanum xanthocarpum* is a spinous xerophytic herb belongs to the family Malvaceae.

    Ans. False

65. Cyathium type of inflorescence is present in *Euphorbia* which appears as a single flower.

    Ans. True

66. Capsule is a dry dehiscent fruit, when ripe opens by two or more valves.

    Ans. True

67. Amaryllidaceae differ from Liliaceae in having inferior ovary.

    Ans. True

68. The petals in Capsella bursa pastoris are replaced with stamens.

    Ans. True

69. Monocarpic plants produce inflorescence only once during their life time and the flowering marks the end of their life.

    Ans. True

70. In Poaceae, the flowers are arranged in spikletes which are the units of the inflorescence.

    Ans. True

71. Berry is a fleshy indehiscent fruit with many seeds and the fruit contains thin epicarp, fleshy mesocarp and stony endocarp.

    Ans. True

72. Generally, the number of stamens in Papilionacea are (9)+1, but in *Dalbergia* it is 9 only.

    Ans. True

73. The flowers are zygomorphic in *Iberis amara*.

    Ans. True

74. The tetralocular ovary in Verbenaceae suggests tetracarpellary condition.

    Ans. False

75. In all taxa of Cyperaceae only one functional microspore is formed in each pollen mother cell.

    Ans. True

76. Family Malvaceae is also known as mallow family.

    Ans. True

77. Carples in a flower free from each other is called apocarpous.

    Ans. True

78. The pentamerous and zygomorphic flowers are present in *Delphinium*.

    Ans. True

79. The species of *Cuscuta* are leafless and rootless total parasite with thread-like twinning stem.

    Ans. True

80. Usually the ovary in the members of family Solanaceae is bilocular, but in genus *Henoonia*, it is unilocular.

    Ans. True

81. The members of Amaranthaceae have sepaloid perianth parts.

    Ans. True

82. Androgynophore is an axis or stalk bearing both stamens and pistil above the point of perianth attachment.

    Ans. True

83. A. Cronquist presented the comprehensive version of his concepts of classification in 'The Evolution and Classification of Flowering Plants'.

    Ans. True

84. All the cereals belong to family Fabaceae.

    Ans. False

85. Androecium is a collective term for all stamens and staminoids.

    Ans. True

86. *Desmodium latifolium* is a xerophyte belongs to family Papilionaceae.

    Ans. False

87. In *Triticum*, the perianth is highly modified structure, represented by two minute structures, known as lodicules.

    Ans. True

88. In members of Poaceae, the basic unit of the inflorescence is called as spikelet.

    Ans. True

89. Culms are present generally in the members of family Poaceae.

    Ans. True

90. Epicalyx in family Malvaceae worked as bracteoles.

    Ans. True

91. Achene is a small, dry, hard, one chambered and one seeded fruit.

    Ans. True

92. Gynandrium is the most important character of Orchid flowers.

    Ans. True

93. Gynandrium of Orchidaceae formed by the adnation of stamens, style and stigma.

    Ans. True

94. Hypanthodium type of inflorescence is present in the members of Asclepiadaceae.

    Ans. False

95. Unisexual and epigynous flowers are present in the members of family Cucurbitaceae.

    Ans. True

96. The condition in which male and female both flowers are on same plant is called monoecious.

    Ans. True

97. Cypsela type of fruits is present in the members of family Asteraceae.

    Ans. True

98. Gynobasic style is present in family Convolvulaceae.

    Ans. False

99. Ligule is a membranous outgrowth at the junction of leaf blade and leaf sheath in grasses.

    Ans. True

100. One seeded pyrenes with drup fruits are present in family Verbenaceae.

    Ans. True

101. Tricapellary gynoecium and inferior ovary present in the members of Orchidaceae.

    Ans. True

102. C. Linnaeus recognized 24 classes in his classification on the basis of the stamens.

    Ans. True

103. Genera Plantarum was written by Bentham and Hooker.

    Ans. True

104. Monocarpellary gynoecium and marginal placentation is present in the members of family Papilionaceae.

    Ans. True

105. Whorl or bracts situated below the flower or inflorescence is called involucres.

    Ans. True

106. Hutchinson divided dicotyledons into two divisions Lignosae and Herbaceae.

    Ans. True

107. C. Linnaeus is also known as father of modern botany.

    Ans. True

108. Chemotaxonomy is the application of chemistry in taxonomy.

    Ans. True

109. Cyathium type of inflorescence is present in the members of family Caryophyllaceae.

    Ans. False

110. When all filaments of the stamens united together to form a bundle, it is called Monadelphous.

    Ans. True

111. Petals are present in *Sterculia*.

    Ans. False

112. Zygomorphic flowers are present in *Ranunculus*.

Ans. False

113. Stipulate, bipinnate and alternate leaves are present in Mimosaceae.

Ans. True

114. Mustard flowers are ebracteate

Ans. True

115. Bicarpellary ovary with parietal placentation is found in Lamiaceae.

Ans. False

116. Pome is the type of fruit in family Cucurbitaceae.

Ans. False

117. Hypogynous flowers are present in the members of family Asteraceae.

Ans. True

118. The book 'Families of Flowering Plants' was written by Engler and Prantl.

Ans. False

119. The oldest known fossils of dicotyldon are from the lower Cretaceous rocks.

Ans. True

120. It is generally believed that the dicotyledons have been originated from gymnosperms.

Ans. True

121. The plant taxonomy begins with the Greeks.

Ans. True

DESCRIPTIVE TYPE QUESTIONS

1. What were the basic criterions to propose the five kingdoms system of classification?

- Whittaker in 1969 proposed this classification, the five kingdoms he divided on the basis of the following criteria:
- Complexity of body organization: it refers to the organization of organism's body (whether the organism is unicellular or mutlicellular).
- Complexity of cell structure: it refers to organization at cellular level (whether the organism shows prokaryotic cellular organization or eukaryotic).

- Mode of nutrition: it refers to the organism whether it is autotrophic or heterotrophic.
- Mode of nutrition: it means the organism is ingestive or absorptive.
- Life cycle: it refers to the status of organisms in the ecosystem (whether it is a producer or a consumer or a decomposer).

2. What are the five kingdoms of classification proposed by Whittaker?
   - Kingdom Monera (includes prokaryotes, e.g. bacteria).
   - Kingdom Protista (unicellular or colonial eukaryotes e.g. *Euglena*, *Volvox*).
   - Kingdom Plantae (includes photosynthetic organisms with cell wall e.g. plants).
   - Kingdom Fungi (non-photosynthetic heterotrophic organisms with cell wall e.g. fungi).
   - Kingdom Animalia (includes multicellular eukaryotic heterotrophic organisms which lack of cell wall e.g. all animals).

3. Write a note on kingdom Monera.

   They are microscopic living organisms showing prokaryotic cell organization. They are regarded as a group of most primitive organisms and generally surrounded by a cell wall. They may be autotrophic or heterotrophic and motile or non-motile. They work as decomposers in the ecosystem and some of them possess ability to fix atmospheric nitrogen in the soil. E.g. Archaebacteria, eubacteria, actinobacteria etc.

4. What are the general characters of kingdom Protista?

   This group evolved from Kingdom Monera. They are unicelled or colonial organisms showing eukaryotic cell organization. Some Protista do possess a well defined cell wall. They exhibit locomotion by cilia or flagella or pseudopodia. The flagella show 9+2 organization. They may be photosynthetic or heterotrophic. E.g. *Euglena*, *Volvox*, protozoans, slime molds etc.

5. Mention the important characters of kingdom Plantae.

   These are multi-celled, photosynthetic organisms showing eukaryotic cell organization. They do not show any locomotion. They may be aquatic or terrestrial with respect to habitat. The aquatic forms may occur in fresh water or marine. The terrestrial plants may or may not show organization of vasculature. The vascular plants may be spore bearing or seed bearing. The seeds may be naked or enclosed within the fruits. E.g. higher algae, bryophytes, pteridophytes, gymnosperms and angiosperms.

6. Write a brief account on kingdom Fungi.

   These are multicelled eukaryotic and heterotrophic organisms. They possess a well defined chitinous cell wall and show absorptive or assimilative type of nutrition. The food is stored in the form of oil droplets and glycogen. They work as decomposers in the ecosystem. E.g. true fungi.

7. Write a note on kingdom Animalia.

   These are multicelled, eukaryotic, heterotrophic organisms which lack a cell wall. They move from one place to other in search of food and partner. They ingest food which is later on digested by enzymatic action, e.g. all animals from sponges to mammals.

8. Whose classifications are considered as phylogenetic?

   The classification proposed by Engler and Prantl, Bessey, Hutchinson, Tippo, Takhtajan, Cronquist etc. are considered phylogenetic.

9. What are the important features of classification proposed by Engler and Prantl?

   Flowers with only stamens or carpels were believed to form the lowest grade of organization. They considered monocots to be primitive than dicots and, placed monocots before dicots. Families in this system are arranged according to increasing complexity of the flower. Primitive forms kept in the beginning have naked flowers. The more advanced families have distinct perianth and the highest evolved families have fused perianth.

10. Write a note on the work of John Hutchinson.

    He put forward his work in his book "Families of Flowering Plants". The seed bearing plants have been divided in two phyla: 1. Gymnospermae and 2. Angiospermae. Angiospermae are further divided into two sub phyla: Dicotyledonae and Monocotyledonae. Trees and shrubs are considered to be more primitive than herbs. Dicotyledons have been considered more primitive than monocotyledons. Polypetaly, actinomorphy and solitary flowers are considered more primitive than Gamopetaly, Zygomorphy and inflory respectively.

11. What is floral diagram?

    It represents a cross-section of the flower; aestivation of sepals or petals; their occurrence in one or more series and their relationship to one another; stamens in one or more series; cross-section of ovary with placentation etc.

12. Define floral formula.

    Symbols are used to avoid verbal descriptions in floral formula. One gets a satisfactory idea of the flower with a floral diagram accompanied by a floral formula. If one gets familiar with these, identification of the natural order becomes very easy.

Tables: Comparisons

Comparison between Natural & Artificial Systems of Classification

Natural Classification	Artificial Classification
Sum total of all the characters taken into consideration.	One or few characters taken into consideration.
Plants groups together, most closely related phylogenetically.	Plants not group together most closely related phylogenetically.
Contains a great deal of information about its members.	Contains only a very limited amount of information about its members.
Easy to incorporate additional information.	Not easy.
High predictive value.	Little or no predictive value.
Identification is easy.	Identification is difficult.
Liable to change with increasing knowledge.	Not possible.

Bentham and Hooker, Engler and Prantl, and Hutchinson's system of classifications

Bentham and Hooker	Engler and Prantl	Hutchinson
Natural system (used for practical purpose).	More or less phylogenetic (proposed after the theory of descent).	Phylogenetic (more close to Bentham & hooker's system).
Gymnosperms in between Dicots & Monocots.	Gymnosperms separate from Dicots & Monocots.	Gymnosperms separate from Dicots & Monocots.
Based on single and mostly artificial characters.	Based on well-defined characters.	Based on definite characters.
202 families.	280 families.	41 families.
Put dicots before Monocots.	Put monocots before Dicots.	Dicots before Monocots.

Poaceae & Cyperaceae

Character	Poaceae	Cyperaceae
Stem	Cylindrical and Hallow with nodes.	Triangular, solid and nodes absent.
Leaves	Simple, alternate, ligulate and sheath split open.	Simple, ligule absent and sheath closed.

(*Contd.*)

Character	Poaceae	Cyperaceae
Flower	Two glumes.	One glume.
Scales	Glumes, Lemma and Palaea.	Pistillate and staminate scales.
Perianth	2 – 3 lodicules.	Absent or scales/bristles.
Stamens	3, (rarely 3+3) versatile.	3 (rarely 1) basifixed.
Pistil	Monocarpellary, one locule, one ovule.	Tricarpellary, Syncarpous, one celled, single, basal ovule.
Fruit	Caryopsis.	Achene or Nut.

Acanthaceae and Lamiaceae

Character	Acanthaceae	Lamiaceae
Stem	With swollen nodes.	Squarish.
Leaves	Opposite, decussate, simple with cystolith	Simple, opposite, decussate.
Inflorescence	Spike or cymose, bract, bracteoles.	Verticillaster.
Flower	Zygomorphic, hermaphrodite, hypogynous	Same as in family Acanthaceae.
Calyx	Sepals five, gamosepalous, unequal size.	Sepals five, fused, bilabinate, persistant.
Corolla	Petals five, gamopetalous, personate.	Petals five, gamopetalous, bilabiate.
Androecium	Stamens 2 or 2+2, epipetalous.	Stamens 2+2 or 2, epipetalous.
Gynoecium	Carpels 2, syncarpous, superior and bilocular ovary, axile placentation.	Carpels 2, syncarpous, superior and tetralocular ovary,, axile placentation.
Style	Simple terminal.	Gynobasic.
Fruit	Capsule.	Schizocarpic carservulous.

Magnoliaceae & Annonaceae

Magnoliaceae	Annonaceae
Leaf thick and stipulate.	Exstipulate with glands.
Flower large & gaudy.	Small flower.
Outer perianth green, in whorl and petaloid.	Outer whorl sepaloid.
Imbricate aestivation.	Valvate aestivation.
Introse anthers with or without filament.	Extrose anthers, linear with elongate connective.
Fruit etaerio/follicles/drupes.	Etaeria of drupes or berries.

Brassicaceae & Capparidaceae

Characters	Brassicaceae (Crucifere)	Capparidaceae
Leaves	Simple, radicle are cauline Pinnatifid or lobed or exstipulate.	Simple or compound, Stipulate.
Inflorescence	Corymb or corymbose Raceme.	Corymbose raceme.
Flowers	Ebracteate, androphore absent.	Bracteate, gynophores or androgynophore present.
Petals	4 and cruciform type.	4, valvate or imbricate aestivation.
Stamens	6, tetradynamous.	6 or indefinite.
Fruit	Siliqua or silicula.	Capsule or berry.
Seeds	Round and oil rich.	Reniform.

Rutaceae & Malvaceae

Characters	Rutaceae	Malvaceae
Habit	Shrub or trees, rarely herbs.	Herbs or shrubs, rarely trees.
Leaves	Compound and alternate.	Simple and alternate.
Inflorescence	Cymose type.	Solitary, axillary or terminal.
Epicalyx	Absent.	Present.

(*Contd.*)

Characters	Rutaceae	Malvaceae
Stamens	10 or 8 or less, polyandrous, obdiblostemonous condition.	Indefinite, monadelphous, monotheous (forming a tube).
Disc	Large cushion like disc below the ovary.	Absent.
Fruit	Capsule/Berry/Hesperidium.	Capsule/schizocarpic/carserulous.

Rutaceae & Caryophyllaceae

Characters	Rutaceae	Caryophyllaceae
Habit	Shrubs or tree	Mostly herbs
Leaves	Compound alternate	Simple opposite
Placentation	Axile type	Free central type
Seed	Non- endospermic	Endospermic

Rutaceae & Anacardiaceae

Characters	Rutaceae	Anacardiaceae
Habit	Herbs, shrubs or trees.	Trees with resin.
Leaves	Alternate or opposite, compound, petioles winged, rarely simple, gland – dotted, exstipulate.	Alternate, Simple, Exstipulate.
Inflorescence	Many types.	Terminal or axillary cyme or panicle.
Flower	Actinomorphic, rarely zygomorphic, bisexual or unisexual, hypogynous, thalamous concave.	Actinomorphic, rarely zygomorphic, hermophrodite or unisexual, hypogynous; thalamus convex, cushion – like, concave or elongated.
Calyx	Sepals five, gamosepalous.	Sepals 5 – 7, free or fused at the base.
Corolla	Petals five, polypetalous, gland – dotted.	Petals 3 – 5 – 7, Free or based at the base.

(*Contd.*)

Characters	Rutaceae	Anacardiaceae
Androecium	Stamen 10 - 8 , free or polyadelphous.	Stamens 10 or 5.
Gynoecium	Carpels 2 - 8 , syncarpous, axile placentation, disc below the ovary.	3-5 carpels, syncarpous or annular disc present.
Fruit	Hespridium or capsule.	A drupe.

Sub-families of Fabaceae (Pappilionaceae, Caesalpinaceae & Mimosaceae)

Characters	Pappilionaceae	Caesalpinaceae	Mimosaceae
Calyx	Sepal 5, gamosephalous odd sepal anterior.	Sepal 5, Free or slightly fused, odd sepal anterior.	Sepal 5, gamosephalous, odd sepal posterior.
Corolla	Petals 5, pappilionaceous, with posterior largest, standard, 2 lateral wings and 2 anterior united keel.	Petals 5, odd petal smallest and innermost ascending imbricate type of aestivation.	Petals 4 or 5, fused at the base valvate aestivation.
Androecium	Stamen 10, Usually diadelphous (9)+1, rarely monodelphous.	Stamen 10 (5+5), with staminodes, rarely free.	Stamens numerous and free.
Gynoecium	Monocarpellary, unilocular and marginal placentation.	Same as in Pappilionaceae.	Same as in Pappilionaceae.
Fruit	Legume.	Legume.	Lomentum.
Habit	Mostly herbs, rarely shrubs, few trees.	Mostly shrubs, trees rarely shrubs.	Mostly shrubs, trees rarely shrubs.
Leaves	Imparipinnate.	Paripinnate.	Bipinnate.
Inflorescence	Racemose.	Raceme or panicle.	Cymose head
Flower	Zygomorphic.	Slightly zygomorphic.	Actinomorphic, regular.

Apocynaceae & Asclepiadaceae

Characters	Apocynaceae	Asclepiadaceae
Habit	Plants herbs or shrubs with latex.	Plants herbs or shrubs or climbers with latex.

(Contd.)

Characters	Apocynaceae	Asclepiadaceae
Leaves	Simple, opposite, deccusate or whorled or exstipulate.	Simple, opposite, deccusate or exstipulate.
Inflorescence	Generally cymose rarely solitary.	Usually cymose umbel, sometimes racemose
Flower	Complete, Hermaphrodite, actinomorphic, hypogynous.	As in Apocynaceae.
Calyx	Sepal 4 – 5, gamosepalous, imbricate aestivation.	Sepals 5, polysepalous, valvate or contorted type of aestivation.
Corolla	Petals 5, gamopetalous – infundibuliform, twisted.	Petals -5, gamopetalous or slightly fused at base.
Corona	Sometimes present.	Present.
Androecium	Stamens 5, Polyandrous, Epipetalous.	Stamens 5, Adnate to gynoecium and forming gynostegium, pollinia and translaters present.
Gynoecium	Carpels 2, syncarpous, placentation marginal or axile, stigma drum shaped.	Carpels 2, syncarpous, placentation marginal, stigmatic 5 angled, united with the anthers to form gynotegium.
Fruit	Capsule, drupe or follicule.	Aggregate, an etaerio of follicles.

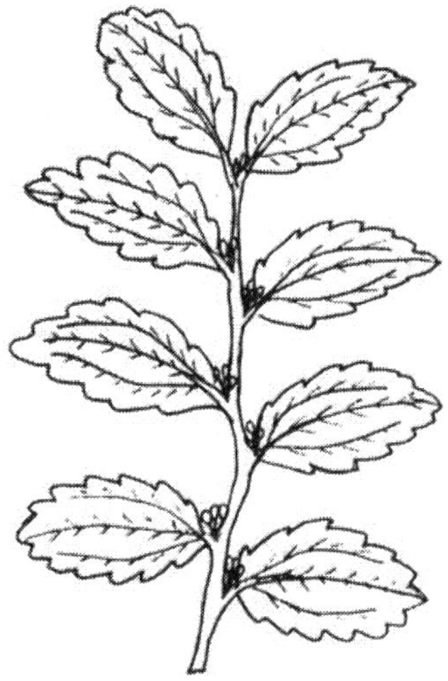

Habit

A single flower

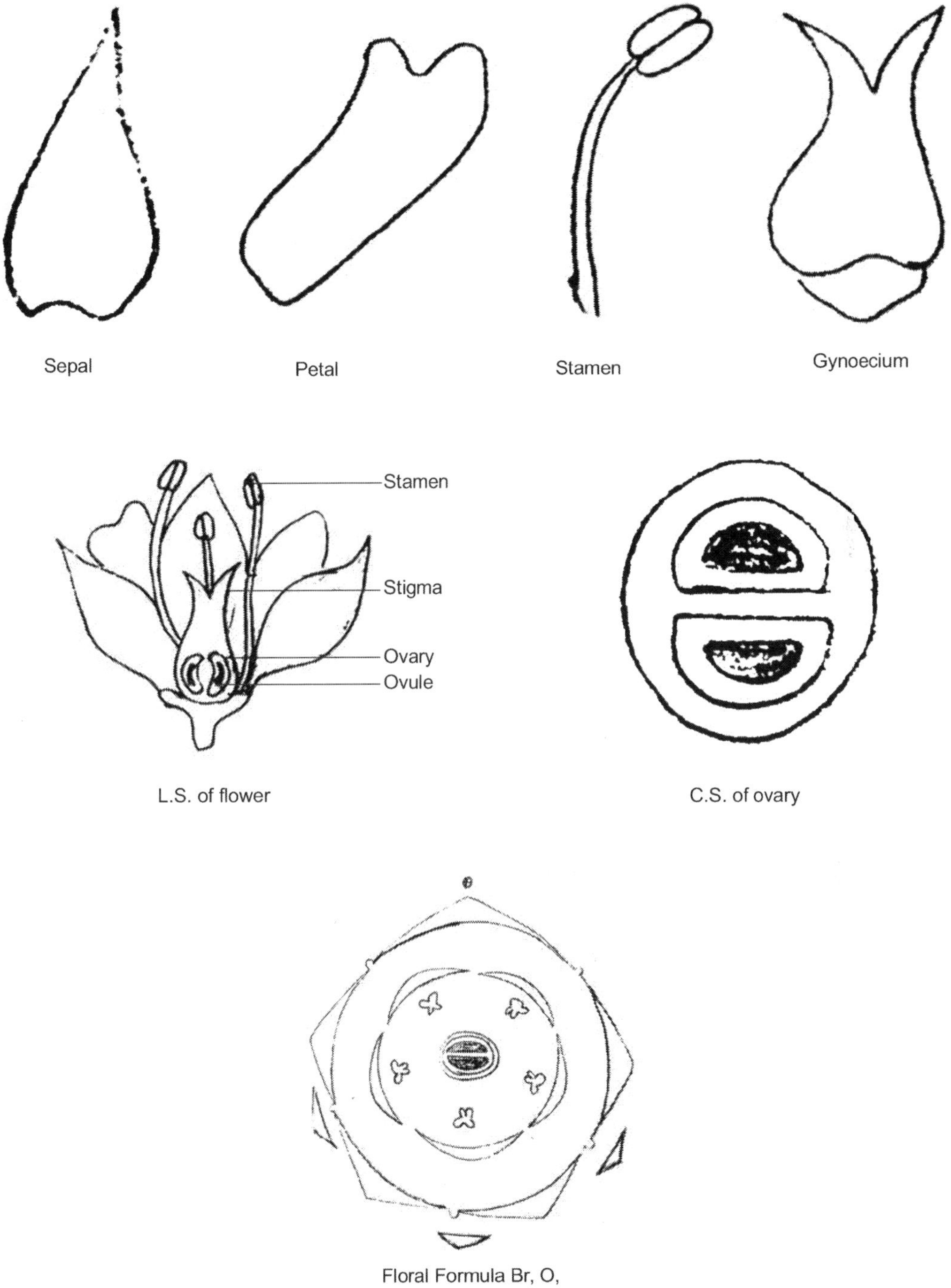

Figure 1.1 Description of family Rhamnaceae

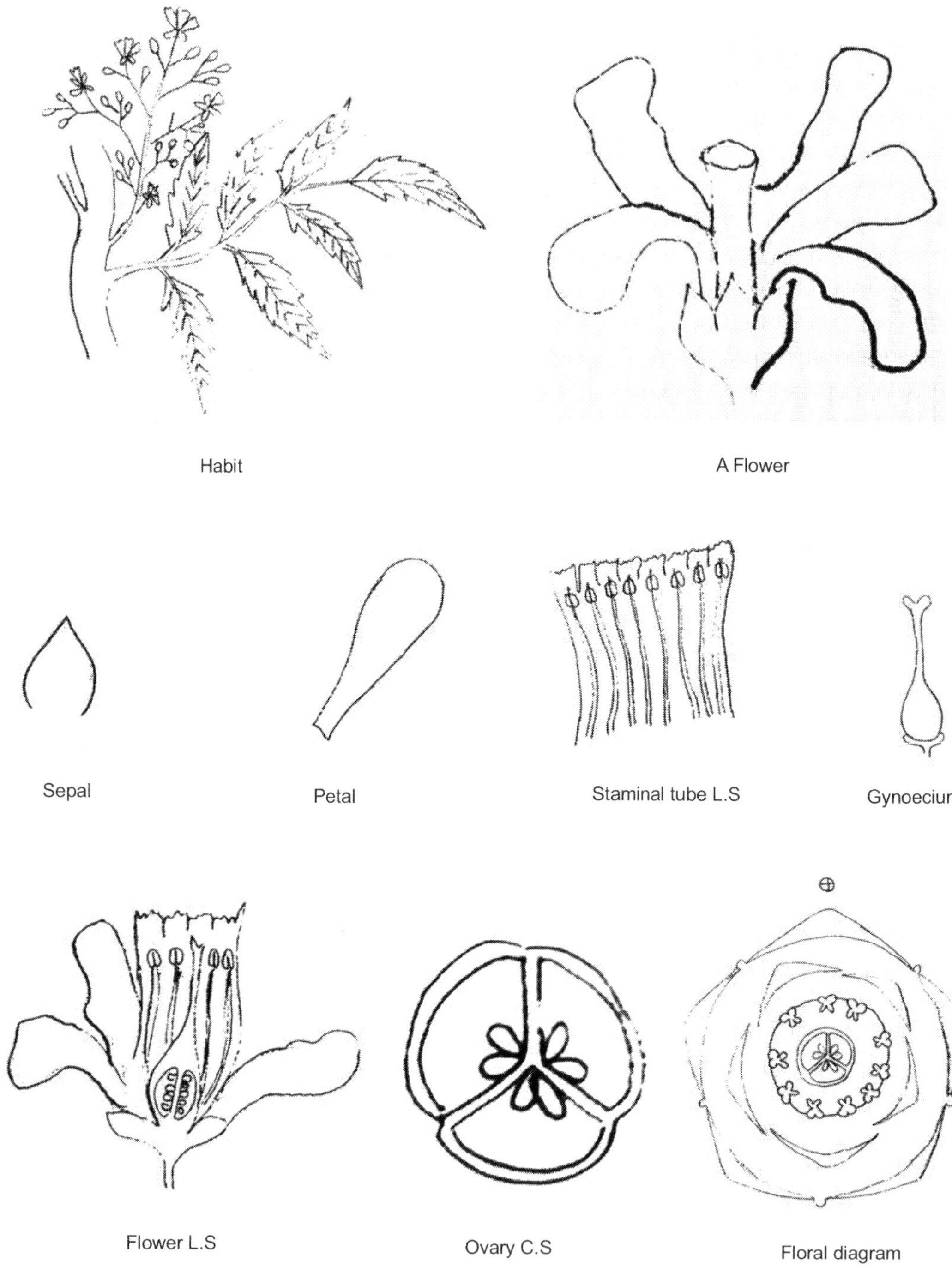

Figure 1.2 Description of family Meliaceae

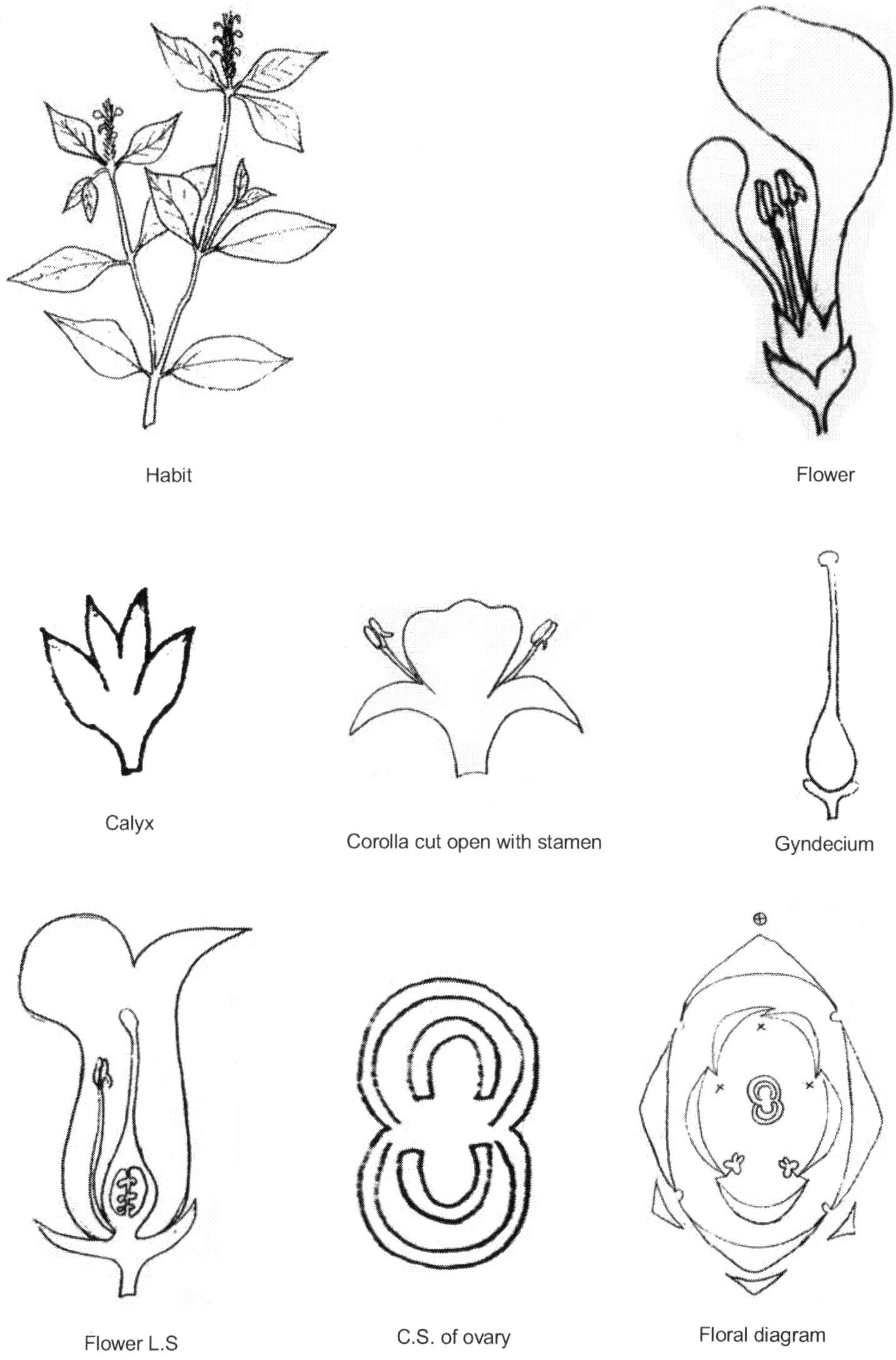

Figure 1.3 Description of family Acanthaceae

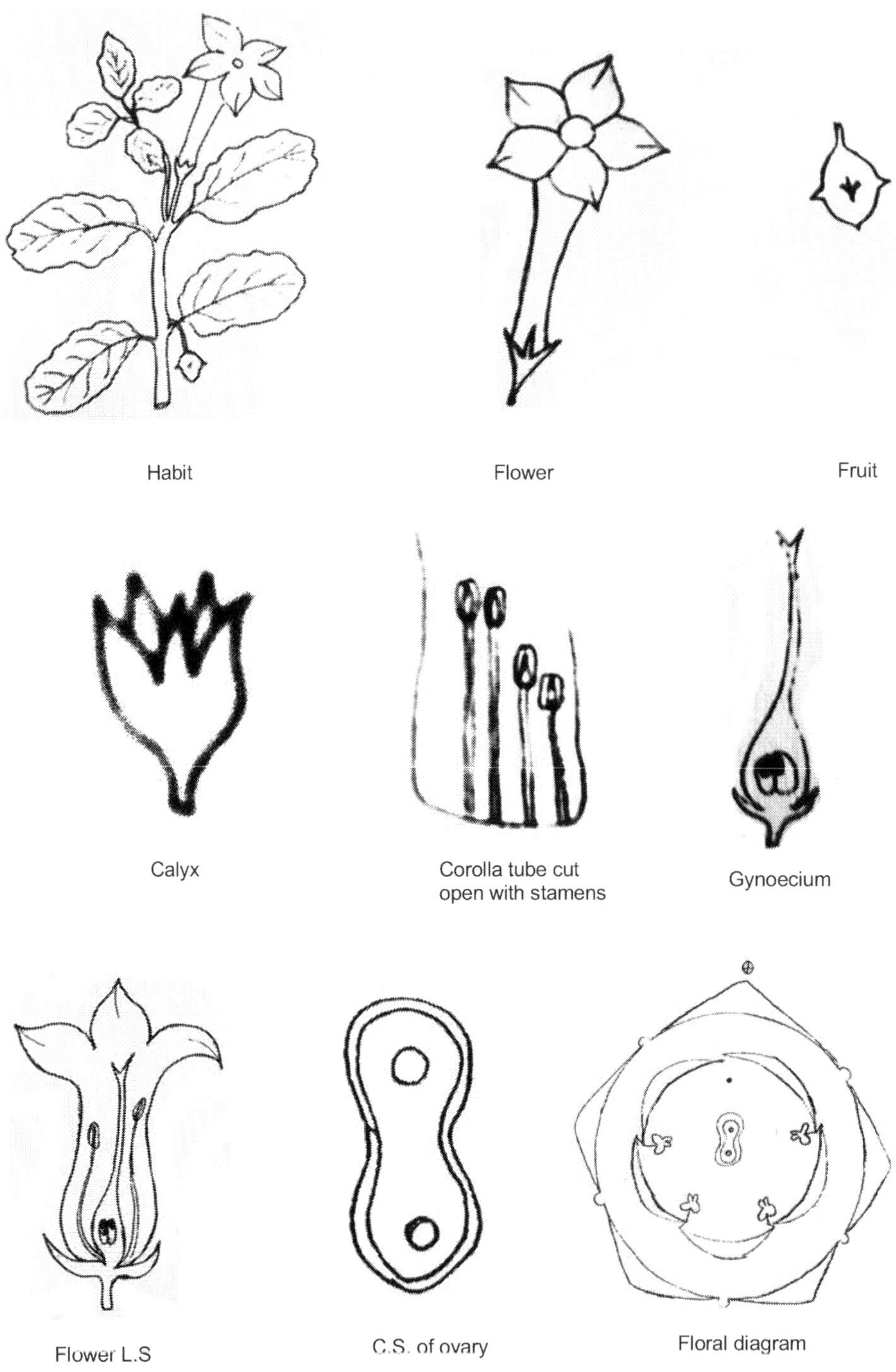

Figure 1.4 Description of family Pedaliaceae

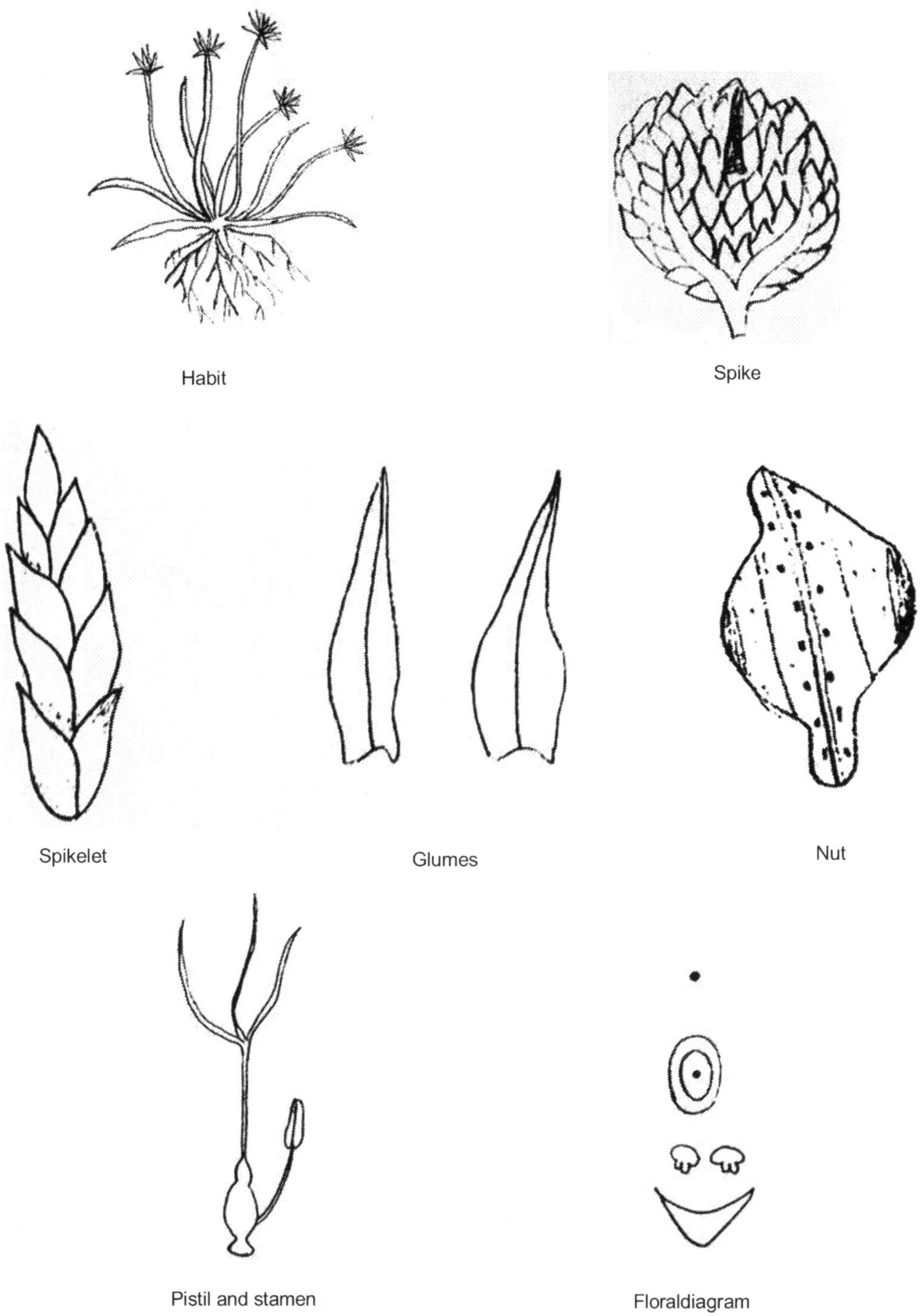

Figure 1.5 Description of family cyperaceae

ANSWERS

1. (c)	2. (d)	3. (d)	4. (a)	5. (b)	6. (d)	7. (d)	8. (c)	9. (c)	10. (c)
11. (c)	12. (c)	13. (c)	14. (d)	15. (a)	16. (d)	17. (b)	18. (c)	19. (b)	20. (d)
21. (b)	22. (c)	23. (a)	24. (d)	25. (d)	26. (b)	27. (d)	28. (b)	29. (b)	30. (b)
31. (b)	32. (c)	33. (b)	34. (c)	35. (d)	36. (c)	37. (c)	38. (b)	39. (b)	40. (b)
41. (a)	42. (d)	43. (c)	44. (d)	45. (a)	46. (d)	47. (d)	48. (c)	49. (a)	50. (a)
51. (b)	52. (b)	53. (b)	54. (d)	55. (a)	56. (c)	57. (b)	58. (a)	59. (b)	60. (a)
61. (b)	62. (d)	63. (c)	64. (d)	65. (c)	66. (a)	67. (d)	68. (c)	69. (c)	70. (b)
71. (c)	72. (a)	73. (b)	74. (c)	75. (a)	76. (d)	77. (b)	78. (b)	79. (d)	80. (c)
81. (b)	82. (c)	83. (b)	84. (a)	85. (a)	86. (c)	87. (d)	88. (c)	89. (a)	90. (a)
91. (b)	92. (b)	93. (d)	94. (b)	95. (c)	96. (d)	97. (b)	98. (b)	99. (a)	100. (d)
101. (a)	102. (a)	103. (d)	104. (a)	105. (c)	106. (b)	107. (a)	108. (c)	109. (a)	110. (d)
111. (b)	112. (d)	113. (b)	114. (b)	115. (c)	116. (a)	117. (d)	118. (a)	119. (d)	120. (c)
121. (b)	122. (a)	123. (b)	124. (b)	125. (c)	126. (d)	127. (b)	128. (d)	129. (a)	130. (a)
131. (c)	132. (a)	133. (b)	134. (c)	135. (d)	136. (d)	137. (b)	138. (d)	139. (c)	140. (d)
141. (b)	142. (c)	143. (d)	144. (d)	145. (c)	146. (d)	147. (b)	148. (b)	149. (d)	150. (b)
151. (c)	152. (b)	153. (d)	154. (d)	155. (a)	156. (b)	157. (c)	158. (d)	159. (a)	160. (a)
161. (c)	162. (b)	163. (d)	164. (d)	165. (c)	166. (a)	167. (a)	168. (c)	169. (c)	170. (c)
171. (c)	172. (d)	173. (d)	174. (c)	175. (b)	176. (c)	177. (c)	178. (c)	179. (d)	180. (b)
181. (b)	182. (a)	183. (d)	184. (a)	185. (c)	186. (c)	187. (c)	188. (b)	189. (b)	190. (c)
191. (c)	192. (a)	193. (b)	194. (c)	195. (b)	196. (b)	197. (d)	198. (b)	199. (c)	

2

ANATOMY

MULTIPLE CHOICE QUESTIONS

1. In a cross section of an old trunk, the outer region of secondary wood was observed lighter in colour. This region of the wood is known as

   (a) spring wood (b) autumn wood (c) sapwood (d) heartwood

2. Biocollateral vascular bundles are common by seen in

   (a) *Helianthus* (b) *Aristolochia* (c) *Cucurbita* (d) *Ricinus*

3. For increase in the girth of plant where should the growth occur?

   (a) Apical meristem (b) Lateral meristem
   (c) Pith (d) Inter-calary meristem

4. The outermost layer of stele in a dicot stem is

   (a) cortex (b) pericycle (c) endodermis (d) hypodermis

5. The function of vessels is

   (a) translocation of water and minerals (b) provide tensile strength
   (c) translocation of food (d) all of these

6. Commercial cork is nothing but a dead tissue with thickened walls formed by the deposition of

   (a) cutin (b) lignin (c) pectin (d) suberin

7. The grittiness in apple and pear is due to

   (a) lignin (b) areolar tissues (c) carotenoids (d) sclerenchyma

8. The cells with excessive length are common in

   (a) sclerenchyma (b) collenchyma (c) epidermis (d) parenchyma

9. A dicot root differs from monocot root by the presence of

   (a) piliferous layer (b) exodermis

(c) radial vascular bundle (d) well developed pith

10. In dicot root the cambium is derived from

(a) epidermis (b) cortex (c) hypodermis (d) pericycle

11. Passage cells are found in

(a) pericycle (b) endodermis (c) epidermis (d) pith

12. Name the special openings seen on a tree trunk through which gaseous exchange is made possible

(a) hydathodes (b) cracks of the bark (c) stomata (d) lenticels

13. Which of the following can be seen in a monocot root?

(a) A large pith (b) Endarch xylem
(c) Medullary ray (d) No or poorly developed pith

14. Multiple epidermis is found in

(a) grass leaf (b) *Nerium* leaf (c) guava leaf (d) grass stem

15. Tyloses are the structures found in xylem vessels which

(a) function is neutral (b) all the time hinders the flow of water
(c) helps in efficient flow of nutrients (d) helps in efficient flow of water

16. Root hairs arise from the

(a) cortex (b) pericycle (c) phloem (d) epiblema

17. The presence of casparian strips is a characteristic feature of

(a) phellem (b) pericycle (c) endodermis (d) endosperm

18. Plerome gives rise to the

(a) stele (b) endodermis (c) hypodermis (d) epidermis

19. Coleoptile is a

(a) covering of root (b) covering of seed
(c) covering of Plumule (d) covering of floral bud

20. In the monocot root we observe

(a) suberized exodermis, casparian strips, passage cells, cambium
(b) suberized exodermis, polyarch xylem, pith
(c) exodermis, endarch, tetrarch closed vascular bundle
(d) conjoint, collateral, open, polyarch vascular bundle

21. The branch of botany which deals with the study of structure and features of different plant organs is known as

(a) Embryology. (b) Anatomy (c) Histology. (d) Morphology.

22. Water secreting glands in plants are called
    (a) Nectaries (b) Digestive glands (c) Hydathodes (d) Epithelium cells
23. One growth ring of plant consists of
    (a) only autumn wood (b) spring wood and early wood
    (c) only spring wood (d) spring wood and autumn wood
24. The formation of distinct annual rings during secondary growth of stem is mainly due to
    (a) formation of unequal quantities of xylem and phloem
    (b) formation of phellogen
    (c) contrasting seasonal variations
    (d) more or less uniform climate throughout the year
25. The commercial cork is obtained from
    (a) Pine (b) Yew (c) Oak (d) Spruce
26. In a young dicot stem, epidermis is covered by a waxy coating called
    (a) suberin (b) periderm (c) cuticle (d) waxylene
27. In the dorsiventral dicot leaf, xylem is found
    (a) below the phloem (b) above the phloem
    (c) around the phloem (d) none of these
28. Grafting is not possible in monocotyledons because they
    (a) lack cambium (b) are herbaceous
    (c) possess scattered vascular bundles (d) have parallel venation
29. Lignin is a component of the secondary cell walls of
    (a) epidermis (b) collenchymas (c) sclerenchyma (d) parenchyma
30. The term used to determine the age of tree by counting the annual rings is called
    (a) Countrochronology (b) Dendrochronology
    (c) ageing and development (d) Chronology
31. Primary growth involves the activity of
    (a) vascular cambium (b) lateral meristem (c) apical meristem (d) phellogen
32. A sieve tube is a living cell, yet it does not possess
    (a) cytoplasm (b) cell wall (c) plasma membrane (d) nucleus
33. Axillary bud and terminal bud are derived from the activity of
    (a) lateral meristem (b) apical meristem
    (c) intercalary meristem (d) parenchyma
34. In old trees, most part of the secondary wood impregnated with extractives is called as
    (a) sapwood (b) soft wood (c) hard wood (d) heartwood

35. In a collateral vascular bundle the phloem is present
    (a) outside the xylem (b) inside the xylem
    (c) on both sides of the xylem (d) all around the xylem
36. Which is called as living mechanical tissue?
    (a) Parenchyma (b) Sclerenchyma (c) Chlorenchyma (d) Collenchyma
37. In angiosperms the leaf gap is found in
    (a) internode (b) node (c) shoot apex (d) petiole
38. The apical meristem of the root is present
    (a) in all the roots (b) only in radicals
    (c) only in tap roots (d) only in adventitious roots
39. Which one of the following is all example of primary meristematic tissue?
    (a) Vascular cambium in dicot roots (b) Phellogen
    (c) Intra-fascicular cambium (d) Inter-fascicular cambium
40. Sieve tubes differ from sieve cells because they
    (a) have sieve pores mainly at the end walls. (b) are dead.
    (c) are shorter. (d) lack nuclei.
41. An example for secondary meristem is
    (a) cork cambium (b) intercalary tissue (c) fascicular cambium (d) phellem.
42. An example for enucleated cell in plants is
    (a) parenchyma cell (b) meristematic cell
    (c) a mature sieve tube cell (d) tracheid
43. Quiescent centre is present in
    (a) root apex (b) shoot apex (c) meristematic tissue (d)all of these
44. A tumour like tissue of thin walled cells developed over the wound is called as
    (a) tyloses (b) callose (c) gall (d) callus
45. Numerous vascular bundles are arranged in a scattered manner in
    (a) monocot stem (b) monocot root (c) monocot leaf (d) dicot stem
46. Pith is well developed in
    (a) monocot root (b) dicot root
    (c) both the above (d) none of these
47. Which are the special openings seen on a tree trunk through which respiration is made possible
    (a) Stomata (b) Cracks of the bark (c) Lenticels (d) None of these
48. Cuticle is secreted by
    (a) hypodermis (b) epidermis (c) endodermis (d) none of these

49. The ground tissue of the leaf is known as
    (a) cataphyll (b) veins (c) chlorophyll (d) mesophyll
50. Tetrarch vascular bundle is characteristic of
    (a) bengal gram root (b) sunflower stem
    (c) maize stem (d) maize root
51. The only living tissue in the xylem is
    (a) vessels (b) wood fibre (c) wood parenchyma (d) tracheids
52. Which is the first layer, formed by dividing plant cells and occurring between the subsequently formed cell walls of daughter cells
    (a) Cellulose layer (b) Middle lamella
    (c) Primary wall (d) Secondary wall
53. Tunica Corpus theory was proposed by
    (a) Hofmeister (b) Nageli (c) Strasburger (d) Schmidt
54. Epidermal tissue generally acts as
    (a) absorbing tissue (b) conducting tissue
    (c) photosynthetic tissue (d) protective tissue
55. Bulliform cells are found in
    (a) dicot stem (b) leaf of monocot
    (c) bulb of onion (d) monocot root
56. Monocots differ from typical dicots in
    (a) absence of secondary growth (b) absence of cambium
    (c) absence of trees (d) being herbaceous
57. What are the parts of periderm?
    (a) Periblem, phellogen, phelloderm (b) Cortex, dermatogens, phelloderm
    (c) Periblem, phellogen, plerome (d) Phellem, phellogen, phelloderm
58. Which is the first wall layer, formed by the dividing plant cells which occurs between the subsequently formed cell walls of daughter cells?
    (a) Primary wall (b) Middle lamella (c) Secondary wall (d) Cellulose layer
59. Which of the following is an non-nucleated live plant cell?
    (a) Xylem (b) Tracheids (c) Vessel (d) Sieve tube
60. Girdling experiments cannot be performed in sugarcane because
    (a) Phloem in it is situated internal to xylem (b) It cannot withstand injury
    (c) Vascular bundles in it are scattered (d) Sugarcane plants are too delicate
61. In which of the following phloem occurs in two patches?
    (a) Maize (b) Sunflower (c) *Cucurbita* (d) *Dracaena*

62. Innumerable vascular bundles, lack of cambium and lack of well demarcated pith are found in

(a) Sugarcane (b) Sunflower (c) Pea (d) Tomato

63. The hypodermis in monocotyledonous stem is

(a) parenchymatous (b) chlorenchymatous
(c) collenchymatous (d) sclerenchymatous

64. Which of the following statement is true?

(a) Vessels are multicellular and with wide lumen
(b) Vessels are unicellular and with narrow lumen
(c) Tracheids are multicellular and with narrow lumen
(d) Tracheids are unicellular and with wide lumen

65. Fibers present in the phloem are

(a) Bast fibers (b) Wood fibers (c) Sclereids (d) Trichomes

66. Radial vascular bundles occur in

(a) leaves (b) monocot stem (c) monocot roots (d) dicot stem

67. The chief function of velamen tissue is to

(a) store food materials (b) absorb moisture
(c) synthesize food (d) conduct food materials

68. The cells of quiescent centre are characterized by

(a) dividing regularly to add to the corpus
(b) dividing regularly to add to tunica
(c) having dense cytoplasm and prominent nuclei
(d) having light cytoplasm and small nuclei

69. A water proofing substance secreted by the cork and endodermis cells is

(a) cutin (b) tannin (c) lignin (d) suberin

70. Periderm is composed of

(a) cork-cambium and bark
(b) cork-cambium and secondary cortex
(c) cork-cambium, secondary cortex and cork
(d) cork, bark and secondary cortex

71. Submerged leaves possess

(a) stomata on the upper surface (b) stomata on the both the surfaces
(c) stomata on the lower surface (d) no stomata on any surface

72. The conductive tissue of roots usually forms a solid mass in the centre instead of being dispersed in bundles around the periphery of the pith and this contributes to

(a) flexibility without loss of tensile strength (b) inflexibility

(c) positive geotropism (d) negative geotropism

73. Diffuse porous woods are characteristic of plants growing in
    (a) Temperate climate (b) Alpine region
    (c) Tropics (d) Cold winter regions
74. In dicot roots
    (a) vascular bundles are scattered and have cambium
    (b) vascular bundles are arranged in a ring and have cambium
    (c) xylem and phloem are radially arranged
    (d) xylem is always endarch
75. Monocot roots differs from dicot roots in having
    (a) open vascular bundles (b) scattered vascular bundles
    (c) well developed pith (d) radially arranged vascular bundles
76. Well developed pith is found in
    (a) monocot root and monocot stem (b) monocot stem and dicot root
    (c) monocot root and dicot stem (d) dicot root and dicot stem
77. Mesophyll is well differentiated into palisade and spongy tissue in
    (a) dicot leaves (b) monocot leaves
    (c) xerophytic leaves (d) hydrophytic leaves
78. Roots show one of the following characters
    (a) exogenous origin (b) presence of leaves and buds
    (c) divisible into nodes and internodes (d) endogenous origin
79. In a dorsiventral leaf location of palisade tissue and phloem respectively are on the
    (a) adaxial and abaxial (b) adaxial and adaxial
    (c) abaxial and adaxial (d) abaxial and abaxial
80. The vessels of heart wood are blocked by
    (a) stomata (b) tyloses (c) latex (d) hydathodes
81. The youngest layer of secondary xylem in a woody stem is located just
    (a) outside the pith (b) outside the cambium
    (c) inside the cambium (d) inside the epidermis
82. The lateral branches in a root arises out of the
    (a) Epidermis (b) Cortex (c) Phloem (d) Pericycle
83. Parenchyma cells are characterized by the
    (a) Presence of thickenings at the corner (b) Presence of intercellular spaces
    (c) Presence of uniform thickening (d) Presence of lignified walls

84. Numerous vascular bundles are arranged in
   (a) monocot root (b) monocot leaf (c) monocot stem (d) dicot root
85. The arrangement of xylem in a stem is said to be
   (a) exarch (b) collateral (c) bicollateral (d) endarch
86. Growth rings are formed due to the activity of
   (a) extrastelar cambium (b) intrastelar cambium
   (c) primary cambium (d) intercalary cambium
87. With increasing secondary wood, which will increase in diameter?
   (a) Sap wood (b) Spring wood (c) Autumn wood (d) Heart wood
88. Secondary growth is anomalous in
   (a) *Pinus* (b) Sunflower (c) *Dracaena* (d) *Annona*
89. Primary xylem is obtained from
   (a) Vascular cambium (b) Procambium
   (c) Cork Cambium (d) Lateral cambium
90. When phloem and cambium are present on both sides of xylem, the vascular bundle is called
   (a) Bicollateral (b) Radial (c) Concentric (d) Collateral
91. In dicot stem the vascular bundles are
   (a) scattered (b) arranged in a ring
   (c) xylem and phloem are situated alternately (d) amphicribal
92. Hydathodes are
   (a) pores through which transpiration takes place
   (b) special openings in the leaf through which liquid water is forced out
   (c) pores in the bark
   (d) stomata in water plants
93. Trapa is a
   (a) Halophyte (b) Mesophyte (c) Xerophytes (d) Hydrophyte
94. Angular collenchyma cells are found in
   (a) *Althaea* sp. (b) *Cucurbita* sp. (c) *Lactuca* sp. (d) *Salvia* sp.
95. Nature of vascular bundle in dicot stem generally is
   (a) Radial (b) Concentric (c) Collateral (d) Ecentric
96. A large water cavity closer to protoxylem elements in the vascular bundles of maize is
   (a) resin canal (b) mucilage duct
   (c) lysogenous canal (d) schizogenous cavity

97. Conjoint, collateral and open vascular bundles are found in
    (a) dicot stem (b) monocot stem (c) dicot root (d) monocot root
98. In leaves the protoxylem elements
    (a) face towards the adaxial surface (b) face towards the abaxial surface
    (c) are surrounded by metaxylem (d) are scattered in the middle
99. Sclerenchyma help in
    (a) photosynthesis (b) support (c) conduction (d) absorption
100. In a dicot stem
    (a) the vascular bundles are usually arranged in a ring and have cambium
    (b) xylem and phloem occur in separate bundles
    (c) xylem is always exarch
    (d) the vascular bundles are scattered and lack cambium
101. Knots are formed in wood because
    (a) branches get buried in the main stem (b) branches fall off and leave scars
    (c) of insect bites (d) of injuries caused by animals

SMALL ANSWER QUESTIONS

1. Name three types of meristems based on position in the plant body.

   Ans. Apical meristem, intercalary meristem and lateral meristem.

2. Who proposed 'Histogen Theory'?

   Ans. Hanstein in 1868.

3. What is the most distinguishing feature of a simple pit?

   Ans. Primary wall is not covered by secondary layers.

4. What is a blind pit?

   Ans. A pit without any complementary pit on the opposite cell wall.

5. What is the most distinguishing feature of a bordered pit?

   Ans. Over arching of the secondary wall over the pit cavity.

6. What is calyptrogen?

   Ans. A meristem in the root apex which gives rise to the root cap.

7. Who introduced the term 'protoplast'?

   Ans. J. E. Purkinje.

8. Name the three histogen layers as proposed in the Histogen Theory.

   Ans. Dermatogens, periblem and plerome.

9. Who proposed 'Anneau Initial' and 'Meristem d' Attente' theory?

   Ans. Buvat.

10. What is closed type of vascular bundle?

    Ans. A vascular bundle without cambium.

11. What is prosenchyma?

    Ans. When parenchyma cells become elongated, pointed and slightly thick walled to provide mechanical support, they are called as prosenchyma.

12. Define bifacial leaf.

    Ans. A leaf with palisade parenchyma on one side of the blade and spongy parenchyma on the other side is called bifacial leaf.

13. What are casparian strips?

    Ans. A band like thickening made up of lignin and suberin, characteristic of endodermal cells.

14. What are osteosclereids?

    Ans. They are bone or barrel shaped sclereids with dilated ends.

15. Who proposed 'Korper-kappe Theory' to explain root apical organization?

    Ans. Schuepp in 1917.

16. What are the types of simple tissues?

    Ans. Parenchyma, collenchyma and sclerenchyma.

17. What is the main function of prosenchyma?

    Ans. Mechanical support.

18. What is collateral type of vascular bundle?

    Ans. A vascular bundle with phloem on the abaxial side of xylem.

19. What are trichosclereids?

    Ans. They are hair like, sometimes branched, and are found in the intercellular spaces in the leaves and stems of certain hydrophytes.

20. Write three types of collenchymas based on the distribution of wall thickenings.

    Ans. Lamellar, lacunar and angular.

21. Which simple tissue is characterized by the absence of living protoplast?

    Ans. Sclerenchyma.

22. Give the example of macrosclereids.

    Ans. Leguminous seeds.

23. What is bast?

    Ans. When sclerenchyma fibres occur in the region of phloem they are called bast.

24. What are complex tissues?

    Ans. Tissues which are made of more than one type of cells and they perform a common function.

25. What is concentric type of vascular bundle?

    Ans. A vascular bundle in which either xylem completely surrounds the phloem or phloem surrounds the xylem.

26. Define xylotomy.

    Ans. The anatomy of xylem is called xylotomy.

27. What is periderm?

    Ans. The phellogen and the tissue produced by it are together known as periderm.

28. Name the types of permanent tissue.

    Ans. Simple and complex tissue.

29. Name a plant which contains aerenchyma.

    Ans. *Hydrilla* stem.

30. Name a plant which contains storage parenchyma.

    Ans. Endosperm of *Coffea arabica.*

31. What are macrosclereids?

    Ans. Rod shaped elongated sclerieds found in the seed coats of pulses are known as macrosclereids.

32. What is protophloem?

    Ans. The part of the primary phloem, which is formed in ontogenetic stages, is termed as protophloem.

33. What is metaphloem?

    Ans. The part of the primary phloem, which is differentiated subsequently, is termed as metaphloem.

34. Name the sieve elements in phloem.

    Ans. Sieve cells and sieve tubes.

35. What is apical meristem?

    Ans. Growth of the shoot occurs at the tip where it constitute a group of actively dividing cells, known as apical meristem.

36. Name two important types of complex tissues.

    Ans. Xylem and phloem.

37. Who gave the Sandwich model of plasma membrane?

    Ans. Danielli and Davson.

38. What are the important tracheary elements present in angiospermic plant body?

    Ans. Tracheids and vessels.

39. What is endarch xylem?

    Ans. Xylem strand in which the cells mature centrifugally.

40. Who gave the fluid mosaic model of cell membrane?

    Ans. Singer and Nicholson.

41. What are open bundles?

    Ans. Vascular bundles with intra-fascicular cambium.

42. What are extrinsic proteins?

    Ans. Proteins which are associated with outer surface of plasma membrane.

43. What are sclereids?

    Ans. These are short or irregular sclerenchyma cells.

44. What is venation?

    Ans. The arrangement of veins in the leaf blade.

45. What are astrosclereids?

    Ans. They are star shaped cells found in the leaves of *Nymphaea*.

46. What are intrinsic proteins?

    Ans. These are associated with plasma membrane and make an internal part of the membrane.

47. What are brachysclereids?

    Ans. They are short, roughly isodiametric sclereids, occurring in the cortex, pith and phloem of stem and pulp of fruits.

48. Define anomalous secondary growth.

    Ans. It is unusual or unfamiliar type of secondary growth.

49. What is a tissue? / Define tissue.

    Ans. When several different kinds of cells are organized into groups they are called as tissue and the cells of a tissue usually have same origin, structure and function.

50. What is an exarch xylem?

    Ans. Xylem strands in which the maturation of the cell progresses centripetally that are the first formed elements are farthest from the centre of the axis.

51. What is the most common example of brachysclereids?

    Ans. *Pyrus*.

52. What is a bicollateral vascular bundle?

    Ans. A vascular bundle with two groups of phloem on either sides of the xylem.

53. What is heart wood?

    Ans. Central region of wood, generally darker in colour, which has closed the functions of storage and conduction.

54. Who proposed the apical cell theory of shoot apical meristem?

    Ans. Nageli in 1878.

55. Write the names of two angiospermic plants which contain cortical bundles.

    Ans. *Nyctanthes* and *Casuarina*

56. What is *velamen*?

    Ans. A mutliseriate spongy tissue present on the aerial roots of some tropical epiphytes such as orchids.

57. What is histogenesis?

    Ans. It is a process of differentiation of tissues.

58. What are simple pits?

    Ans. These are the areas with only primary walls without any secondary thickening.

59. What are annual rings?

    Ans. Growth rings seen in transaction of the stem depicting the growth in a year.

60. What is torus?

    Ans. The primary wall present between a pit pair develops a thickening in its central part is known as torus.

61. What are fibers?

    Ans. These are long, narrow, thick and lignified slcerenchyma cells, with pointed or blunt ends.

62. Write one angiospermic plant name which possesses two rings of medullary bundles.

    Ans. *Boerhaavia.*

63. What are lithocyst?

    Ans. A cell containing cystolith is called lithocyst.

64. What is apical cell theory?

    Ans. A single apical cell is the structural and functional unit of the apical meristem and it governs the whole process of growth.

65. What are extra-xylary fibres?

    Ans. These fibers occur in various tissues outside the xylem such as cortex, pericycle or in the phloem.

66. What is the source of commercial cork?

    Ans. *Quercus suber*

67. What are stone cells?

    Ans. Sclereids are also called stone cells.

68. In which monocot plant, secondary growth can be seen.

    Ans. *Dracaena* sp.

69. Who proposed Histogen theory of shoot apical meristem?

    Ans. Hanstein in 1868.

70. Name three angiosperms which lacks vessels.

    Ans. *Drimys*, *Pseudowintera* and *Trochodendron.*

71. In which angiospermic plant, interxylary phloem and two large medullary bundles are present.

    Ans. *Achyranthus* sp.

72. Name the kinds of tracheary elements of xylem.

    Ans. Tracheids and vessels.

73. Name an angiospermic plant which possesses inter and intra-xylary phloem in stem.

    Ans. *Leptadenia* sp.

74. What is Histogen theory of shoot apical meristem?

    Ans. Three distinct meristematic zones are present; these layers are termed as histogen.

75. Name the dicotyledonous plants where scattered vascular bundles are present.

    Ans. *Thalictrum*, *Piper*, *Peperomia* etc.

76. Define open type of vascular bundle?

    Ans. A bundle which retains a portion of cambium.

77. What are tracheids?

    Ans. They originat from single cells and perforated cells having only pit pair on their common walls.

78. What are transfer cells?

    Ans. The phloem cells which are responsible for short distance transport of metabolites are called transfer cells.

79. Name the monocot plants which have vascular bundles arranged in rings.

    Ans. *Triticum*, *Oryza* etc.

80. What are vessels?

    Ans. These are joined into long continuous tubes of varying lengths, originate from a longitudinal file of cells. They have perforations on their end walls.

81. Name some secretory products of plants?

    Ans. Rubber, gums, oils, resins, mucilage etc.

82. What is a passage cell?

    Ans. Thin walled cells of the exodermis and endodermis which permit passage of solutions.

83. What are cystoliths?

    Ans. Calcium carbonate crystals that occur in the *Ficus* leaves hanging from the epidermal cells are called cystoliths.

84. Name the plant where glandular tricomes are present.

    Ans. *Pinguicula.*

85. Name the three distinct layer of histogen theory.

    Ans. Dermatogens, periblem and plerome.

86. What are laticifers?

    Ans. These are specialized parenchyma cells which secrete a viscous fluid, known as latex.

87. What is internal or intraxylary phloem?

    Ans. The primary phloem strands are found in pith, inner to the xylem.

88. What are Alluring glands?

    Ans. These are present on the under surface of the lid of pitcher, secrete nectar.

89. Name the families where internal phloem is present.

    Ans. Apocynaceae, Asclepiadaceae, Myrtaceae, Solanaceae etc.

90. Name the main components of latex.

    Ans. It is a suspension which contains various substances like waxes, resins, proteins, essential oils and sugars.

91. What is anatomy?

    Ans. The branch of morphology that deals with the internal structure of organs.

92. What are fibre tracheids?

    Ans. They are xylary fibre with comparatively thin walls and usually with bordered pits.

93. Chicle is obtained from which plant?

    Ans. *Achras*.

94. What is the role of companion cells?

    Ans. They maintain pressure gradient in sieve tubes.

95. What are phelloid cells?

    Ans. A cork cell without suberin deposition in the walls.

96. Papain is obtained from which plant?

    Ans. *Carica papaya*.

97. What is secondary growth?

    Ans. Increase in thickness of a stem or root due to the activity of vascular cambium and cork cambium.

98. What is tylosis?

    Ans. An outgrowth from a ray or axil parenchyma cell into the lumen of a vessel which partially or completely blocks the cavity.

99. What is the source of commercial cork?

    Ans. *Quercus suber*.

100. Name the enzymes secreted by the inner glands of Pitcher wall.

    Ans. Phosphatase and esterase.

101. Who proposed Korper-Kappe theory of root apical meristem?

    Ans. Schuepp in 1917.

102. Name the plant which produce brazilin.

    Ans. *Caesalpinia sappan*.

103. What is included or interxylary phloem?

Ans. When the phloem patches or strands are included in the secondary xylem they are called so.

104. Who has given concept of quiescent centre?

Ans. Clowes.

105. What are plasmodesmata?

Ans. Fine cytoplasmic strands interconnecting adjacent cells in plant tissues.

106. Name a monocot which contains laticifers.

Ans. *Dracaena*.

107. Name the plants where interxylary phloem is present.

Ans. *Leptadenia*, *Salvadora* etc.

108. What are amphicribral vascular bundles?

Ans. Concentric vascular bundle in which xylem is surrounded by phloem.

109. What is bark?

Ans. The tissue present outside the cambium.

110. Name the plant which produces santalin.

Ans. *Pterocarpus santalinus*.

111. What is sap wood?

Ans. Outer portion of the wood that contains living cells and reserve materials and in which conduction of water takes place.

112. What are the products of phellogen?

Ans. Phellum or cork and Phelloderm or secondary cortex.

113. How many types of hydathodes are present in plants?

Ans. Unicellular, trichome and hydathodes with water-conducting elements.

114. What types of thickenings are present on protoxylem?

Ans. Annular and spiral.

115. Name the families where, included or interxylary phloem is present.

Ans. Amaranthaceae, Chenopodiaceae and Nyctaginaceae.

116. What types of thickenings are present on metaxylem?

Ans. Scalariform and reticulate.

117. What is protoxylem?

Ans. The first formed xylem elements.

118. What are raphides?

Ans. Needle shaped crystals of calcium oxalate occurring in bundles of certain plant cells.

119. Name the families which have medulary and cortical vascular bundles.

Ans. Amaranthaceae, Chenopodiaceae, Nyctaginaceae etc.

120. What are the major components of nectar?

Ans. Sucrose, glucose and fructose.

121. Name the monocot which has secondary growth in root.

Ans. *Dracaena*.

122. Name the genera which contain medullary vascular bundles.

Ans. *Achyranthes*, *Boerhaavia*.

123. What are libriform fibres?

Ans. Xylery fibres with very thick secondary walls and simple pits.

124. What is rhytidome?

Ans. The outer bark, consisting of periderm and tissues isolated by it.

125. Who introduced the term plasmodesmata

Ans. E. Strasburger

TRUE OR FALSE QUESTIONS

1. Intra-fascicular cambium is primary in origin.

Ans. True

2. In the vascular bundle of leaf, xylem is located towards the adaxial side of the leaf.

Ans. True

3. Shoot apex is sub-terminal in position because it is covered by leaf primordial.

Ans. False

4. Callose is found in sieve elements in association with sieve plate and lateral sieve areas.

Ans. True

5. The presence of palisade tissue is a characteristic feature of dicot leaf.

Ans. True

6. Meristem occurs only at the apices of roots and shoots.

   Ans. False

7. The xylem is exarch in roots.

   Ans. True

8. Hardwood means wood with both tracheids and vessels in it.

   Ans. True

9. The age of tree growing in temperate regions can be determined by counting the primary xylem.

   Ans. True

10. Radially elongated epidermal cells present at margins, is a xerophytic character.

    Ans. True

11. A non-articulated laticifer develops from a single cell.

    Ans. True

12. Hydathodes are pores through which transpiration takes place.

    Ans. False

13. Bulliform cells are present in the epidermis of monocot leaf.

    Ans. True

14. The presence of sunken stomata is a hydrophytic character.

    Ans. False

15. Resin ducts are the best example of schizogenous ducts.

    Ans. True

16. Branches in root arise from cortex.

    Ans. False

17. In the land plants, guard cells differ from other epidermal cells in the possession of chloroplast.

    Ans. True

18. The innermost layer of cortex is endodermis.

    Ans. True

19. Well developed cuticle is a xerophytic character.

    Ans. True

20. Secondary growth involves in the activity of lateral meristem.

    Ans. True

21. Polystelic condition is present in Pteridophytes.

    Ans. True

22. Resin ducts are present in conifers.

    Ans. True

23. The presence of casparian strips is a characteristic feature of pericycle.

    Ans. False

24. Dorsiventral leaves are generally present in dicot plants.

    Ans. True

25. Isobilateral leaves are present in monocot plants.

    Ans. True

26. A potato tuber contains ecentric starch grains.

    Ans. True

27. Cork cambium mostly arises from cortical cells.

    Ans. True

28. Primary phloem originates from procambium.

    Ans. True

29. Laticiferous tissue helps in conduction of food material.

    Ans. False

30. A leaf is described as isobilateral when the two surfaces cannot be distinguished anatomically.

    Ans. True

31. Only mesophyll tissue is present in dicot leaf.

    Ans. False

32. The youngest layer of secondary xylem in a woody stem is located just outside of the pith.

    Ans. False

33. Porous wood is also known as soft wood.

    Ans. False

34. In *Betula* (Bhojpatra) alternating layers of suberized and non-suberized cells are present.

    Ans. True

35. Phloem parenchyma functions as storage tissue.

    Ans. True

36. Meristematic tissues divide indefinitely.

    Ans. True

37. The lateral branches in a root arise out of the cortex.

    Ans. False

38. Articulated laticifers are made up of a number of cells.

    Ans. True

39. Non-porous wood is also known as hard wood.

    Ans. False

40. Lenticels possess complementary cells.

    Ans. True

41. Movement of water and minerals into the vascular column of a root is controlled by the endodermis.

    Ans. True

42. Softwood means wood with only tracheids in it.

    Ans. True

43. A collateral vascular bundle is one in which either phloem strand or xylem strand present.

    Ans. False

44. Medullary vascular bundles are present in *Amarnthus*.

    Ans. True

45. The secondary phloem originates from vascular cambium.

    Ans. True

46. Mesophyll is differentiated into palisade and spongy layers in isobilateral leaf.

    Ans. False

47. Permanent tissues lose their power of division either permanently or temporarily.

    Ans. True

48. Torus is very well developed in gymnosperms.

    Ans. True

49. Extensive air spaces are found in various plant parts of hydrophytes.

    Ans. True

50. Anomalous secondary growth is present in *Achyranthes* stem.

    Ans. True

51. The important constituent of cork cell wall is lignin.

    Ans. False

52. Epiphytes are characterized by the presence of a specialized tissue in their roots called as velamen.

    Ans. True

53. Collenchyma cells are dead cells.

    Ans. False

54. Xylem is a water conducting complex tissue.

    Ans. True

55. Heartwood means wood in the center of the tree with no longer conducting property.

    Ans. True

56. The secondary xylem is not differentiated into protoxylem and metaxylem.

    Ans. True

57. Sieve tubes are found in xylem.

    Ans. False

58. Phloem is a principal food conducting tissue.

    Ans. True

59. The meristems which have been delayed in its differentiation are called as secondary meristem.

    Ans. True

60. Annual rings are present in primary xylem strands.

    Ans. False

61. Foliar roots are present in *Bryophyllum*.

    Ans. True

62. Concept of quiescent centre to explain the root apical organization was proposed by Schuepp.

    Ans. False

63. Sclerenchyma tissue is composed of dead cells.

    Ans. True

64. A leaf in which stomata are restricted to the lower surface only is known as hypostomatic.

    Ans. True

65. Plerome gives rise to stele.

    Ans. True

66. Vessels containing tyloses only in secondary xylem.

    Ans. True

67. Heart wood is made up of living cells.

    Ans. False

68. Sieve cells are present generally in pteridophytes and gymnosperms.

    Ans. True

69. Ray parenchyma is only present in secondary xylem.

    Ans. True

70. With increasing secondary wood, sap wood will increase in diameter.

    Ans. False

71. Xylary fibers are also known as wood fibres.

    Ans. True

72. The protoxylem elements are less prominent.

    Ans. True

73. Short and irregular sclerenchyma cells are known as stone cells.

    Ans. True

74. Sap wood is made up of dead cells.

    Ans. False

75. The shell of walnut is made up of stone cells.

    Ans. True

76. Radial types of vascular bundles are present in the roots.

    Ans. True

77. Starch sheath is present in dicot stem at the place of endodermis.

    Ans. True

78. The phellogen originates from the outer layer of pericycle.

    Ans. True

79. Root hairs are formed from extensions of the epidermal cells.

    Ans. True

80. Hypodermis is collenchymatous in dicot stem.

    Ans. True

81. Pith cavity is present in *Triticum*.

    Ans. True

82. Laticifers are specialized parenchyma cells which secrete a viscous fluid, known as resin.

    Ans. False

83. Sieve tubes are present in gymnosperms.

    Ans. False

84. A thin walled parenchymatous tissue with large intercellular space is known as aerenchyma.

    Ans. True

85. Collenchyma and parenchyma belong to simple tissue.

    Ans. True

86. Hydathodes are water secreting glands in plants.

    Ans. True

87. A colourless plastid meant for storing starch is called amyloplast.

    Ans. True

88. In strict technical terms, periderm is a part of the bark.

    Ans. True

89. Sieve elements are living cells but without nucleus.

    Ans. True

90. Inversely oriented cortical bundles are seen in the stems of *Nyctanthus*.

    Ans. True

91. Aerial roots are present in *Fucus* and *Oxalis*.

    Ans. True

92. Hypodermis is sclerenchymatous in monocot stem.

    Ans. True

93. In plants like *Leptadenia* and *Salvadora*, interxylary phloem is of common occurrence.

    Ans. True

94. Lateral roots are formed from the pericycle.

    Ans. True

95. Tylosoids are the proliferations from contiguous parenchyma cells.

    Ans. True

96. Concentric vascular bundles in which phloem is surrounded by xylem are called amphivasal vascular bundles.

    Ans. True

97. In pteridophytes and gymnosperm albuminous cells are present at the place of companion cells.

    Ans. True

98. Hematoxylin is obtained from *Hematoxylon campechianum*.

    Ans. True

99. Wood means xylem tissue.

    Ans. True

100. Sapwood means wood at the periphery of the stem and active in conduction.

     Ans. True

101. In the stem of *Dianthera americana*, polystelic condition is seen.

     Ans. True

DESCRIPTIVE TYPE QUESTIONS

1. What are the types of underground modifications of stem?

   The following modifications are present in the underground stems:

   - Rhizome: Sub-terranean, grows horizontally. E.g. Ginger, turmeric etc.
   - Tuber: Swollen stem tip, axillary buds present. E.g. Potato.
   - Corm: Sub-terranean, grows vertically. E.g. Gladiolus, saffron.
   - Bulb: Stem condensed, leaves scaly and fleshy. E.g. Onion, garlic etc.

2. What are the sub-aerial modifications of stem?

   The following modifications are present in the sub-aereial stems:

   - Runner: Prostrate, creeping, rooting at nodes. E.g. Doobgrass, Oxalis etc.
   - Sucker: Arises from basal underground stem, grows obliquely upwards. E.g. Mentha, *Chrysanthemum* etc.
   - Stolon: Sub-terranean, roots at nodes and gives off aerial branches. E.g. *Colocasia.*
   - Offset: Internodes short and thick, group of leaves at nodes. E.g. Water hyacinth.

3. What are the aerial modifications of stem?

   The following modifications are present in the aerial stems:

   Stem tendril: Thin, wiry, leafless, spirally coiled structures. E.g. Passiflora, grape vine etc.

   - Bulbils: Modified vegetative or floral buds. E.g. *Dioscorea.*
   - Thorn: Hard, straight, pointed axillary branch. E.g. *Duranta*, *Citrus* etc.
   - Phyllocladode: Stem modified into leaf-like structures. E.g. *Opuntia.*
   - Cladode: Phylloclade of one or two internodes. E.g. *Asparagus*, *Ruscus* etc.

4. Name the types of roots present in plant community?

   - Tap roots: Roots directly growing from radicle.
   - Adventitious roots: Roots growing from any place other than radicle.

5. What are the types of tap roots?

   Tap roots can be following types:

   - Pneumatophores: These are vertically growing, negatively geotropic, with breathing pores. E.g. *Rhizophora* , *Avicenia* etc.
   - Fusiform roots: Swollen in the middle and tappers towards both ends. E.g. Radish.
   - Napiform roots: Swollen at base, suddenly tapering towards apex. E.g. Turnip, beet root etc.

- Conical roots: Broad at base, gradually tapering towards apex. E.g. Carrot.
- Tuberous root: Without definite shape. E.g. Four O'clock plant.

6. Write the types of adventitious roots.

Following types of adventitious roots are generally present in plants:

- Fibrous roots: Cluster of roots arising from the base of stem. E.g. Wheat, rice.
- Foliar roots: Roots arise from leaves. E.g. *Begonia*, *Bryophyllum* etc.
- True adventitious roots: Arising from nodes, internodes etc. E.g. *Ficus*, money plant etc.
- Tuberous roots: Without definite shape. E.g. Sweet potato.
- Fasciculated roots: Tuberous roots in clusters. E.g. Dahlia, *Asparagus* etc.
- Moniliform or beaded roots: Swollen at regular intervals. E.g. *Portulaca*, *Momordica* etc.
- Nodulus roots: Swollen at tips. E.g. Arrowroot, mango, ginger etc.
- Annulated roots: Rings at regular intervals. E.g. Ipecac.
- Prop roots: Grow vertically downwards from branches E.g. Banyan, rubber plant etc.
- Stilt roots: Grow from basal nodes, E.g., maize, sugarcane etc.
- Climbing roots: Attach by roots from nodes and internodes. E.g. Betel.
- Floating roots: Spongy roots arising from nodes, on upper side of stem. E.g. *Jussiaea.*
- Epiphytic roots: Cling to the support, E.g. Vanda, orchids etc.
- Assimilatory roots: Green and photosynthetic. E.g. *Trapa*, *Tinospora* etc.
- Parasitic roots: Enter the host; suck food, water and mineral from vascular tissue of the host. E.g. *Cuscuta.*

7. What are the different types of meristems?

The classification of different types of meristems are as follows:

A. On the basis of plane of divisions, meristem can be classified into:

- Mass meristem: Present in young embryos, endosperm etc.
- Plate meristem: Present in flat angiospermic leaf.
- Rib meristem: Epidermis, cortex of root and stem are made up of this meristem.

B. On the basis of development, meristem can be classified into:

- Primary meristem: It is present on the apices of stem and root, leaf primordia etc.
- Secondary meristem: Parts of vascular cambium and cork cambium made up of this type of meristem.

C. On the basis of the position of meristem, it can be classified into:

- Apical meristem: Present on shoot apex and root apex.

- Lateral meristem: Vascular cambium, cork cambium etc are made up of this meristem.
- Intercalary meristem: Bases of internodes, leaves of grasses, peduncle, gynophores of ground nut **(Figure 2.1)**.

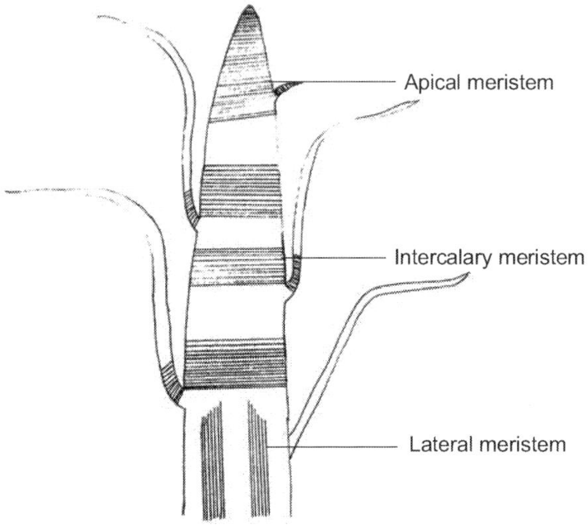

Figure 2.1 Types of meristem on the basis of their position

D. On the basis of the functions of meristem, it can be classified into:

- Protoderm: Formation of epidermis.
- Procambium: Formation of primary vascular tissues.
- Ground meristem: Formation of cortex, meristem, pith etc.

8. Explain following types of inflorescences with examples.
   - Catkin: It is also known as amentum type of inflorescence. It is spike consisting of unisexual flowers. The inflorescence axis is long and pendulous. E.g. *Morus*, *Salix* etc.
   - Spadix: It is spike with thick and fleshy inflorescence axis surrounded by one or several brightly coloured bracts, called spathe. E.g. Banana, *Colocasia*.
   - Capitate: In this type of inflorescence a large number of sessile flowers arise from a suppressed axis forming a globose structure. E.g. *Acacia*, *Mimosa*.
   - Hypanthodium: The main axis forms a fleshy cup-shaped receptacle. It is almost closed and has an apical opening (ostiole) guarded by scales. Flowers are born on the inner wall of the cup on cymose groups. They are unisexual, male flowers are near the ostiole and fertile female flowers at the base of the cup. Besides these two types of flowers, few sterile female flowers are also present. E.g. *Ficus*.
9. What are the main functions of xylem and phloem elements?

   Functions of xylem elements:

   - Tracheary elements: a. Tracheids b. Vessels: Conduction of water.

- Fibres: a. Fibre tracheids b. Libriform fibres: Provide mechanical support.
- Parenchyma cells: Storage and translocation of ergastic substances.

Functions of phloem elements:

- Sieve elements: a. Sieve cells b. Sieve tubes: Conduction of organic food.
- Companion cells: Control sieve elements.
- Sclerenchyma cells: a. Fibres b. Sclereids: Provide mechanical support.
- Parenchyma cells: Storage and translocation of food material

10. What are the general characters of tracheids and vessels?

General characters of tracheids:

- The cells are long but narrow.
- Each tracheid originates from a single cell.
- The transverse walls are septate and are without perforations.
- The contact between the neighbouring tracheids is maintained through the pits on the lateral walls.

General characters of vessels:

- The cells are short but broad.
- Vessel constitute a series of vessel members originating from many cells.
- The transverse walls are dissolved and hence show distinct perforation plates.
- The contact between the neighbouring vessels is maintained through perforated transverse walls, and also through the pits on the lateral walls.

11. Write about the structure of a simple pit.

- Pits are depressions in the cell wall that are circular to oval where the secondary cell wall is the portion of the cell wall that is hollowed out to make the pit.
- Pits are adjacent to one another on either side of the cell wall. Thus, pits come in pairs and are called pit pairs.
- The pit membrane is composed of the primary wall and middle lamella of the adjacent cell walls of the pit pairs.
- U-shaped in morphology.
- Found in parenchyma, collenchyma, fibers, and sclerids.

12. Write notes on the structure of bordered pit.

- Similar to the simple pit.
- The secondary cell wall forms a border over the pit membrane leaving a small opening called pore.
- In the center of the pit membrane, knob of secondary wall is present called torus.

- The torus can swell to the extent of plugging the pore in the border and prevent water loss during water stress in the plant.
- Bordered pits are generally present in tracheids, vessels, and xylem fibers.

13. What are vessel elements? Write about the morphology of vessels.
    - It is one of the two conducting cells of the xylem, the other one is the tracheid.
    - Dead when mature and contains lignin.
    - Thick and cross cell walls.
    - Many times are missing in the mature cell.
    - At most will contain large holes or just bars of tissue to keep the end of the cell from collapsing.
    - It is almost horizontal in some to slanted in others, but not as slanted as in tracheids.
    - Cells are much shorter than tracheids.
    - Cells are much wider than tracheids.
    - Contain characteristic secondary wall thickenings (annular, reticulate, pitted, scleriform etc.).
14. What are the differences between pith and pith rays?

    1. Pith:
    - Large parenchymatous cells in the center of the vascular bundle ring.
    - Used for food storage.
    - Used for waste storage.

    2. Pith rays:
    - Large parenchyma cells between the vascular bundles.
    - Used for food storage.
    - Used for lateral transport of food and water.
15. Explain in details about the primary and secondary growth.

    1. Primary growth is responsible for:
    - Growth in length of the stem.
    - Produces the basic tissue pattern in the stem.

    2. Secondary growth is responsible for:
    - Growth in girth of the stem.
    - Produces secondary (new and more) vascular tissues.
    - Provides a continuous connection of meristem cells between the primary tissues of the roots and the primary tissues of the shoots.
    - This is why some trees (California Redwoods and Bristlecone Pines) to live to such a great age (several thousand years).

3. Development of the vascular cambium:

- ✦ The function of the vascular cambium is to produce secondary growth, thus the vascular cambium must be formed before secondary growth can occur.
- ✦ Two regions of the primary stem contribute to the vascular cambium:

  a. Fasciscular cambium- the meristem cells within the vascular bundle.

  b. Interfascicular cambium- the meristem cells between the vascular bundles.

4. The differentiation of the fascicular cambium:

- ✦ All of the procambium in the vascular bundle do not differentiate into xylem or phloem.
- ✦ This undifferentiated procambium is called as the residual procambium.
- ✦ The residual procambium is 2-4 cells wide and remains meristematic.
- ✦ The residual procambium will be called vascular cambium when the vascular cambium begins to divide to form secondary tissue.
- ✦ Thus, fascicular cambium is the regions of the vascular cambium that originates within the vascular bundle.

5. The differentiation of the interfascicular cambium:

- ✦ This portion of the vascular cambium originates in the pith rays between the vascular bundles.
- ✦ A band of parenchyma cells, about 2-4 cells wide de-differentiate and become meristematic.
- ✦ When these cells begin to divide they are called vascular cambium and are referred to as interfascicular cambium because of where they originated.

6. The vascular cambium:

- ✦ When the fascicular and interfascicular cambium join to form a complete cylinder around the stem it becomes vascular cambium.
- ✦ As soon as the cylinder is formed the vascular cambium becomes active by dividing on both the inner and outer surface of the vascular cambium.
- ✦ New xylem cells are formed inwardly and attached to the previously produced xylem.
- ✦ New phloem cells are formed outwardly and are attached to the previously produced phloem.
- ✦ Two kinds of vascular cambium cells exist:

(1) Ray initials:

- ✦ Produces the radial (lateral) transport system cells.
- ✦ In xylem and phloem these are the parenchyma cells.

(2) Fusiform initials:

- Produces the axial (vertical) transport system cells.
- In xylem and phloem these are the sieve-tube members, companion cells, tracheids, vessels, and fibers.

7. Generally the xylem-producing cells are more active than the phloem-producing cells.

8. The yearly activity of the cambium produces annual rings in the xylem.

16. What are the functions of periderm (cork)?

- Increase in diameter of the stem occurs with the activity of the vascular cambium.
- This causes the protective epidermis to crack and split open.
- Thus, a need for a meristematic layer at the outer edge of the phloem for the internal protection of the stem.
- Thus a layer of cork cambium forms outside the phloem.
- The cork cambium forms a layer of waxy cork cells.
- The cylinder of cork cambium increase in diameter as the stem increases in diameter.

17. How does the cork form?

1. In the young stem (1 year old or less):

- Cortical cells just under the epidermis become meristematic.
- Produces a layer 1-2 cells thick of cork cambium called phellogen.
- Phellogen produces a layer of cork cells 4-6 cells thick external (toward the epidermis) to the phellogen.
- Phellogen produces a single layer of cells, phelloderm, internal (toward the xylem and phloem).

2. Structure of the cork cells:

- Cells are flattened.
- Cell walls contain suberin, a waxy substance.

3. In old stems (more than 1 year old, generally 3-4 years):

- A new phellogen forms because the former phellogen dies.
- The new phellogen forms in the outer region of the still-living phloem.
- New phellogens will form about every one to four years depending upon the species of tree.

18. What is bark?

- As the layers of cells outside the vascular cambium die, they are sloughed off as bark.
- In the young stem the bark contains: Epidermis, cork, cork cambium, phelloderm, cortex and phloem.
- In the old stem the bark contains: Cork, cork cambium, phelloderm and phloem.

Tables:

Differences between the internal structures of monocot and dicot stems.

Character	Dicot stem	Monocot stem
1. Hypodermis	May be present or absent; if present generally collenchymatous.	Generally present and sclerenchymatous.
2. Cortex	Well differentiated, generally parenchymatous.	The tissue present is not differentiated. It is mostly parenchymatous and extends up to the centre of axis and is called ground tissue.
3. Endodermis	True endodermis, casparian strips are generally absent; instead starch sheath is usually present.	Absent.
4. Pericycle	Present between cortex and vascular bundles in the form of patches of sclerenchyma and parenchyma.	Absent.
5. Medullary rays	Parenchymatous, present between vascular bundles.	Absent.
6. Vascular bundles	Conjoint, collateral or bicollateral, endarch and open, arranged in a ring, all the vascular bundles almost of the same size, phloem parenchyma present and bundle sheath generally absent.	Scattered in the ground tissue, larger towards the centre and smaller near the periphery, phloem parenchyma absent, bundle sheath generally well developed and distinct.
7. Pith	Distinct in centre, parenchymatous or sclerenchymatous.	Pith not differentiated.
8. Secondary growth	By vascular cambium, by secondary vascular tissues.	Vascular cambium absent, no secondary growth.

Differences between dicot and monocot roots.

Character	Dicot stem	Monocot stem
1. Vascular bundles	The number of vascular bundles varies from 2 to 6.	The number of vascular bundles is generally more than six.
2. Pericycle	Lateral roots, vascular cambium and cork cambium originate from this layer.	Only lateral roots are produced from this layer.
3. Cambium	Develops later during secondary growth.	Does not develop at any stage.
4. Pith	Poorly developed or absent	Large and well develop.

Anamolous primary structures:

In dicot stem	Examples
Medullary vascular bundles	*Amaranthus* sp.
	Acyranthus sp.
	Mirabilis sp.
	Raphanus sp.
	Chenopodium sp.
	Bougainvillea sp.
	Boerhavia sp.
	Pepromia sp.
Polystele	*Dianthera* sp.
	Primula sp.
Interxylary phloem or included phloem	*Leptadaenea* sp.
	Salvadora sp.
Exclusive phloem bundles	Cuscuta (total stem parasite)
Exclusive xylem bundles	*Paeonia* sp.
In Monocot stem	
Vascular bundles arranged in ring	*Triticum* sp.
	Oryza sp.
	Secale sp.
	Avena sp.
	Hordaeum sp.
Eustele in monocots	*Tamus cummunis* sp.

Anomalous secondary growth:

Anomalous secondary growth	Examples
Secondary growth in monocots	*Yucca* sp.
	Dracena sp.
	Aloe arborescenae
	Sansi viera
	Agave Americana
	Lomendra sp.
	Kingia sp.
Successive rings of vascular cambium	*Gnetum* sp.
	Mirabilis sp.
	Boerrhavia sp.
	Bougainvillea sp.
	Cycas sp.
Interxylary phloem or included phloem	*Leptadaenea* sp.
	Salvadora sp.

(*Contd.*)

Anomalous secondary growth	Examples
Origin of cork cambium from epidermis	*Quercus suber* sp. *Solanum dulcamara* sp
Origin of cork cambium from pericycle	*Clemalis* sp. *Thuja* sp.
Interxyllary cork or included cork	*Artemesia tridentate* sp. *Epiboliam*

Differences in stem and root.

Stem	Root
It develops from plumule.	It develops from the radicle.
Upper part of the plant axis.	Lower part of the plant axis.
Generally positive phototropic and negatively geotropic.	Generally positively geotropic and negatively phototropic.
May be green (at juvenile stage) or brown at later stage.	Underground generally, so generally non green, exposed to light may appear green.
Stem hairs are unicellular or multicellular.	Root hairs are unicellular.
Differentiated into nodes, internodes, leaves, flowers and fruits.	Not differentiated.

Differences in parenchyma, collenchymas and sclerenchyma.

Parenchyma	Collenchymas	Sclerenchyma
1. Cells are living.	Cells are living.	Cells are dead.
2. The cell walls are thin and do not have secondary depositions.	Pectin deposits are present outside the primary wall.	The secondary cell walls is made of lignin.
3. Pits are absent from the wall.	Pits are absent from the wall.	Simple pits are present in the cell wall.
4.Intercellular spaces are present.	Intercellular spaces are absent	Intercellular spaces are present.
5. Functions: Synthesis of food and its storage, packing tissue, also acts as meristematic tissue.	Functions: Being a mechanical tissue, provides rigidity with flexibility, and also carries out vital processes like photosynthesis if required.	Functions: Being a major mechanical tissue, provides rigidity with hardness.

Differences between aggregate and multiple fruits.

Aggregate fruits	Multiple fruits
1. This type of fruits develops from a single flower having a number of free carples.	This type of fruits develops from inflorescence.
2. Examples: Rose , strawberry, lotus etc	Examples: Jack fruit , pineapple, mulberry etc.

Differences between drupe, berry and schizocarpic fruits:

Drupe	Berry	Schizocarpic fruits
1. Pericarp is differentiated into epicarp, mesocarp and endocarp.	As like in drupe fruits.	The fruit wall does not show differentiation into different layers.
2. The endocarp is hard.	The endocarp, if present, is thin and membranous.	The endocarp is absent.
3. Seed is enclosed by endocarp.	The seed is enclosed by mesocarp.	The seed is enclosed by endocarp.
4. Examples: Mango, almond, coconut etc.	Examples: Banana, tomato, chilies etc.	Examples: Ground nut, coriander etc.

Differences between legume and follicle fruits.

Legumes	Follicle
1. These fruits open along both dorsal and ventral sutures at maturity.	They open along ventral sutures only.
2. Legumes do not cluster to form etaerio.	Follicles are clustered together to form etaerio.
3. Examples: Gram, beans etc.	Examples: *Aconitum* , *Calotropis* etc.

Differences between sap wood and heart wood.

Sapwood	Heartwood
The outer region of the old trees is sapwood.	The central region of the old trees is heartwood.
Also known as alburnum.	Also known as duramen.
Consists of newly formed xylem elements.	Old xylem elements filled with resins, gums etc.
It is soft and not durable.	It is hard and durable.

(*Contd.*)

Sapwood	Heartwood
Vessels are not plugged with tyloses.	Vessels are plugged with tyloses.
It conducts water, store food and nutrients.	Provide mechanical support.

ANATOMY—A QUICK GLANCE

1. The founders of plant anatomy were Marcello Malpighi (1628-94) and Nehemiah Grew (1641-1712).
2. The branch of botany dealing with the internal organization of plants is called anatomy.
3. The structural and functional unit of body is called cell and the group or mass of the cells that are alike in origin, structure and function are called tissue.
4. The term 'tissue' was coined by Grew.
5. The different tissues are often grouped together into larger functional units called organs.
6. Tissues are three types in general: 1. Meristematic tissue, 2. Permanent tissue and 3. Secretory tissue.
7. The meristematic cells are spherical, oval or polygonal in shape without any intercellular spaces, their walls are thin and homogenous, densely filled with protoplasm and have large nuclei.
8. Vacuoles are small or absent in meristematic cells, and these are actively metabolizing and dividing cells.
9. The term meristem was given by Nageli (Meristo means divided).
10. The small meristem which occupy a small area at the tip of the stem or root and which gives rise to primary meristem is called promeristem or primordial meristem.
11. The meristems which arise from amongst the non-meristematic tissues by de-differentiation of permanent cells at later stages in the life are called secondary meristems.
12. Usually the epidermal cells are colourless, but in hydrophytes and sciophytes these are green in colour due to presence of chloroplasts.
13. The secondary meristems are usually lateral in position in plant body and they give rise to secondary tissues; the growth by these types of tissue are called secondary growth.
14. In *Aloe*, *Agave*, *Hakea* and *Mesembryanthemum;* chlorophyll-free, thin walled and water-turgid cells are found which represent water storage parenchymatous tissue.
15. As regard the position of meristem in plant body, the meristems can be classified into, apical meristem, intercalary meristem and lateral meristem.

16. The vascular cambium and cork cambium are the examples of lateral meristem.
17. When the parenchyma cells are exposed to light they develop chloroplasts in them, and such tissue is called chlorenchyma.
18. On the basis of function, meristem can be classified as protoderm, procambium and ground or fundamental meristem.
19. Protoderm is the outer most layer of the apical meristem which develops into epidermis later on.
20. Procambium is located inside the protoderm and gives rise to primary vascular tissues.
21. Ground meristem constitutes the major part of the apical meristems and is the precursor of ground tissue such as pith, cortex, hypodermis, pericycle etc.
22. On the basis of plane of division, the meristems are classified into three types: 1. Mass meristem, 2. Plate meristem 3. Rib meristem or file meristem.
23. In mass meristem cell division occurs in all planes leading to increase in volume, e.g. development of endosperm.
24. In plate meristem, cell division occurs only in two planes showing increase in area. E.g. epidermis.
25. In rib meristem or file meristem cell divide only in one plane increasing in length. E.g. development of cortex and pith.
26. Permanent tissues are those in which growth has stopped completely or for the time being and these can be simple and complex permanent tissues.
27. In aquatic plants, the parenchyma cells in the cortex posses well developed air spaces (intercellular spaces) and such tissues are known as aerenchyma.
28. The latex cells are also known as non-articulate latex ducts and the latex vessels are also known as articulate latex ducts.
29. Arborescent is opposite to herbaceous where secondary growth is present, e.g. trees.
30. The term 'xylem' was introduced by *Nageli* and 'tracheid' was given by Sanio.
31. In most of the monocots and herbaceous plants primary growth is the only type of growth.
32. The sclerenchyma cells are grouped in to fibres and sclereids; fibres are elongated sclerenchyma cells, usually with pointed ends and sclereids are usually not much longer than they are broad, occurring singly or in groups.
33. Vascular bundles are of three types: 1. Collateral, 2. Radial and 3. Concentric.
34. When xylem and phloem are present together on the same radius, the vascular bundles are called collateral.
35. When xylem and phloem are present as separate strands on different radii, the bundles are described as separate and radial vascular bundles.

36. The concentric type of vascular bundles are of two types: 1. Leptocentric or amphivasal (phloem surrounded by xylem), and 2. Hadrocentric or amphicribal (xylem surrounded by phloem).
37. The simple tissue is made up of one type of cells forming a uniform system of cells (e.g. parenchyma, collenchymas and sclerenchyma).
38. The complex tissue is made up of more than one type of cells working together as a unit (e.g. the xylem and phloem).
39. Glandular epidermal hairs, nectaries etc. are the examples of common external glands which occur on the epidermis.
40. The first formed plant body is known as the primary plant body, since it is built up by means of primary tissues and growth.
41. The first meristems appear in the embryonic shoot and embryonic root. All primary tissues of the body originate either from shoot apical meristem or from root apical meristem.
42. Bicollateral vascular bundles are present in the stem of Cucurbits. The outer cambium in Cucurbit vascular bundle is many-layered and more or less flat, while the inner cambium is few-layered and curved.
43. The formation of shoot apex is two types: 1. At the plumular end of the embryonic axis or 2. In the axil of a leaf.
44. Xylem is composed of tracheids, vessels, xylem fibres and wood parenchyma and, phloem is composed of sieve elements, companion cells, phloem fibres and phloem parenchyma.
45. Oil glands, mucilage secreting glands, glands secreting gums, resin and tannin etc. are the examples of internal glands.
46. The fibres present outside of xylem are called extraxylary fibres, these are of three types: 1. Cortical fibres, 2. Pericyclic fibres (also called perivascular fibres) and 3. Phloem fibres (also called bast fibres).
47. The stele in dicot stem is called eustele and in monocot stem it is called as atactostele, monocot stele is most advanced type among the steles.
48. In the Pteridophytes and Gymnosperms, sieve cells, phloem parenchyma are present but companion cells are absent.
49. According to the tunica carpus theory (proposed by Schmidt, 1924) of shoot apex is dome shaped at its tip and the outer layer is called tunica and the inner is called corpus.
50. The tunica constitutes one or more layers; they divide anticlinally and increase the surface area (useful in the formation of epidermis).
51. Tunica surrounds the inner core called corpus and the cells of the corpus divide in various planes, resulting in the increase in volume (useful in the formation of cortex and vascular tissues).

52. All the dead tissues lying outside the cork cambium together constitute the bark of the plant. It is two types: 1. Outer bark or rhytidome (it consists of outer dead portions of bark outside the cork cambium) and 2. Inner bark (located between vascular cambium and cork cambium).
53. Bark is of two types: 1. Ring bark as in *Betula* where cork cambium appears in the form of a complete ring so that bark is formed as a sheet and 2. Scale bark as in Guava where cork cambium appears in strips so that the resulting bark is formed as scales.
54. According to the histogen theory of shoot apex, the apical initials are differentiated into three histogens: 1. Dermatogens, 2. Periblem, and 3. Plerome.
55. Tracheids are most primitive type of conducting elements in xylem, and are present in all group of plants, the xylem of ancient vascular plants consists of tracheids only.
56. Vessels are advanced type of conducting elements and are characteristically found in Angiosperms.
57. The examples of vessel-less Angiosperms are the members of Winteraceae, Trochodendraceae and Tetracentraceae.
58. Vessel elements are also present in some Pteridophytes, e.g. *Selaginella*, *Equisetum*, *Actinopteris*, *Pteridium*, *Regnellidium*, *Marsilea* etc.
59. Dermatogens divides anticlinally to form the epidermis, plerome is the central mass of cells which gives rise to central cylinder and periblem is present in between dermatogens and plerome form cortex.
60. On the basis of rate of cell division, shoot apices can be distinguished in two regions: 1. The summit and 2. The flanks.
61. The summit shows slow mitotic activity but flanks exhibit high meristematic activity and gives rise to leaf primordial.
62. At the time of flower formation, the vegetative apex is converted into a reproductive apex and the vegetative apex stops producing leaf primordial.
63. The tissues formed by means of secondary growth are called secondary tissues and the growth in the diameter of stem and roots due to secondary tissues is called secondary growth.
64. The tracheids are short and are generally up to 1 mm in length but vessels are comparatively longer and may reach up to 10 cm in length.
65. In *Pinus*, the resin ducts are schizogenous in nature, and when they mature, they have the structure of a tube with an epithelial lining.
66. In rare case vessels attain the length up to 2-6 meters as in *Eucalyptus* and *Quercus*.
67. The digestive glands in *Drosera* are the example of internal glands.
68. According to Sachs, the plant body of a vascular plant is composed of three systems of tissues: 1. The dermal, 2. The vascular and 3. The fundamental or ground system.

69. In leaf the ground tissue consists of photosynthetic parenchyma, and is known as mesophyll.
70. The special structures which exudates water under conditions of low transpiration and abundant soil moisture are called hydathodes or water stomata.
71. The vascular tissue system enlarges by secondary growth which takes place in the vascular cambium and the periderm may be derived from a separate meristem, the phellogen or cork cambium.
72. Tracheids consist of a single elongated cell which possesses tapering end walls and the vessels consists of a row of cells placed one above the other and the intervening walls are absent.
73. The root apex differentiates from the radicle end of the embryonal axis and it is enclosed by a fully differentiated mature structure called the root cap.
74. The apical meristem in roots is not terminal, but is sub-terminal in position; the meristem produces cells not only towards the axis as in shoot apex but also away from the axis to form the root cap.
75. In the apical meristem of root of *Zea mays*, and in some other plants with a regular arrangement cells in the apical meristem, it is possible to conclude from the study of cell lineages that there is a central region of cells which divide rarely or not at all, these inactive or passive cells constitute the quiescent centre.
76. The root apex is differentiated into three regions: 1. Protoderm (produce epidermis), 2. Cortical initials (produced cortex) and 3. Vascular cylinder.
77. Vessels are tubular and have no cross walls but tracheids are not tubular and these are separated by cross walls which bear bordered pits.
78. The periderm consists of phellem (cork), phellogen (cork cambium) and phelloderm.
79. The phellogen develops in the epidermis, the cortex, the phloem or the root pericycle and produces phellem toward the outside and phelloderm toward the inside.
80. The parenchyma cells are characteristically living cells capable of growth and division; they are concerned with photosynthesis, storage, wound healing, and origin of adventitious structures.
81. Apical cell theory was given by Nageli in 1858 and according to this theory the cell is a constant structural and functional unit of apical meristems governing the whole process of growth.
82. Starch sheath of stem are also called endodermoid because it is homologous to endodermis but is not so specialized.
83. The elongated parenchyma cells are called prosenchyma.
84. The leaf primordial is formed on the flanks of shoot apical meristem.
85. The quiescent centre in root apex acts as a reservoir of relatively resistant cells which constitute a permanent source of active initials.
86. The histogen theory was introduced by Hanstein in 1870 who considered that the primordial meristem was sharply separable into three distinct zones or Histogens.

87. The conducting elements of the phloem are collectively called as sieve elements.
88. Parenchyma cells are the most basic type of differentiated cells from which other tissues have evolved.
89. In some stems, pericycle opposite to vascular bundles is sclerenchymatic alternating with parenchymatic patches; the sclerenchymatic patches are called bundle caps.
90. The parenchyma cells which store ergastic substances are called idioblasts.
91. The sieve elements may be segregated into the less specialized sieve cells and the more specialized sieve tubes.
92. Collenchyma is absent generally in roots and monocot plants.
93. Collenchyma provides support, flexibility and elasticity to the organ wherein it is present.
94. Sclerenchyma also provide support and mechanical strength to the plant, these cells are dead with thick lignified cell walls.
95. The lateral roots are endogenous roots and arise from pericyle.
96. Sclerenchyma is of two types: 1. Long, flexible with tapered ends called fibres and 2. Usually rough and spherical sclereids called stone cells.
97. Sclereids are usually scattered singly or in groups in the plant body especially in phloem, fruit walls and in the pulp of fruits.
98. Complementary cells are rounded cells located in lenticels and which have intercellular spaces and permit diffusion of oxygen into the organs.
99. The grittiness of the fruits like guava and pears is due to the presence of stone cells in groups in their pulp.
100. Sclereids occurring singly are called idioblastic sclereids or sclerites and isodiametric sclereids resembling parenchyma in shape, found in soft parts of plants, called brachysclereids.
101. Bone like sclereids are columnar in shape with dilated ends called osteosclereids or prop cells. E. g. Present in leaves and seed coat of many monocots.
102. Macroslereids are columnar in shape, forming palisade-like epidermal layer in seed coats of peas and beans and epidermis of onion.
103. Trichosclereids are hair like sclereids, sometimes branched, present in leaves of *Olea* and aerial roots of *Monstera*.
104. Medullary vascular bundles are found in the scars of *Boerhaavia*, *Mirabilis*, *Amaranthus* etc.
105. In *Cannabis* phloem fibres are lignified, whereas fibres of *Linum* are of cellulose and without lignin.
106. The phloem fibres has been known since early times as bast or bass, used in the manufacture of cords, mats and cloth and these fibres are known as bast fibres.

107. Companion cell is a specialized type of parenchyma cell which is closely associated in origin, position and function with sieve tube elements.

108. Abscission is the organized shedding of a plant part from the plant without causing any injury (e.g. falling of leaf, flower or its parts, fruits, branch etc.).

109. Phloem parenchyma cells are concerned with storage of starch, fat, tannins, resins and other organic substances.

110. Collenchyma cells make a living tissue closely related to parenchyma but the presence of uneven thickened primary wall is the most distinctive character of these cells. It is regarded as a form of parenchyma specialized as supporting tissue of young organs.

111. Sclerenchyma cells possess thick, secondary lignified walls and lack protoplast at maturity.

112. The tissues which are concerned with the secretion of gums, resins, volatile oils, nectar, latex etc. are called secretory tissue. These are two types: 1. Laticiferous tissue and 2. Glandular tissue.

113. Xylem cells make a complex tissue which, in association with the phloem is continuous throughout the plant body and concerned with water conduction, storage and support.

114. Phloem cells also make a complex tissue which is concerned with translocation of solutes and storage of food.

115. The articulated laticifers are formed through union of cells in which parts of walls are dissolved and may be primary or secondary in origin.

116. The non-articulated laticefers are single cells and much branched and primary in origin.

117. The members of Euphorbiaceae contain latex cells generally but *Hevea brasiliensis* and *Manihot glaziovii* contain latex vessels.

118. Latex cells are present in *Cannabis*, *Vinca minor* and *Urtica dioica.*

119. Para rubber obtained from *Hevea brasiliensis* and Indian rubber from *Ficus elastica.*

120. P-proteins are proteinaceous structures present in sieve tubes and are believed to be responsible for movement of materials through the cell, and sealing of pores after wounding.

121. The word 'leptom' is used for soft-walled conducting part of the phloem which includes sieve elements, companion cells and parenchyma cells.

122. *Thalictrum*, *Nymphea* and *Papaver orientale* are the examples of scattered vascular bundles in dicot stem.

123. Sieve cells differ from sieve tube members in that the openings of sieve areas are small and sieve areas occur throughout the end walls and lateral walls.

124. Internal phloem or medullary phloem or intraxylary phloem or vascular phloem on inner border is present in *Capsicum*, *Calotropis*, *Leptadenia*, *Punica granatum* etc.

125. The enzyme papain is obtained from the latex of *Carica papaya.*
126. Slime plugs are dense funnel-shaped structures formed by coagulation of slime bodies on sieve plates.
127. The albuminous cells of Conifers are analogous to companion cells.
128. The companion cells and sieve tube arise from same mother cell, so these are called sister cells.
129. Cortical vascular bundles are found in Casuarina, *Lathyrus*, *Nyctanthus* etc.
130. Vascular bundles in Cucurbitaceous members are present in two rings, five in outer ring and five in inner ring.
131. Due to unusual position of vascular cambium several vascular cylinders are found in, *Bauhinia*, *Serjania* etc.
132. The complete ring of vascular cambium is originated from pericycle in the members of Chinopodiaceae and Amaranthaceae.
133. Epidermal cells contain cystoliths are called lithocysts.
134. The increase in the diameter of monocot root and stem is due to primary thickening meristem, this type of growth is called primary growth.
135. Callose pad which is deposited on sieve is areas only for limited unfavorable and period is called seasonal, while that formed permanently in functionless old sieve tube is called definitive callose.
136. Generally root is positive geotropic but in mangroves the pneumatophore roots are negative geotropic.

Figures:

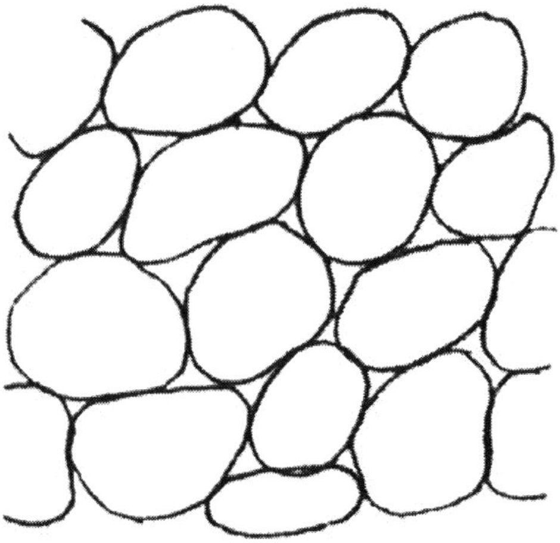

Parenchyma cells

(a)

Palisade Paranchyma

(b)

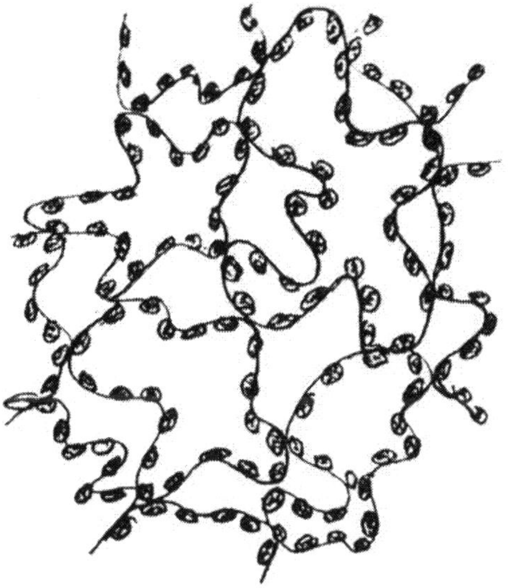

Spongy chlorenchyma

(c)

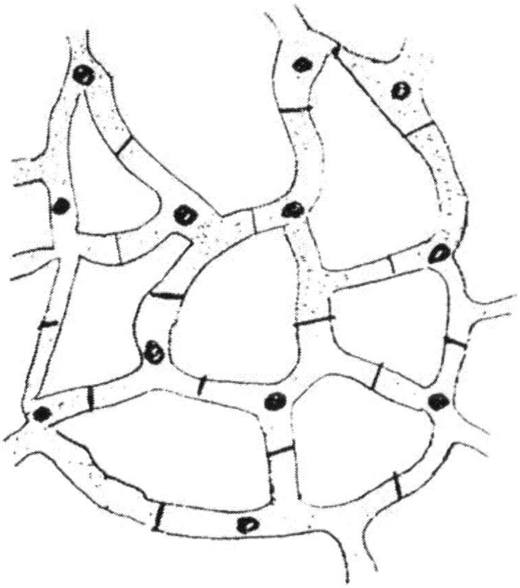

Aerenchyma in the petiole of canna

(d)

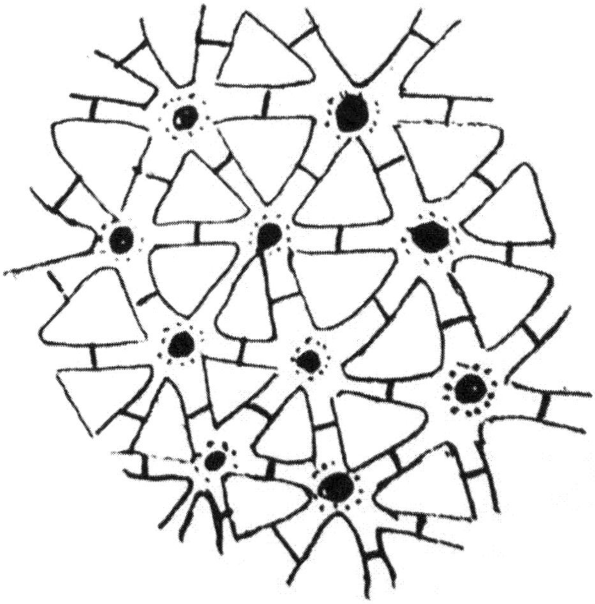

Arenchyma in the petiole of banana

(e)

Figure 2.2 Types of parenchyma

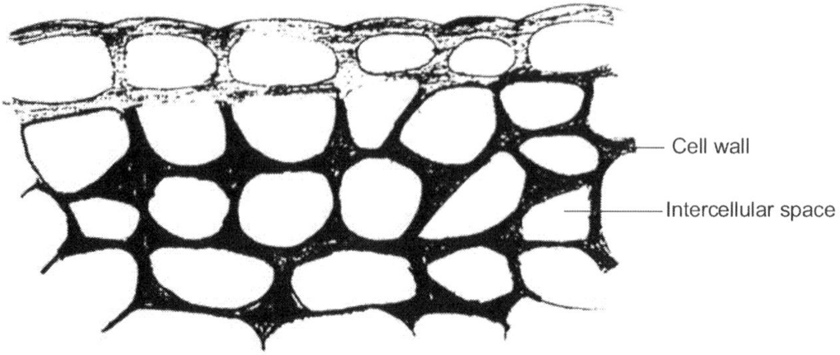

Angular collenchyma

(a)

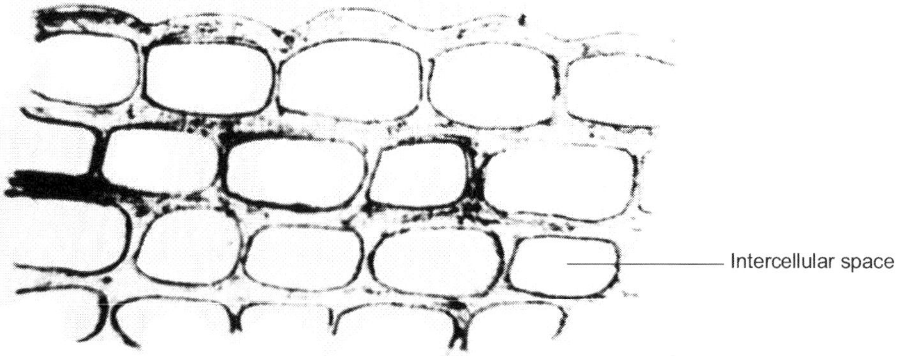

Lamellar collenchyma

(b)

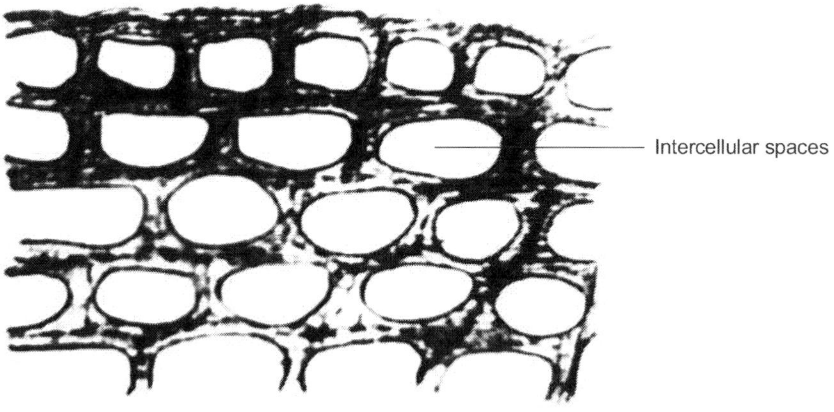

(c)

Figure 2.3 Types of collenchymas

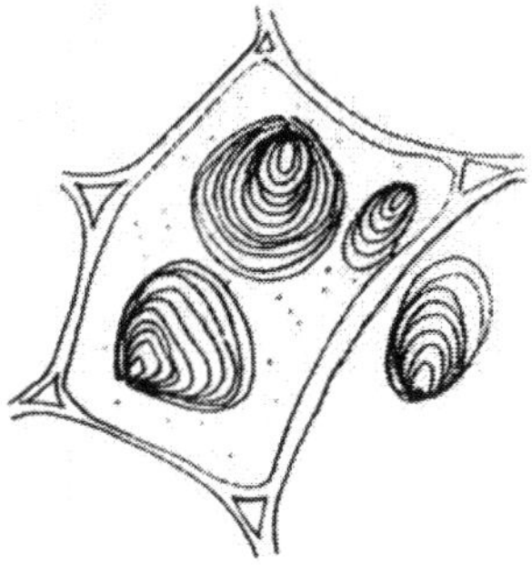

Excentric grains
(*Solanum toberosum*)

(a)

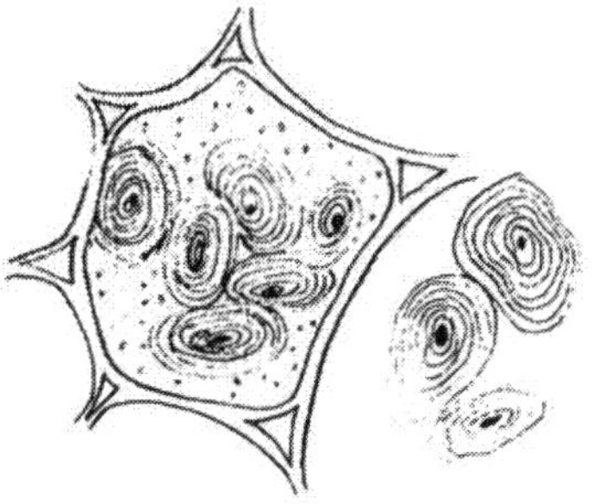

Concentric grains
Pea seed

(b)

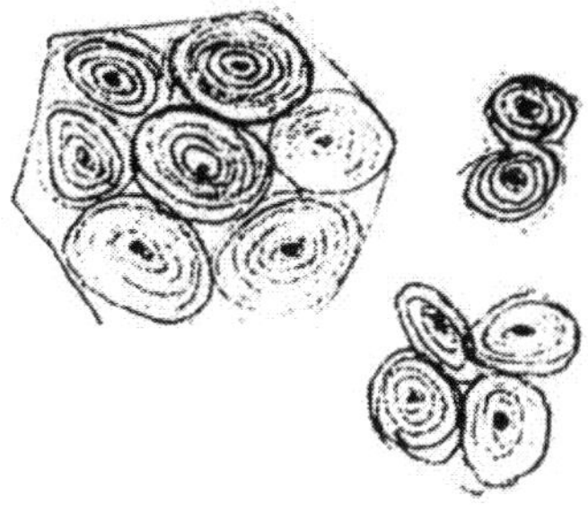

Compound starch grains
Enclosperm of *Oryza*

(c)

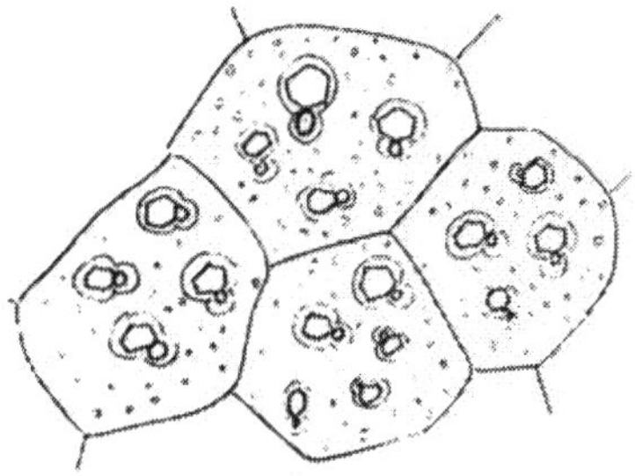

Aleurone grains *Ricinus*

(d)

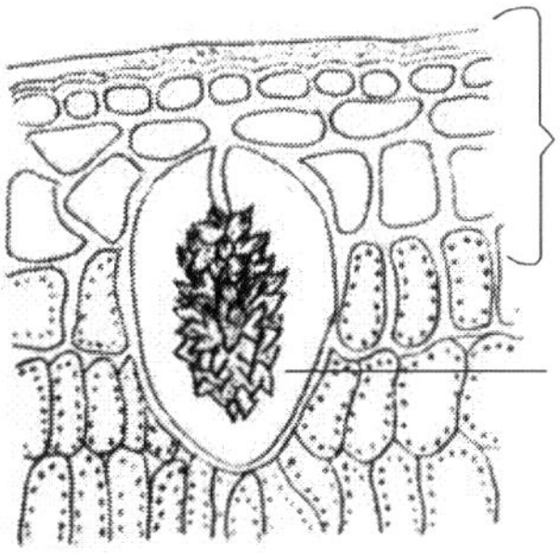

Cystolity-caco$_3$ crystals (*Ficus*)

(e)

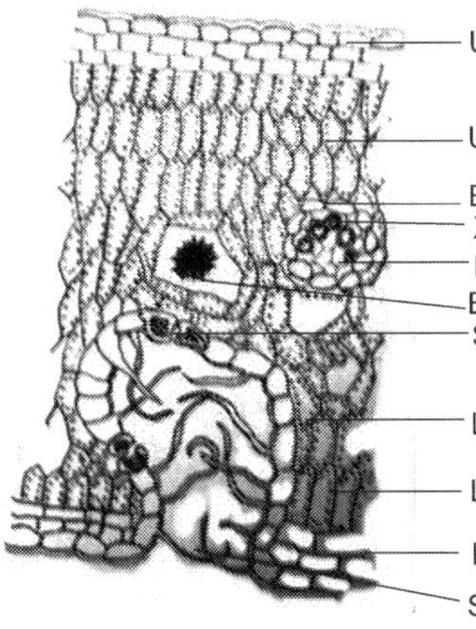

A portion of nerium leaf

(f)

Figure 2.4 Types of ergastic substances in plant cells

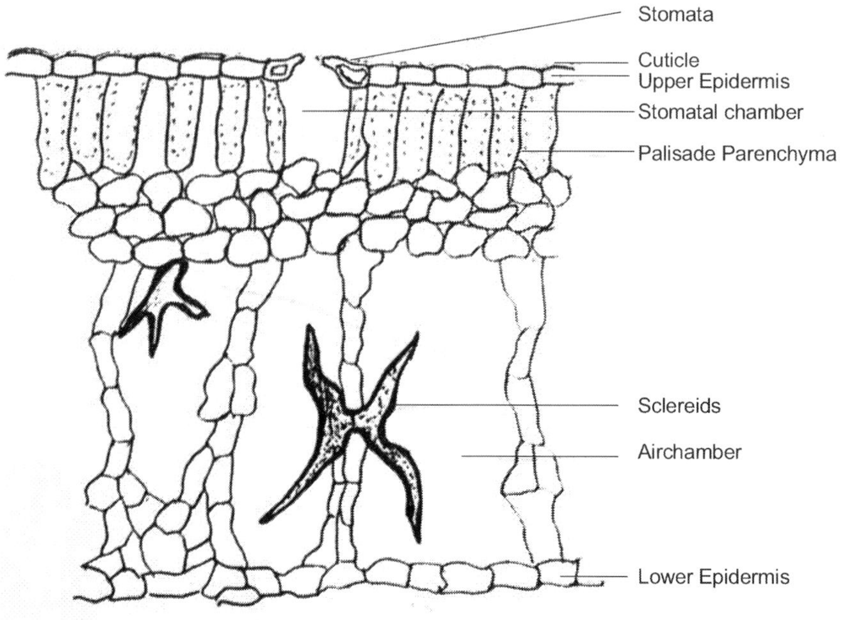

Figure 2.5 T.S. of *Nymphaea* leaf showing presence of sclereids in the cells

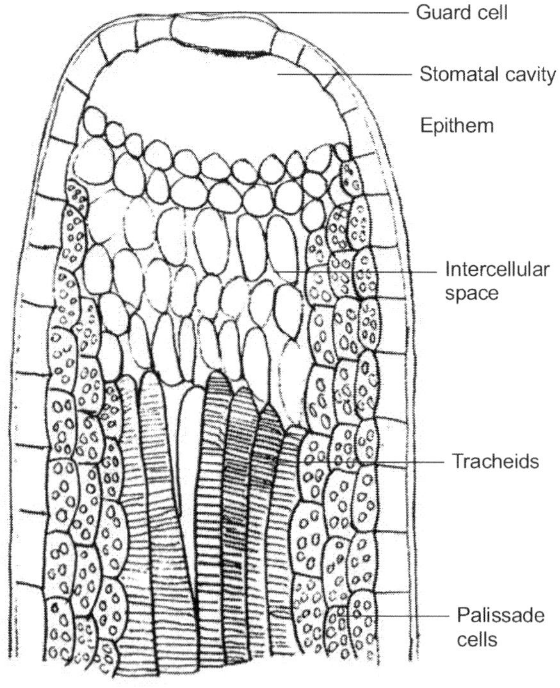

Figure 2.6 Structure of water hydathode

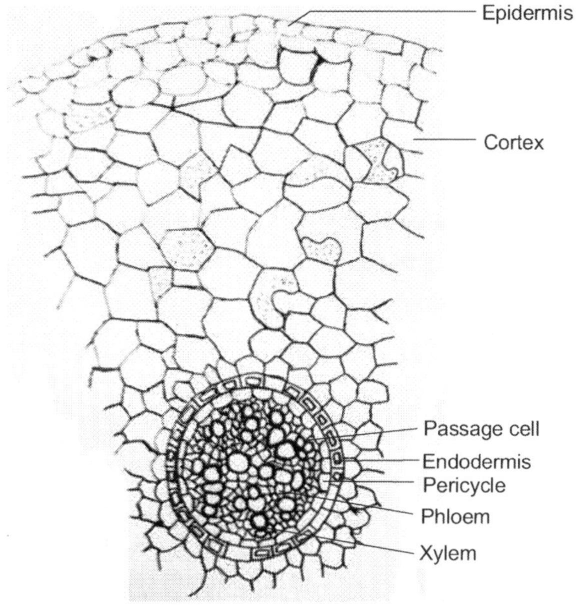

Figure 2.7 Internal structure of monocot root

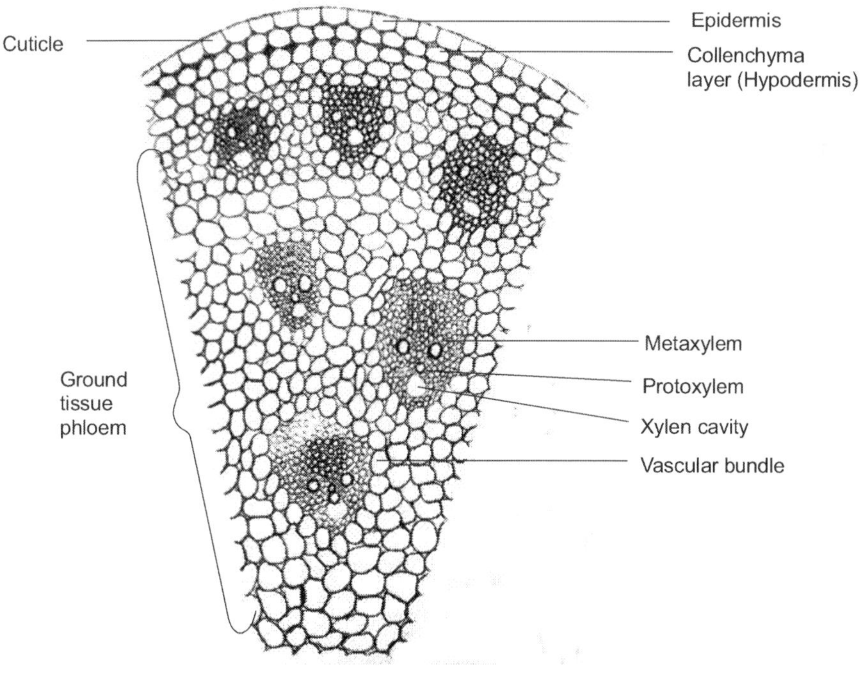

Figure 2.8 Internal structure of monocot stem

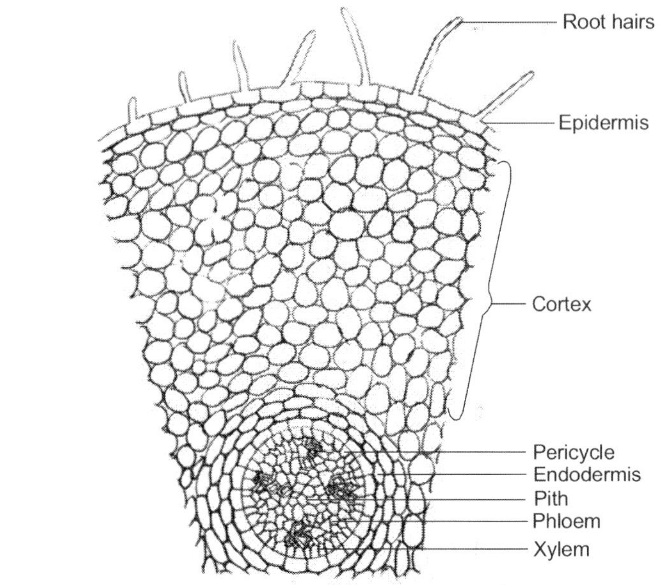

Figure 2.9 Internal structure of dicot root

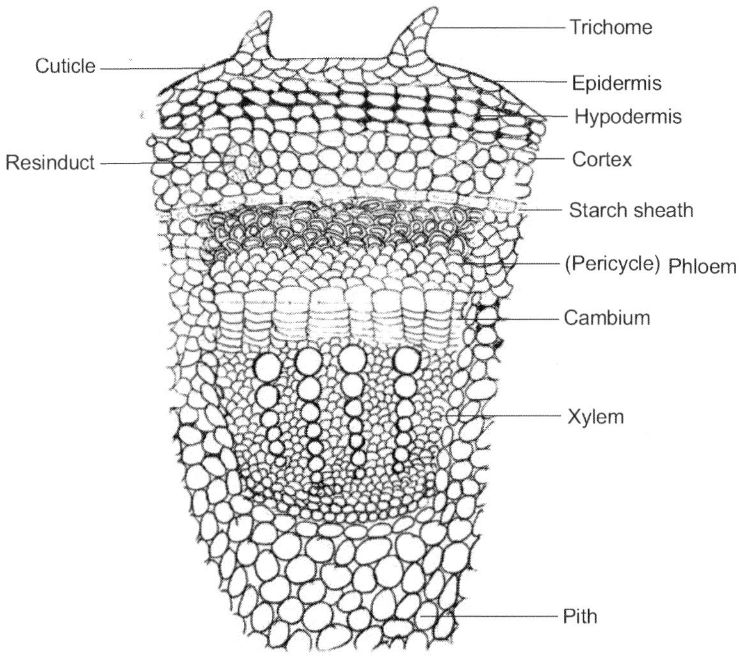

Figure 2.10 Internal structure of dicot stem

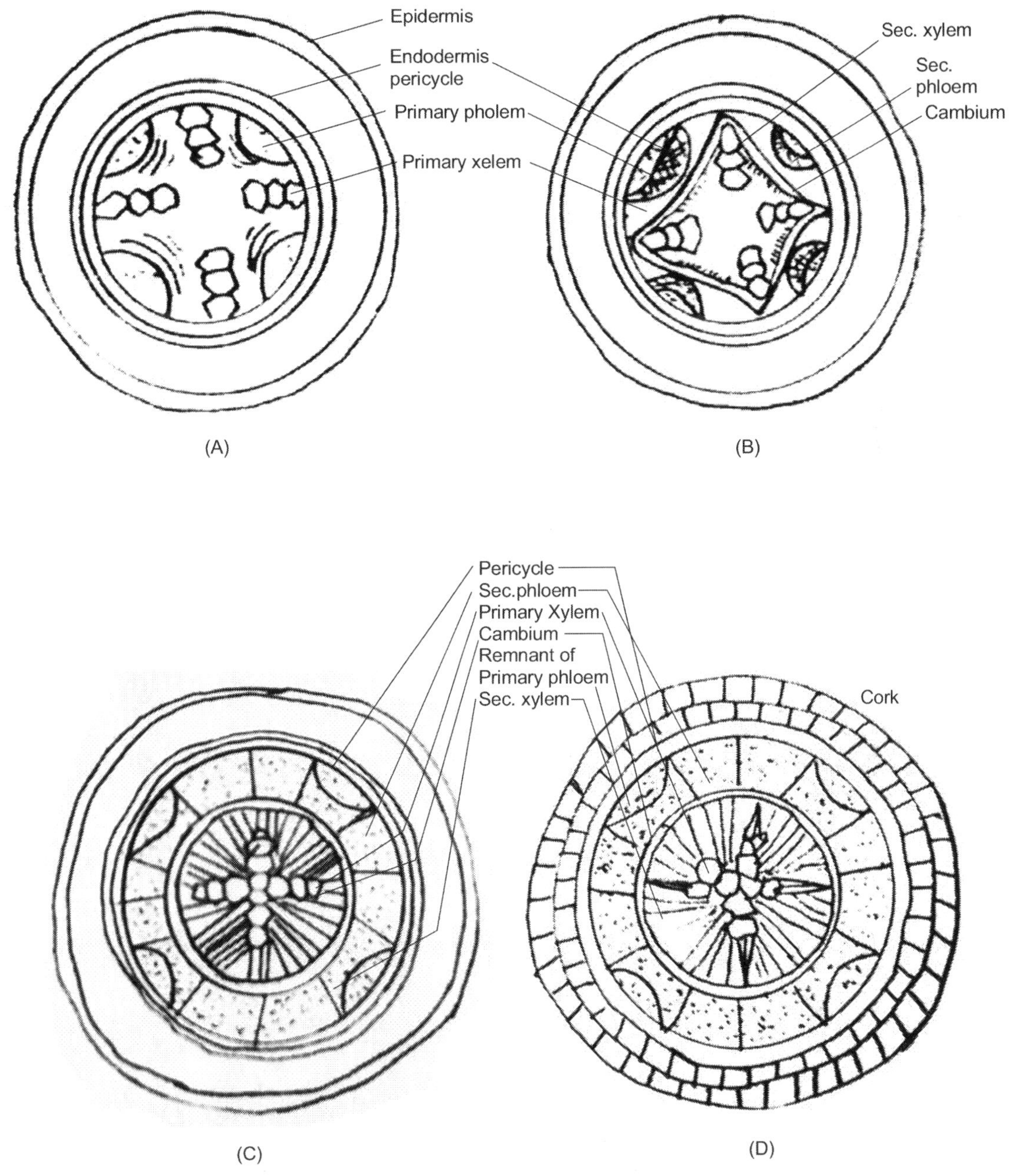

Figure 2.11 Secondary growth in dicot root

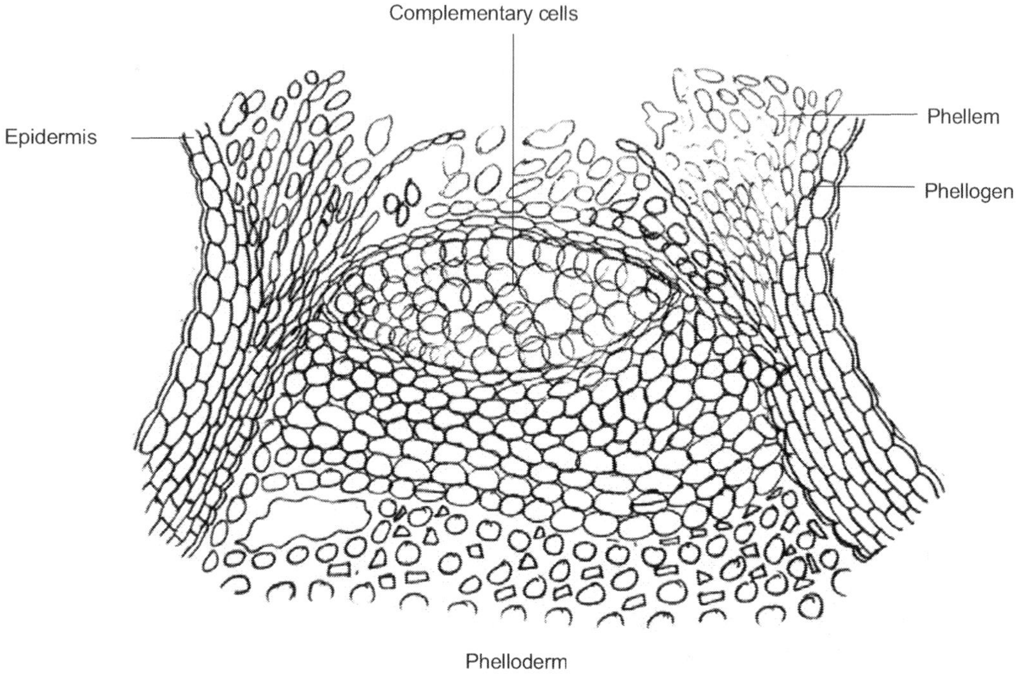

Figure 2.12 Structure of a mature lenticel

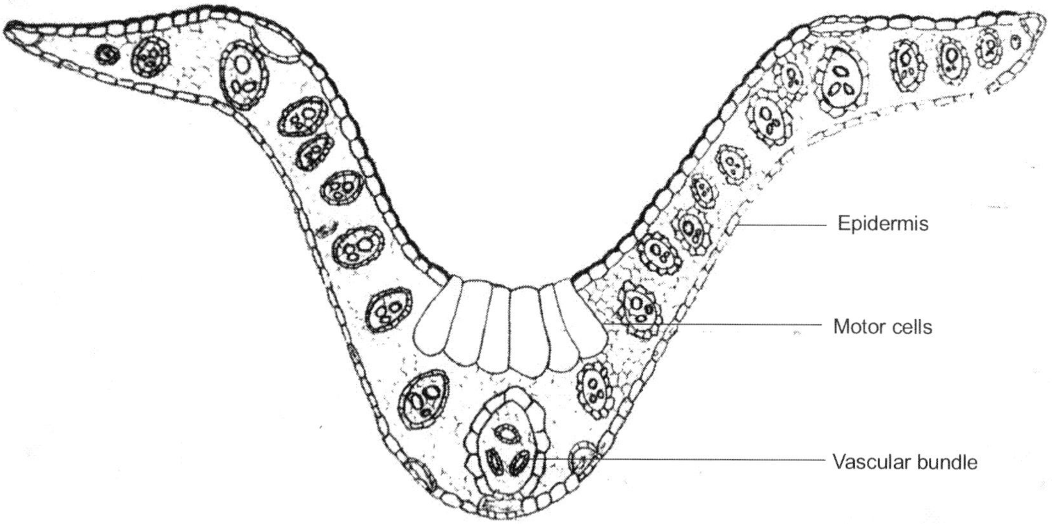

Figure 2.13 T.S. of grass leaf showing motar cells

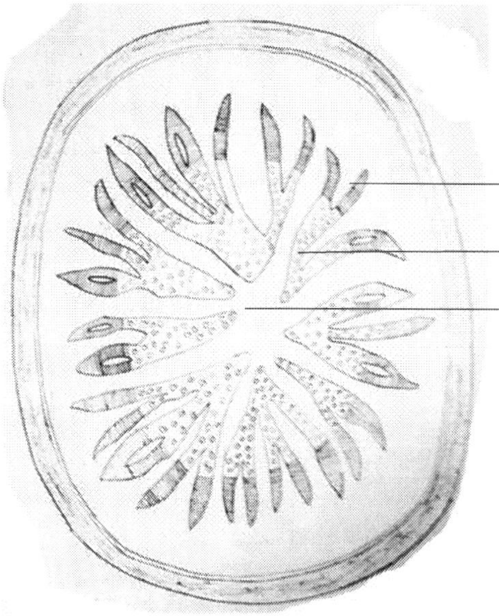

Figure 2.14 T.S. of *Aristolochia* stem showing anomalous secondary growth

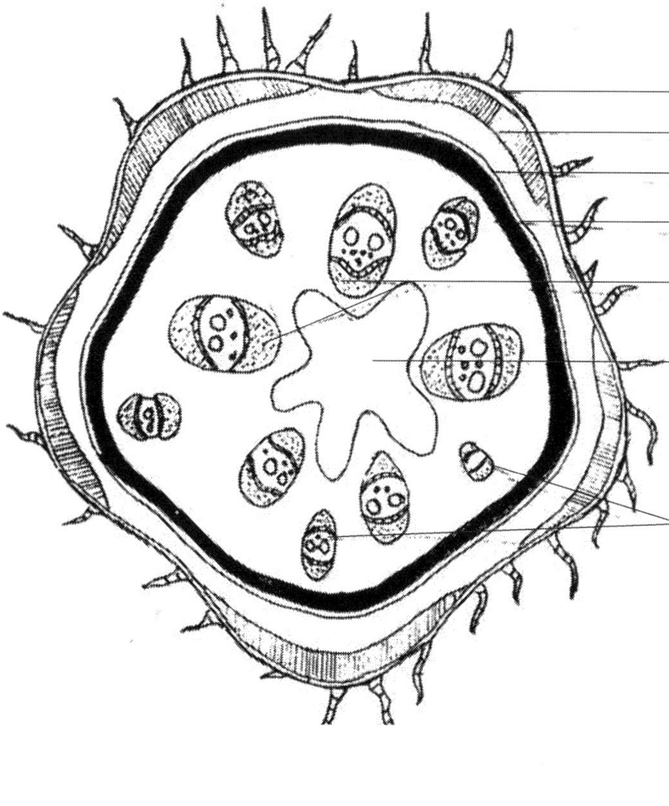

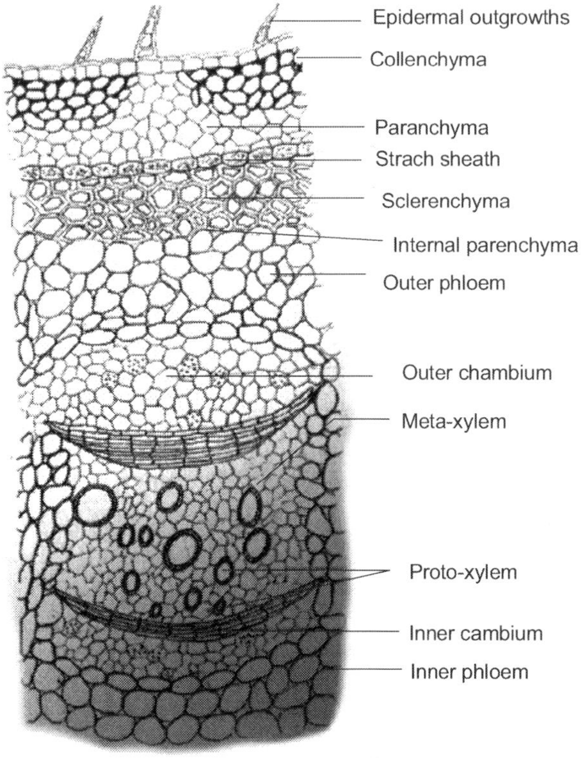

Figure 2.15 T.S. of *cucurbita* stem

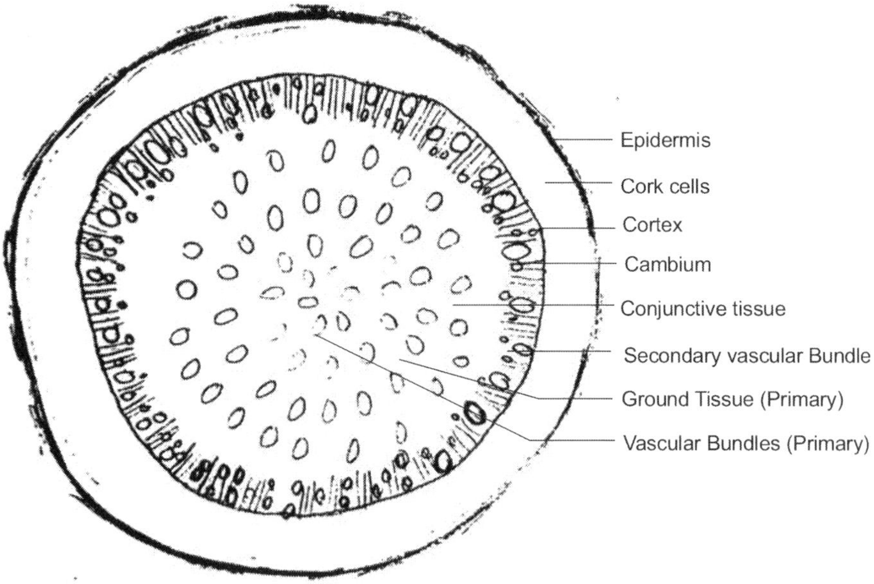

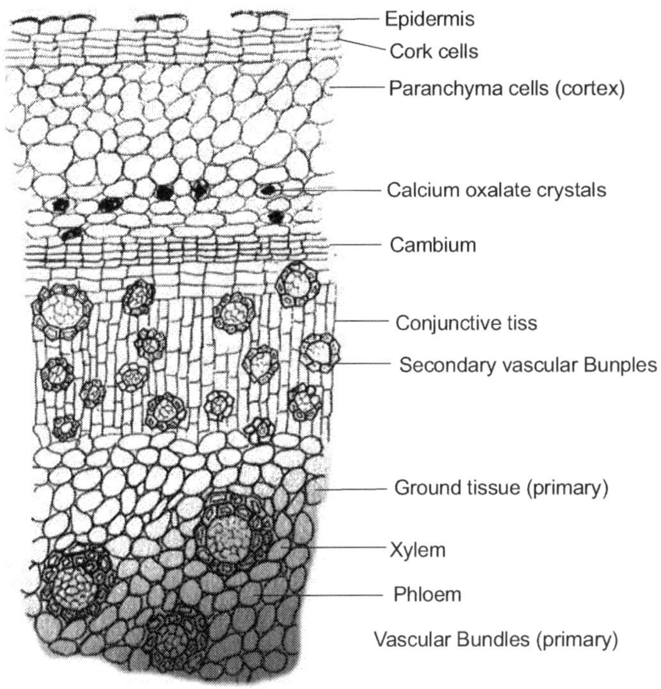

Figure 2.16 T.S. *Draceana* stem

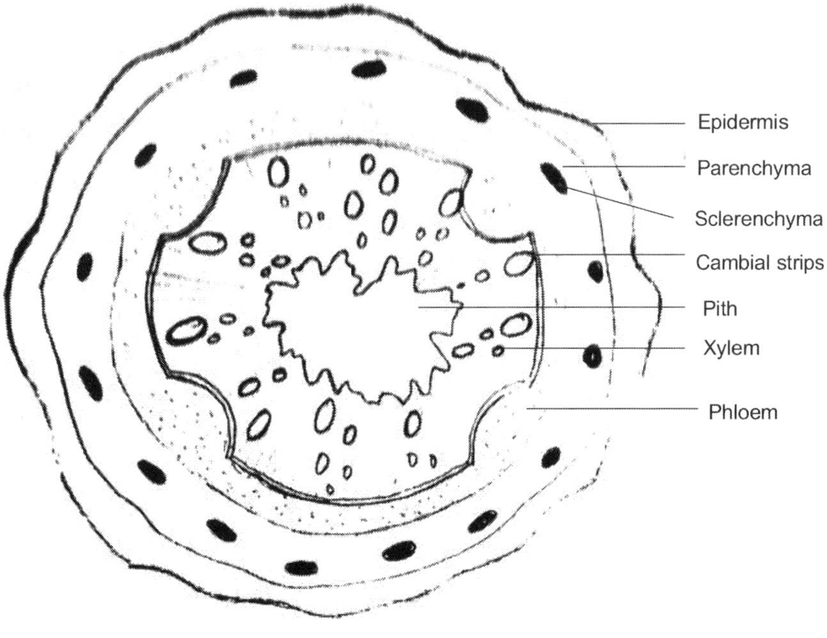

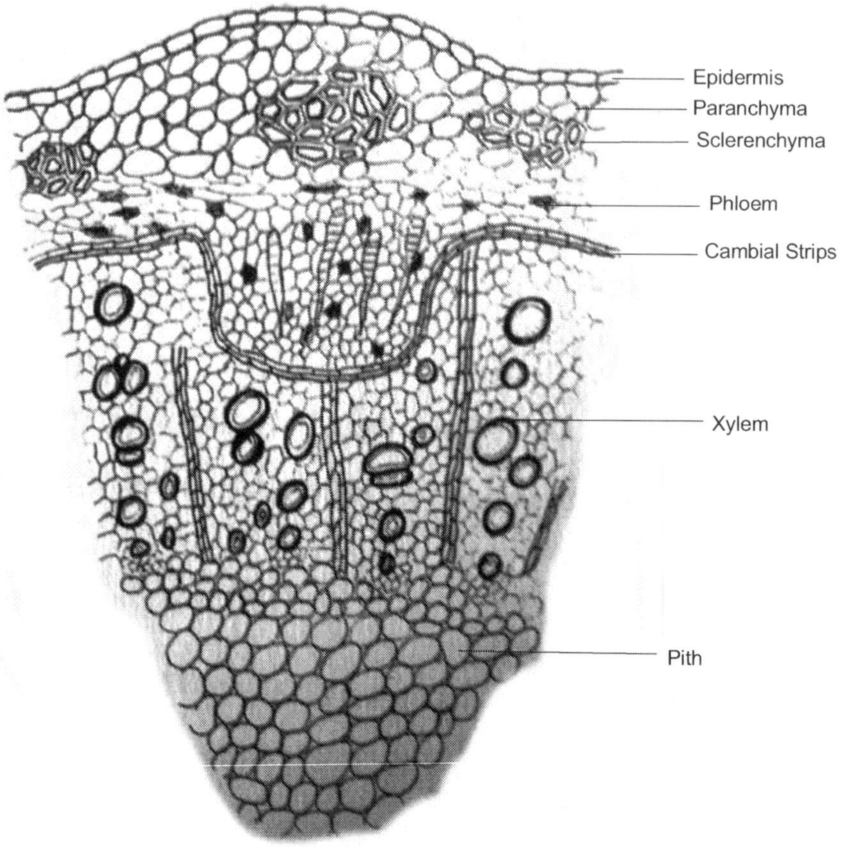

Figure 2.17 T.S. of *Bignonia* stem

ANSWER

1. (c)	2. (c)	3. (b)	4. (c)	5. (a)	6. (d)	7. (d)	8. (a)	9. (d)	10. (d)
11. (b)	12. (d)	13. (a)	14. (b)	15. (b)	16. (d)	17. (c)	18. (a)	19. (c)	20. (b)
21. (d)	22. (c)	23. (d)	24. (c)	25. (c)	26. (c)	27. (b)	28. (c)	29. (c)	30. (b)
31. (c)	32. (d)	33. (b)	34. (d)	35. (a)	36. (d)	37. (b)	38. (a)	39. (c)	40. (a)
41. (a)	42. (c)	43. (a)	44. (d)	45. (a)	46. (b)	47. (c)	48. (b)	49. (d)	50. (a)
51. (c)	52. (c)	53. (d)	54. (d)	55. (b)	56. (b)	57. (d)	58. (c)	59. (d)	60. (c)
61. (c)	62. (a)	63. (d)	64. (a)	65. (a)	66. (c)	67. (b)	68. (d)	69. (d)	70. (c)

71. (d)	72. (a)	73. (c)	74. (c)	75. (c)	76. (d)	77. (a)	78. (d)	79. (a)	80. (b)
81. (b)	82. (d)	83. (b)	84. (c)	85. (d)	86. (d)	87. (d)	88. (c)	89. (b)	90. (a)
91. (b)	92. (b)	93. (d)	94. (b)	95. (c)	96. (c)	97. (a)	98. (a)	99. (b)	100. (a)
101. (a)									

3

ECONOMIC BOTANY

MULTIPLE CHOICE QUESTIONS

1. The product of commercial importance, obtained from coconut palm are
   (a) Latex and oil (b) Oil and fibre (c) Latex and fiber (d) Rubber and drug
2. Potatoes are usually propagated by vegetative means because
   (a) Potato seeds have long dormancy period
   (b) By this method incidence of diseases may be reduced
   (c) They do not produce seeds
   (d) By this method it is possible to maintain genetic quality
3. The true stem of banana is
   (a) Bulb (b) Rhizome (c) Corm (d) Runner
4. The artificial ripening of banana is done by
   (a) Keeping them in a room where the temperature is high
   (b) Keeping them in a room having temperature below 0°C
   (c) Keeping them at room temperature
   (d) None of these
5. What is the source of chewing gum
   (a) Fluid from *Musa* sp (b) Gum from *Acacia* sp
   (c) Sugar from sugarcane (d) Latex from *Achras sapota*
6. Botanical name of jute plant is
   (a) *Musa textilis* (b) *Cannabis sativa*
   (c) *Gossypium arboreum* (d) *Corchorus capsularis*
7. Teak wood is obtained from *Tectona grandis* which belongs to the family
   (a) Verbinaceae (b) Lamiaceae

(c) Fabaceae (d) Sterculariaceae

8. Paddy can grow well on
   (a) Sandy soil (b) Clay to clay loam soil
   (c) Alluvial soil (d) Stony soil
9. Reserpine, a drug is extracted from
   (a) *Papaver somniferum* (b) *Withania somnifera*
   (c) *Rauwolfia serpentina* (d) *Colchicum autumnale*
10. The centre of origin of wheat plant is
    (a) Africa (b) Brazil (c) South-West Asia (d) South America
11. One of the following is bio-fuel
    (a) Wood (b) Coal
    (c) LPG gas (d) Fuel obtained from biological sources
12. The cotton fiber from cotton plant is obtained from
    (a) seed coat (b) leaves (c) stems (d) root
13. The commercial jute fibres are
    (a) inter-xylary fibres (b) bast fibres
    (c) surface fibres (d) xylem fibres
14. Black soil is good for the cultivation of
    (a) Wheat (b) Tea (c) Cotton (d) Rice
15. The quality of tobacco depends on
    (a) Its variety (b) Curing process
    (c) Agro-climatic conditions in which plant grows (d) All of these
16. The principal crop of India is
    (a) Wheat (b) Rice (c) Maize (d) Sorghum
17. To get food containing starch we grow
    (a) Grapes and orange (b) Cabbage and cauliflower
    (c) Rice and wheat (d) Groundnut and coconut
18. Latex containing petroleum like substances are present in the families
    (a) Euphorbiaceae and Apocynaceae (b) Euphorbiaceae and Asclepiadaceae
    (c) Both (a) and (b) (d) None of these
19. Half of the world population is dependent for food on
    (a) Wheat (b) Potato (c) Rice (d) Bread

20. Which is the most important cereal of temperate region?
    (a) *Sorghum vulgare* (b) *Avena sativa*
    (c) Maize (d) Barley
21. Which of the following state leads in the production of wheat?
    (a) Punjab (b) Uttar Pradesh (c) Haryana (d) Bihar
22. Which acid is present in lemon
    (a) Lactic acid (b) Ascorbic acid (c) Citric acid (d) None of these
23. The richest source of vitamin A is
    (a) Radish (b) Carrot (c) Lemon (d) Orange
24. Asafoetida is a gum resin obtained from
    (a) Stem (b) Root (c) Leaves (d) Bark
25. Which vitamin was first time produced by microorganism
    (a) Vitamin- A (b) Vitamin- C (c) Vitamin- B (d) Vitamin- K
26. Which is the highest rubber producing country
    (a) India (b) Malaysia (c) China (d) Australia
27. Maximum amount of carbohydrate is present in
    (a) Rice (b) Barley (c) Maize (d) Wheat
28. Tea and coffee is classified as
    (a) Distilled beverage (b) Non-alcoholic beverage
    (c) Fermented beverage (d) Alcoholic beverage
29. Coconut oil is classified as
    (a) semi-drying oil (b) essential oil (c) drying oil (d) vegetable oil
30. Which is included in cereal crop
    (a) Rice (b) Wheat (c) Maize (d) All of these
31. The botanical name of pearl millet is
    (a) *Hordeum vulgare* (b) *Setaria italica*
    (c) *Pennisetum typhoides* (d) *Sorghum bicolor*
32. Which of the following is not a true fruit
    (a) Tomato (b) Mango (c) Apple (d) Grape
33. Pulses belong to the family
    (a) Rosaceae (b) Poaceae (c) Fabaceae (d) Cucurbitaceae
34. The percentage of proteins in wheat is
    (a) 14% to 18% (b) 2% to 6% (c) 10% to 14% (d) 50%

35. The edible part of litchi is
    (a) Aril (b) Mesocarp (c) Thalamus (d) Seed coat
36. The fatty oils which do not absorb atmospheric oxygen are known as
    (a) drying oils (b) semi-drying oils
    (c) essential oils (d) non-drying oils
37. The composition of cotton fibre is
    (a) Cellulose (b) Callose (c) Pectin (d) Chitin
38. Tea can be grown
    (a) At sea level (b) In elevated areas (c) Sea shore (d) In dry places
39. Which is the best source of protein among these
    (a) wheat (b) bread (c) apple (d) soybean
40. Edible part of cucumber is
    (a) pericarp and thalamus (b) placenta, pericarp and thalamus
    (c) mesocarp (d) endocarp
41. If tea leaves are kept in hot water for longer period, the liquid becomes bitter because
    (a) of the defect in the tea leaves (b) the tannin dissolve out
    (c) the volatile oil in the leaves dissolve out (d) the thein dissolve out
42. The sticky yellow resin which is exuded from the female flowers of Indian hemp plant is commercially known as
    (a) Hashish (b) Bhang (c) Ganja (d) Opium
43. One of the following is an oil crop
    (a) Rose (b) Marigold
    (c) Sunflower (d) *Chrysanthemum*
44. Which of the following wood has highest calorific value
    (a) Teak (b) *Acacia* (c) Neem (d) Mango
45. Para rubber is obtained from
    (a) *Ficus elastica* (b) *Castilla elastica*
    (c) *Palaquium gutta* (d) *Hevea brasiliensis*
46. Morphine is obtained from
    (a) *Withania somnifera* (b) *Adhatoda vasica*
    (c) *Rauwolfia serpentina* (d) *Papaver somniferum*
47. Wheat is
    (a) Achene (b) Caryopsis (c) Samara (d) Sorosis

48. Khus –khus oil is obtained from
    (a) Roots of grass *Vetiveria zizanioides* (b) Roots of *Mentha* sp.
    (c) *Cocos nucifera* (d) *Salix* sp.
49. *Triticum aestivum* is
    (a) Diploid (b) Triploid (c) Tetraploid (d) Hexaploid
50. Bast fibres are obtained from
    (a) *Linum usitatissimum* (b) *Corchorus*
    (c) *Crotolaria juncea* (d) All of these
51. Which of the following crop is not sown by seeds
    (a) Brinjal (b) Mosses (c) Potato (d) Wheat
52. The major source of sugar in the world today is
    (a) Dates (b) Sugarcane (c) Water melon (d) Beet root
53. Sandal wood oil is extracted from which part of the plant
    (a) Wood (b) Bark (c) Leaf (d) Flower
54. Palm oil is obtained from
    (a) *Glycine* (b) *Olea* (c) *Arachis* (d) *Elaeis*
55. Which of the following yields fiber, oil and animal food?
    (a) Jute (b) Cotton (c) Sunflower (d) Sun hemp
56. In India hockey sticks are generally made from
    (a) Melia (b) Salix (c) *Morus alba* (d) *Morus nigra*
57. The edible part in coconut is
    (a) Endosperm (b) Epicarp (c) Endocarp (d) Mesocarp
58. In cabbage, the edible part is
    (a) Axillary bud (b) Foliar bud (c) Apical bud (d) Floral bud
59. Food grains which provide the most important staple food for man are
    (a) Millets (b) Cereals (c) Legumes (d) Oil seeds
60. Cultivation of honey bee is known as
    (a) Laciculture (b) Apiculture (c) Sericulture (d) Bee culture
61. Gifts of the new world to old world are
    (a) paddy and cotton (b) beet root, tobacco and maize
    (c) potato, tobacco and maize (d) wheat and barley
62. One of the major 'Basmati Rice' producing state in our country is
    (a) Andhra Pradesh (b) Uttar Pradesh (c) Kerala (d) Karnataka

63. Opium is obtained from
    (a) *Aconitum napals* (b) *Argemone maxicana*
    (c) *Papaver sominiferum* (d) *Withania sominiferum*
64. Seeds of one of the following are the richest source of protein
    (a) Soybean (b) Lentils (c) Hyacinth bean (d) Groundnut
65. Saffron is produced from
    (a) Stamens (b) Styles and stigmas of *Crocus* plant
    (c) Roots (d) Petals
66. One of the major foreign exchange earning crop of India is
    (a) Tea (b) Rice (c) Coffee (d) Wheat
67. Sal wood is obtained from
    (a) *Shorea robusta* (b) *Salmalia malabarica*
    (c) *Dalbergia sissoo* (d) *Tectona grandis*
68. Which of the following wood is used in making cricket bats
    (a) *Cedrus deodara* (b) *Morus alba*
    (c) *Tectona grandis* (d) *Salix purpurea*
69. Black wood is obtained from
    (a) *Acacia* (b) *Dalbergia* (c) *Albizzia* (d) *Pongamia*
70. The major source of food and fodder is
    (a) Fungi (b) Cereals (c) Algae (d) Lichens
71. Tendu leaf 'Bidi Patta' is obtained from
    (a) *Depterocarpus* (b) *Piper betel*
    (c) *Diospyros melanoxylon* (d) *Diospyros ebenum*
72. Medicine to check high blood pressure is obtained from
    (a) *Digitalis* (b) *Cinchona*
    (c) *Rauwolfia* (d) None of the above
73. The legume which is cultivated for both proteins and oils is
    (a) *Cicer arietinum* (b) *Cajanus cajan*
    (c) *Vigna radiata* (d) *Glycine max*
74. Marijuana, a drug, is obtained from
    (a) *Canabis indica* (b) Opium
    (c) Fungus (d) *Canabis sativa*
75. The fibres obtained from Jute is
    (a) Bark (b) Pith (c) Xylem (d) Endodermis

76. Essential oils are those which are
    (a) used in soap manufacture (b) used in perfumes
    (c) essential for plants producing them (d) essential for human kind
77. Groundnut in India is mainly cultivated in
    (a) Bihar (b) Assam (c) Kerala (d) Gujarat
78. Which state in India is a major producer of saffron
    (a) Jammu and Kashmir (b) Rajasthan
    (c) Uttar Pradesh (d) Tamil Nadu
79. Which of the following is used to prepare a milk-like preparation?
    (a) Barley (b) Pigeon pea (c) Soybean (d) Mustard
80. The major Rabi crop in India is
    (a) Maize (b) Barley (c) Wheat (d) None of these
81. From which part of *Atropa belladonna* the drug belladonna is obtained?
    (a) Root (b) Stem
    (c) Leaves (d) All parts of the plant
82. Which medicinal plant is becoming extinct in India?
    (a) *Rauwolfia* (b) *Aconitum* (c) *Trapa* (d) All the above
83. *Digitalis purpurea* belongs to the family
    (a) Papilionaceae (b) Acanthaceae
    (c) Solanaceae (d) Scrophulariaceae
84. Cloves are obtained from *Eugenia aromatica* and the morphology of the product being
    (a) dried pedicels (b) dried seeds
    (c) product of petioles (d) dried flower buds
85. Plant parts useful for extraction of opium from *Papaver somniferum* are
    (a) Young seedlings (b) Ripened seeds
    (c) Old leave (d) Unripe fruits
86. The commercial jute fibres are
    (a) Phloem fibres (b) Xylem fibres
    (c) Interxylary fibres (c) None of these
87. The economic product of tobacco plant is
    (a) Flowers (b) Roots (c) Leaves (d) Stems
88. Short fibres are known as
    (a) Flint (b) Fluff (c) Lint (d) Fuzz

89. Long fibres are known as

(a) Lint (b) Fuzz (c) Fluff (d) Flint

90. The bark of *Cinnamomum zeylanicum* is used as spice because of

(a) etheral oils secreted (b) aromatic oil secreted

(c) a sweet flavor (d) none of these

91. Nicotine is obtained from which part of the plant

(a) root (b) leaf (c) stem (d) flowers

92. Corms and root tubers of Arecaceae have a pungent acrid taste because they contain

(a) latex (b) alkaloid

(c) aromatic oil (d) oxalate crystals

93. Rubber is collected from

(a) crashing the stem of *Hevea brasiliensis*

(b) tapping the stem of *Hevea brasiliensis*

(c) crushing the fruits and collecting the latex of *Achras sapota*

(d) tapping the stem of *Carica papaya*

94. Which of the following is the man made cereal?

(a) *Hordeum vulgare* (b) *Oryza sativa*

(c) Triticale (d) *Triticum aestivum*

95. Major cotton producer state in India

(a) Tamil Nadu (b) Andhra Pradesh (c) Maharashtra (d) Punjab

96. Food grains among the following are

(a) Caryopsis (b) Cypsela (c) Siliqua (d) Sorosis

97. Opium yields

(a) Alkaloids (b) Latex (c) Resin (d) Tannin

98. Which state of India, is the largest producer of jute?

(a) Andhra Pradesh (b) Gujarat (c) West Bengal (d) Kerala

GENERAL HEALTH RELATED QUESTION BANK RELATED TO PLANTS

1. The most commonly used preservative added to jams, jellies etc is
   (a) Sodium oxalate (b) Calcium oxalate
   (c) Sodium benzoate (d) Sodium thiobenzoate
2. If you are anemic, you must take
   (a) beet root (b) spinach (c) cabbage (d) radish
3. Which of the following properties have been recognized for Neem tree
   (a) Pesticidal (b) Blood purifier (c) Antiseptic (d) All of these
4. Which one of the following vitamins alleviate asthma
   (a) Vitamin B (b) Vitamin C (c) Vitamin K (d) Vitamin A
5. Tobacco has been reported to cure lung cancers, the active principle responsible for this effect is
   (a) Tobacine (b) Tobaccinine (c) Nicotine (d) None of these
6. The most widely used drug in the world is
   (a) Wine (b) Cocaine (c) Caffeine (d) Tobacco
7. Avoid taking over-ripe fruits as it may cause
   (a) Sugar loss to body (b) Food poisoning
   (c) Indigestion (d) Sleepiness
8. If you wish to impart more vitality and resistance to disease; better have
   (a) a tonic (b) milk (c) wine (d) raw juice
9. Which of the following vegetables has highest content of proteins
   (a) Pea (b) Cauliflower (c) Cabbage (d) Drumstick
10. While playing, an average tennis player burns about
    (a) 3.35 cal/min (b) 4.00 cal/ min (c) 5 cal/min (d) 5.5 cal/min
11. How much calories of energy are required by an urban person
    (a) 3000 kcal/day (b) 2000 kcal/day (c) 2500 kcal/day (d) 2200 kcal/day
12. Alcohol, a stimulant, 100 ml of it contains
    (a) 500 calories (b) 1500 calories (c) 1000 calories (d) 700 calories
13. When you have a bad tooth-ache better take
    (a) Hot tea (b) Aspirin
    (c) Cotton soaked with olive oil/long oil and place over the aching tooth
    (d) Place a hot-pack of Vicks over it

14. Complete the famous saying,——————— a day keeps doctors away.
    (a) An orange (b) An amala (c) An apple (d) An egg
15. Herbal formulations have come to the international screen during last one decade and owes a brilliant future prospects, because they are
    (a) Cheap/eco friendly (b) Easily available
    (c) With least side effects (d) All of these
16. Which of the following solid foods can be easily digested by a beginner-infant
    (a) Fruits (b) Wheat (c) Pulses (d) Rice
17. Retinol or carotene, abundantly present in carrot is converted into vitamin—— in the human body
    (a) C (b) B (c) A (d) K
18. Which of the following vitamin is poorly found in fruits and vegetables, but can be synthesized in presence of sun light
    (a) Calciferol (b) Vitamin C (c) Niacin (d) Kampferol
19. A mixture of lemon juice with ginger that prevents dehydration and cold, is called as
    (a) Gingerin (b) Gingerale (c) Lemonin (d) Lemonale
20. The stem bark extract of mango tree has been recently shown to suppress
    (a) Cancer (b) AIDS (c) Dengue (d) Hepatitis
21. Cabbage, carrots and sprouted beans have recently been looked upon as
    (a) Anti-AIDS agents (b) Anti-cancer agents
    (c) Anti-diabetic agent (d) Anti-pyretic agents
22. For proper health and body development, the amount of vitamins and minerals can be compensated by taking
    (a) A glass of milk and one fruit (b) Two orange per day
    (c) Any one fruit per day (d) One apple per day
23. Isolates from the leaves of the tree checks the development of AIDS, name the tree?
    (a) Papaya (b) Mango (c) Apple (d) Fig
24. Coffee has been recently shown to prevent 'Pellagra' a deficiency disease due to the presence of
    (a) Vit- C (b) Vit- B (c) Vit- B3 (d) Vit- K
25. Which of the following vegetables is good for bones, teeth and skin
    (a) Cabbage (b) Pumpkin (c) Carrot (d) Spinach
26. A daily dose of 150 mgs of vitamin C can be compensated by taking
    (a) A lemon (b) An orange (c) Two orange (d) An amala

27. Normal cold can be effectively treated and cured with
    (a) Garlic juice + oil
    (b) Garlic + onion juice + water
    (c) Onion juice + oil
    (d) All of these
28. Highest content of dietary fibres can be found in
    (a) Rice bran (b) Oats bran (c) Wheat bran (d) Corn bran
29. Which of the following vegetables contains highest water content
    (a) Cucumber (b) Carrot (c) Potato (d) Radish
30. Which of the following vegetables is iron-rich
    (a) Spinach (b) Cabbage (c) Beans (d) Cauliflower
31. Retinol- which is good for eyes can be found in
    (a) Carrot (b) Green leafy leaves(c) Yellow fruits (d) All of these

SMALL ANSWER QUESTIONS

1. All cereals belong to which family of plants?

   Ans. Poaceae (grass family).

2. Name any pseudocereal.

   Ans. *Fagopyrum esculentum* (buckwheat).

3. Name the famous Russian botanist who worked a lot on the origin of cultivated plants.

   Ans. Nikolai Ivanovich Vavilov.

4. Name a man made cereal?

   Ans. Triticale.

5. Triticale produced by the crossing of

   Ans. *Triticum* and *Secale*.

6. *Apium* belongs to which family?

   Ans. Apiaceae.

7. What is golden tip in tea industry?

   Ans. The terminal buds of tea plant are the most valued in the trade and commonly known as golden tips.

8. Who put forward the concept of mega gene centre for the origin of cultivated plants?

   Ans. Zhukovsky (1968).

9. Name a tall variety of wheat which is highly resistant to rusts.

   Ans. N.P. 846.

10. Name an artificial textile fiber, which is prepared from zein.

    Ans. Vicara.

11. Which tree is to be considered the man's most useful tree?

    Ans. *Cocus nucifera.*

12. What is the botanical name of rye?

    Ans. *Secale cereale*

13. Name the state which is leading in coffee production in India.

    Ans. Karnataka.

14. Write the botanical names of three important millets.

    Ans. *Pennisetum typhoides, Sorghum bicolor, Eleusine corocana.*

15. Name the protein obtained from maize which is used as a foaming agent in the fire extinguishers.

    Ans. Zein.

16. Name the native place of chickpea?

    Ans. Southern Asia.

17. Which fruit plant is also known as 'Adam's fig'?

    Ans. *Musa paradisiaca.*

18. Name two important sugar yielding plants of India.

    Ans. *Saccharum officinarum,* and *Beta vulgaris.*

19. Where is Sugarcane Breeding Institute located in India?

    Ans. Coimbatore.

20. Write the alkaloids present in sesame oil.

    Ans. Sesamin, sesamolin.

21. What is the use of sesamol?

    Ans. It is a powerful antioxidant.

22. Name the semi-drying oil.

    Ans. *Sesamum indicum* oil.

23. Name the source of cocoa butter.

    Ans. *Theobroma cocoa* seeds.

24. Name the part of plant from which palm oil is extracted.

    Ans. Fibrous mesocarp of *Elaeis guineensis.*

25. Which country is known as 'The Home of Spices'?

    Ans. India.

26. Name a spice which is obtained from the bark.

    Ans. *Cinnamomum zeylanicum.*

27. Name the location of Coir Board in India.

    Ans. Cochin.

28. Which non-alcoholic beverage is commonly used in India?

    Ans. *Camellia sinensis.*

29. Coffee plant belongs to which family?

    Ans. Rubiaceae.

30. What is Tossa jute?

    Ans. The fibres of *Corchorus olitorius* are yellowish-red or brown in colour are known as tossa jute.

31. Cocoa, is obtained from which part of plant?

    Ans. Roasted kernels of ripe seeds.

32. Name the major coffee plantation state in India.

    Ans. Karnataka state.

33. Which spice is considered as 'Queen of spice'?

    Ans. Cardomom.

34. Name the metal utilized in the hydrogenation of oils.

    Ans. Nickel.

35. What are fumitories?

    Ans. Substances which are used for smoking.

36. Which spice is known as king of Indian spices?

    Ans. *Piper nigrum.*

37. What kind of wood is present in gymnosperm?

    Ans. Non-porous.

38. What is the meaning of 'CTC' tea?

    Ans. The term 'CTC' refers to Crush, Tear and Cure.

39. What is the source of sun hemp fibre?

    Ans. It is obtained from the stem of *Crotolaria juncea*.

40. Which plant produce quinine drug, write botanical name of the plant?

    Ans. *Cinchona* sp.

41. Cardamom of commerce are dried aromatic fruits belongs to which family?

    Ans. Zingibraceae.

42. What is the botanical name of Bhojpatra?

    Ans. *Betula utilis*.

43. Name the drug and its source which is used in ophthalmology to dilate pupil.

    Ans. Belladonna, from the leaves of *Atropa belladonna.*

44. What is the name of cork which is removed first time from plant?

    Ans. Virgin cork.

45. What is retting?

    Ans. The jute fibres are separated from the stems by the process of retting, which is usually done in gently flowing, deep, clear and tepid water.

46. Name two important surface fibers obtained from plants.

    Ans. Cotton and kapok.

47. Name the botanical name, family and source of coir.

    Ans. *Cocos nucifera*, Arecaceae and mesocarp.

48. Name the characteristic fatty acid present in mustard oil.

    Ans. Erucid acid.

49. Name the alkaloids present in areca nut.

    Ans. Arecoline, arecaidine, guvacoline etc.

50. The pungency of mustard oil is due to the presence of

    Ans. Allyl-iso-thiocyanate.

51. What is charas?

    Ans. The pure and undiluted, sticky yellow resin which is exuded from the female flowers of *Canabis sativa* is known as charas.

52. What is 'toughness' in woods?

    Ans. This is the ability of the wood not to tear off or split even after continuous hard use.

53. Name the Indian state which produces highest quantity of *Piper nigrum*.

    Ans. Kerala.

54. What are the important techniques used for the seasoning of woods?

    Ans. Open air seasoning and kiln seasoning.

55. Write the name of three important timber yielding plants of India.

    Ans. *Tectona grandis* (teak). *Cedrus deodara* (deodar) and *Dalbergia sissoo* (shisham).

56. Name the two most important alkaloids in coffee.

    Ans. Caffeine and caffetonic acid.

57. Name the source of citronella oil.

    Ans. Leaves of *Cymbopogon nardus*.

58. What are fumitories and masticatories?

    Ans. The substances used for smoking are known as fumitories and those for chewing are called masticatories.

59. Write the plant name which belongs to Euphorbiaceae family and produce rubber.

    Ans. *Hevea brasiliensis*.

60. What is 'Bhojpatra' ?

    Ans. Thin sheets of the bark of *Betula utilis*.

61. Savoury seeds are highly aromatic fruits belonging to which family?

    Ans. Apiaceae.

62. What is the source of vetiver oil?

    Ans. Roots of *Vetiveria zizanioides*.

63. Where is the Potato Research Institute located in India?

    Ans. Shimla.

64. Name the centre of origin of coffee?

    Ans. Ethiopia.

65. Write the examples of fermented alcoholic beverages.

    Ans. Beers and wines.

66. Name the two largest tea producing countries.

    Ans. Ceylon and India are the two largest tea producing countries.

67. What are non-drying oils?

    Ans. The oils do not absorb atmospheric oxygen and as such are incapable of forming elastic film after long exposure to air.

68. Give the example of some non-drying oils.

    Ans. Ground nut oil, castor oil, olive oil etc.

69. What are drying oils?

    Ans. They readily absorb atmospheric oxygen when exposed and form a dry thin elastic film.

70. The aroma and flavour of the tea is due to the presence of

    Ans. Ethereal oil and theol.

71. Which spice is known as 'Black Gold of India'.

    Ans. *Piper nigrum.*

72. Give example of drying oil.

    Ans. Linseed oil, Safflower oil etc.

73. What is hard wood and soft wood?

    Ans. Technically, porous and non-porous woods are known as hard wood and soft wood.

74. What is ethnology?

    Ans. Studies of medicinal plants based on traditional literature and its investigation in the modern light is known as ethnology.

75. Who wrote the book 'Origin of Cultivated Plants'?

    Ans. De Candolle in 1883.

76. Who proposed diffusion origin theory for the origin of cultivated plants?

    Ans. Harlan in 1971.

77. Name the native place of mango plant.

    Ans. South East Asia.

78. The special qualities of castor oil is due to the presence of

    Ans. Ricinoleic acid.

79. Name the place where Central Tobacco Research Institute is located.

    Ans. Rajahmundry.

80. What is the centre of origin of clove?

    Ans. Volcanic islands of Moluccas.

81. Give an example of vegetable fats.

    Ans. Coconut oil.

82. Coffee is usually blended with the powder of roots of which plant?

    Ans. Chicory (*Cichorium intybus*).

83. What is caffelite?

    Ans. It is a plastic like material which is obtained from coffee seeds.

84. What is the use of vulcanization?

    Ans. Vulcanization increases elasticity, tensile strength and abrasion resistance of rubber.

85. What are the methods for the extraction of oils from plants?

    Ans. Mechanical expression and solvent extraction.

86. Name the plant where flower bearing branches is pushed underground after fertilization.

    Ans. Ground nut.

87. Name the principal fatty acid present in castor oil.

    Ans. Ricinoleic acid.

88. What are semi-drying oils?

    Ans. Intermediate between drying and non-drying oil and absorb atmospheric oxygen slowly and produce a soft film even after long exposure.

89. Write the examples of semi-drying oils.

    Ans. Cotton seed oil, sesame oil, rape oil etc.

90. Write down the types of mechanical compression of oil.

    Ans. Hot and cold.

91. Which plant is mostly used for the making of musical instruments?

    Ans. Bottle gourd (*Lagenaria vulgaris*).

92. What are the solvents used for the extraction of vegetable oils?

    Ans. Gasoline, benzene, carbon disulphide and petroleum ether etc.

93. Which plant produce aconite?

    Ans. *Aconitum ferox.*

TRUE OR FALSE QUESTIONS

1. Rice is essentially a temperate crop.

   Ans. False

2. Zhukovsky put forward the concept of mega gene centre of the origin of cultivated plants.

   Ans. True

3. N. I. Vavilov made a milestone contribution on the origin of cultivated plants.

   Ans. True

4. Olong tea is a partially fermented product, intermediate between green and black tea.

   Ans. True

5. Wheat is the principal cereal crop of the temperate regions of the world.

   Ans. True

6. Wheat is grown as Rabi crop in India.

   Ans. True

7. Curing and fermentation of tobacco leaves impart required texture and aroma to them.

   Ans. True

8. N. I. Vavilov also recognized secondary centre of origin.

   Ans. True

9. Ten center of origin of cultivated plant was recognized by Vavilov.

   Ans. False

10. The rice grain is a cypsela which remains enclosed by golden or brown husk.

    Ans. False

11. The botanical name of emmer wheat is *Triticum dicoccum.*

    Ans. True

12. Ganga 5 is a wheat variety, which is resistant to downy mildew infection and lodging.

    Ans. False

13. High temperature with high humidity is harmful for the wheat crop.

    Ans. True

14. Baggase is the byproduct of sugar industries.

    Ans. True

15. Corn oil is a drying oil, which is obtained from dried germs of maize.

    Ans. False

16. Assam and West Bengal are the highest tea producing states in India.

    Ans. True

17. The main credit to develop improved varieties of wheat in India goes to Dr. B.P. Pal and M.S. Swaminathan.

    Ans. True

18. On hydrolysis, sesamolin produce sesamol.

    Ans. True

19. Potato was introduced from new world to the old world.

    Ans. True

20. *Hordium vulgare* is the botanical name of the oat.

    Ans. False

21. Oat plant can be distinguished from other cereal plants by bluish tint of the foliage.

    Ans. True

22. Rye is essentially a hot weather crop.

    Ans. False

23. Banana, mango, citrus are indigenous to India.

    Ans. True

24. Small seeded cereals and forage grasses are known as millets.

    Ans. True

25. Apple and pear are considered as false fruits.

    Ans. True

26. The botanical name of ragi is *Secale cereale.*

    Ans. False

27. Due to the resemblance of the spike to finger, *Eleusine corocana* is known as finger millet.

    Ans. True

28. Castor oil has fungicidal and antibacterial properties.

    Ans. True

29. Ground nut is a rain fed Kharif crop.

    Ans. True

30. Drying oils are used in paint and varnish industries.

    Ans. True

31. Semi-drying oils are used in making soaps and candles.

    Ans. True

32. Non-drying oils are used as food, lubricants etc.

    Ans. True

33. Vegetable ghee can be prepared by the hydrogenation of vegetable oils.

    Ans. True

34. Date palm was perhaps the earliest fruit cultivated by man.

    Ans. True

35. Tobacco is essentially a tropical crop but can be grown in almost all agroclimatic conditions.

    Ans. True

36. The flavour of coffee is due to the presence of a volatile oil, caffeol.

    Ans. True

37. Fenugreek is a cold seasoned crop and is fairly resistant to frost.

    Ans. True

38. Simple fruits are developed from a single ovary.

    Ans. True

39. Saffron is the dried stigmas and tops of styles of *Crocus sativa.*

    Ans. True

40. Mustard oil is an example of drying oil.

    Ans. False

41. Dried unopened flower buds of *Syzyzium aromaticum* are known as cloves.

    Ans. True

42. Dasheri, Langra, Alphonso, chausa are mango varieties.

    Ans. True

43. The botanical name of pigeon pea is *Cicer arietinum* and Africa is its native place.

    Ans. False

44. *Allium cepa* is the botanical name of garlic and belongs to family Liliaceae.

    Ans. False

45. Soybean oil is an example of non-drying oils.

    Ans. False

46. The fruit is a mature and ripened ovary.

    Ans. True

47. Banana is also known as 'Apple of Paradise'.

    Ans. True

48. Banana is also known as 'Adams fig'.

    Ans. True

49. Guava, the rich source of vitamin C, is the native of tropical America.

    Ans. True

50. The thalamus part of apple is edible.

    Ans. True

51. Non-drying oils readily absorb atmospheric oxygen when exposed and from a thin dry elastic film.

    Ans. False

52. Sugar can also be obtained from sugar maple (*Acer saccharum*).

    Ans. True.

53. In India barley is grown as rabi crop.

    Ans. True

54. Spices are natural plant products which are used for imparting flavour, aroma and piquancy and for seasoning of foods.

    Ans. True

55. Saffron is the dried unopened floral bud of a tree, *Crocus sativus*.

    Ans. False

56. Sugarcane Breeding Station is located at Coimbatore.

    Ans. True

57. Lerma Rojo 64 is a dwarf wheat variety.

    Ans. True

58. Cedar wood oil, which has a high refractive index, is used as mounting medium with oil-immersion lenses in microscopy.

    Ans. True

59. Macerization of cotton is done by treatment of fabric with caustic soda solution.

    Ans. True

60. Central Rice Research Institute is located in Cuttack.

    Ans. True

61. The center of origin of rice plant is India.

    Ans. True

62. Maize is a subtropical crop.

    Ans. True

63. Red wood is obtained from *Shorea robusta.*

    Ans. False

64. The cereal obtained by the crossing of wheat and rye is named as triticale.

    Ans. True

65. Studies on the medicinal plants, based on old literature and modern investigation is known as ethnology.

    Ans. True

66. Cocoa butter is an example of vegetable fat.

    Ans. True

67. Bala, Jamuna, Karishna are some improved rice varieties.

    Ans. True

68. Olive oil is a semi-drying type of oil.

    Ans. False

69. Alluvial and loam soils are most suitable for maize cultivation.

    Ans. True

70. Potato is an introduced crop in India.

    Ans. True

71. Multiple fruits are developed from the entire inflorescence.

    Ans. True

72. Pharmacology deals with the action of drugs.

    Ans. True

73. Ganga, Shkati and Amber are the improved maize varieties.

    Ans. True

74. Coffee and tea are alcoholic beverages.

    Ans. False

75. Red pepper belongs to the family Piperaceae.

    Ans. False

76. An artificial textile fibre, known as Vicara, is prepared from Zein.

    Ans. True

77. Cloves are directly obtained from the seeds.

    Ans. False

78. The aerial portion of banana is a pseudo-stem.

    Ans. True

79. Charas is the dried juice that exudes from the male flowers of *Cannabis sativa.*

    Ans. False

80. Barley is the world's fourth most important cereal crop.

    Ans. True

81. Cotton fibres are epidermal prolongations of seed coat of cotton plant belonging to the family Crusiferae.

    Ans. False

82. Barley grows best in cool climate but does not prefer hot and humid climate.

    Ans. True

83. All cereals belong to family Liliaceae.

    Ans. False

84. The native of Brinjal crop is India.

    Ans. True

85. In India bajra is grown as a rabi crop.

    Ans. False

86. Indian Institute of Sugarcane Technology is situated at Kanpur.

    Ans. True

87. The characteristic of cereals fruit is caryopsis in which pericarp remains completely fused with seed coat.

    Ans. True

88. India is known as the 'Home of Spices.'

    Ans. True

89. Hilly areas are most suitable for rye cultivation.

    Ans. True

90. Proso millet is also known as common millet or hog millet.

    Ans. True

91. Coffee Board is responsible for maintaining the quality of coffee.

    Ans. True

92. Spices are generally highly nutritious.

    Ans. False

93. Lettuce produces lactucarium used as hypnotonic in bronchitis and asthma.

    Ans. True

94. Ginger is the underground stem of a perennial herb, *Zinziber officinale*.

    Ans. True

95. Seeds of *Papaver somniferum* are free from narcotics.

    Ans. True

96. The true stem of banana plant is greatly reduced to underground corm.

    Ans. True

97. The branch of medicine dealing with the history, botany, identification, preservation and commerce of crude drugs is known as pharmacognosy.

    Ans. True

DESCRIPTIVE TYPES QUESTIONS

1. What are the grades of tea available in market?

   The following grades/classes of tea are available in Indian market:

   - Broken orange pekoe
   - Flowery broken orange pekoe
   - Broken pekoe
   - Orange pekoe
   - Flowery pekoe
   - Broken orange pekoe fannings
   - Fannings and dust

2. Write down the types or groups of fibres.

   The fibres are grouped in following categories:

   - Textile fibres: Fibres utilized for the manufacture of fabrics netting, cordage etc.
   - Brush fibres: Fibres utilized in the manufacture of brushes and brooms.
   - Rough weaving fibres: Fibres utilized in making baskets, chairs, mats etc.
   - Natural fibres: Natural fibres used as cloth without weaving.
   - Filling fibres: Fibres used for stuffing mattresses, cushions, pillows etc.
   - Paper-making fibres: Wood fibres utilized for making of paper in paper industry.

3. Write a note on the various uses of coconut palm.

   Each and every part of coconut palm is useful, some parts and their use are mentioned below:

   - Coir: Coir is a natural fibre extracted from the husk of the coconut. It is rot proof, sound proof, water damp and pest resistant. The items like fibres, yarn, mats, mattings, rugs, carpets, rope, rubberized coir etc. are made by coir.
   - Coir dust: A waste product in the coir industry. It can be used as manure. It is excellent surface soil mulch. It can also be used for building slabs, or for hard boards.
   - Shell powder: It is used as compound filler for synthetic resin glues and surface smoothing of mould articles.
   - Leaf blade: Dried leaf blades, cut into pieces and can be used as packing material.
   - Tender husk: Orange dwarf, tender nut husk, and its juice are extensively used in Ayurvedic medicine.
   - Coconut water: Used as medicine.
   - Copra: It is the dried meat and is a highly valued commodity for oils and fats.

- Coconut oil: It is an essential ingredient in manufacture of soap, margarine and other industries. This oil is extensively edible as a substitute of ghee.
- Vinegar: Coconut water is fermented, kept in the sun light and decanted. After adding acetic acid, the clear solution is vinegar.
- Toddy: It is a sugar containing juice, obtained by tapping the unopened spadix of the coconut palm. It is an excellent beverage, and a rich source of sugar and of baker's yeast. When it undergoes fermentation it forms an alcoholic drink.

4. Name the important gum producing plants.
   - *Acacia nilotica*
   - *A. catechu*
   - *A. modesta*
   - A. senegel
   - *Anogeissus latifolia*
   - *Bauhinia retusa*
   - *Lannea coromandelica*
   - *Sterculia urens*
   - *S. villosa*
   - *Pterocarpus marsupium*

5. What is fenny?

   It is a famous liquor of Goa prepared by the Cashew apple. It is also known as Cashew apple liquor of Goa. It is unique liquor obtained from the excessive distillation of cashew apple juice. It is liquor perfect by itself without the addition of any extraction ingredients and having its own unique flavour.

6. What are the special things to remember before energy plantations?
   - The plant species should perfectly be local as they are better adapted to climatic and soil conditions of the area.
   - The plant should show rapid growth and high coppicing ability.
   - The plant species should be hard.
   - Besides, providing fuel wood, the species should preferably produce additional products like fruits, seeds, fodder, green manure, medicines, tannins etc.
   - The species selected should have high calorific value of wood.
   - The species should have low requirement of water and fertilizer and must have the ability to improve soil quality.

7. What are the quality determining factors of Jute fibre? What is white and tossa jute?

The quality of jute fibres is determined by their fineness, colour, lusture, length and strength. The fibres obtained from *Corchorus capsularis* are better quality than those of *C. olitorius*. The fibres obtained from *C. capsularis* are usually white in colour and hence known as white jute. The fibres of *C. olitorius* are yellowish-red or brown in colour and known as tossa jute.

8. What are the different types of cotton (*Gossypium*)?

The genus *Gossypium* is divided into four main groups on the basis of geographical distribution and chromosome number:

- Cultivated cottons of the old world (2n =13)
- Cultivated cottons of the new world (2 =26)
- Wild cottons of the old world and new world (2 =13)
- Wild cottons of Polynesia (n =26)

All cultivated cottons belong to the four species, *G. arboretum*, *G. herbaceum*, *G. hirsutum*, *G. barbadense*.

9. What are the alkaloids obtained from Sarpagandha?

The total alkaloid content of the Sarpagandha root varies from 1.7 to 3%. The bark alone accounts for nearly 90% of alkaloid contents. Pharmacological activity is due to the presence of several alkaloids, of which reserpine is the most potent. Other important alkaloids present are serpentine, serentinine, reserpinine, rauwolfinine, deserpidine, ajmaline, ajmalicine, yohimbine, isoaimaline etc.

10. What are different grades of commercial Betel leaves?

Bleached leaves are cleaned and sorted into different grades, according to their size, colour, texture, position on the vines, maturity and of plucking. The different grades are Mahoba, Maghai, Bangla, Jagannathi, Banarasi, Mandava, Poona, Kapoori, Desi etc.

11. Write a note on the properties of cocoa?

The roasted seeds of cocoa contain:

- 3% theobromine
- 30-50% fatty oils
- 15% starch
- 155 proteins
- Minerals like iron, phosphorus, manganese, zinc, aluminium etc.

The stimulating qualities of cocoa are due to the presence of theobromine and small quantities of caffeine, and the characteristic aroma due to the essential oil, 'Cacool'.

Tables

Plants with good quality fuel wood.

Common name of the plant	Botanical name
1. Babool	*Acacia nilotica*
2. Haldu	*Adina cordifolia*
3. Jamun	*Syzygium cuminii*
4. Shisam	*Delbergia sissoo*
5. Neem	*Azadirachta indica*

Plant which are used for paper production.

Name of the plant	Botanical name
Bamboo	*Bambusa arundinacea*
Pycea	*Picea smithiana*
Bamboo grass	*Eulaliopsis binata*
Bamboo	*Bambusa polymorpha*
Kagaji Bamboo	*Dendrocalamus humiltooni*

Important timber yielding plants.

Common name of the plant	Botanical name
Babul	*Acacia arabica*
Paludar	*Abies pindrow*
Toon	*Cedrella toona*
Deodar	*Cedrus deodara*
Shisam	*Delbergia sissoo*
Mahua	*Madhuca indica*
Mango	*Mangifera indica*
Pinus	*Pinus rouxburghii*
Bellow	*Salix alba*
Shal	*Shorea robusta*
Teak	*Tectona grandis*

Rubber yielding plants.

S.No.	Name of the Plant	Family
1.	*Hevea sp.*	Euphorbiaceae
2.	*Castilla elastic*	Moraceae
3.	*Ficus elastic*	Moraceae
4.	*Achras sapota*	Sapotaceae
5.	*Alstonia elastic*	Apocynaceae
6.	*Euphorbia tirucalli*	Euphorbiaceae
7.	*Funtumia elastic*	Apocynaceae

Major oil yielding crops.

Common name	Botanical name	Source	Family
Ground nut	*Arachis hypogaea*	Seeds	Fabaceae
Mustard	*Brassica campestris*	Seeds	Brassicaceae
Sesame	*Sesamum indicum*	Seeds	Pedaliaceae
Coconut	*Cocos nucifera*	Dried copra	Arecaceae
Castor	*Ricinus communis*	Seeds	Euphorbiaceae
Palm	*Elaeis guinensis*	Fruit	Arecaceae
Olive	*Olea europaea*	Mesocarp	Arecaceae
Cotton	*Gossypium* sp.	Fruit pulp	Oleaceae
Sunflower	*Helianthus annus*	Seeds	Malvaceae
Linseed	*Linum usitatissimum*	Seeds	Asteraceae
Safflower	*Carthamus tinctorius*	Seeds	Linaceae
Soybean	*Glycine max*	Seeds	Asteraceae
Clove	*Syzygium aromaticum*	Seeds	Fabaceae
		Flower buds	Myrtaceae

Centre of origin of some crop plants.

Centre of origin	Name of the crop
New world	
Peruvian Andes	Potato Tamato

(*Contd.*)

Centre of origin	Name of the crop
Brazil	Pine apple *Hevea*
United States of America	Sunflower
Maxico and central America	Maize
Old world	
Asia minor and Afganistan	Lentil, rye, apple, pear, almond, pomegranate
Ethiopia	Barley, coffee, sorghum,
Mediterranean sea	Beet, lettuce, oat, cabbage, olive
China	Soyabean, tea, onion, wheat
South-West Asia	Rice, Banana, orange, brinjal, pigion pea,
South-East Asia	Sugarcane, mango
Europian Siberia	Cherry, black pepper

Plants of medicinal importance.

Name of the medicine	Name of the plant	Useful part of the plant
Quinine	*Cinchona sp.*	Bark
Black pepper	*Piper nigrum*	Fruit
Sentonine	*Artimisia cina*	Bud
Aconite	*Aconitum napellus*	Tuberous root
Licorine	*Glycyrrhiza glabra*	Root
Oak	*Calotropis procera*	Stem
Colchicum	*Colchicum autumnale*	Root
Ephedrine	*Ephedra equisetina*	Wood
Neem	*Azadirachta indica*	Leaves
Sun hemp	*Cannabis sativa*	Leaves
Sandal	*Santalum album*	Stem
Opium	*Papaver somniferum*	Fruits
Husk of isabgol	*Plantago* sp.	Seeds
Kaiphal	*Myrica esculenta*	Bark
Belladonna	*Atropa belladonna*	Leaves
Sweet flag	*Acorus calamus*	Root

Important fibre yielding plants.

Common name	Botanical name	Source	Family
Cotton	*Gossypium* sp.	Seed hair	Malvaceae
Jute	*Corchorus capsularis*	Bast	Tiliaceae

(*Contd.*)

Common name	Botanical name	Source	Family
Flax	*Linum usitatissimum*	Bast	Linaceae
Hemp	*Cannabis sativa*	Bast	Cannabinaceae
Manila hemp	*Musa textilis*	Outer part of leaf stalk	Musaceae
Coir	*Cocos nucifera*	Mesocarp	Arecaceae
Sisal hemp	*Agave sisalana*	Leaves	Agavaceae

Important Petro-crop plant species:

Name of the plant	Family
Copaifera langsdorfii	Euphorbiaceae
Euphorbia antquorum	,,
Hevea brasiliensis	,,
Jatropha curcas	,,
Asclepias curassavia	Asclepiadaceae
Calotropis gigantic	,,
Ceropegia tuberosa	,,
Pergularia daemia	,,
Allamanda cathartica	Apocynaceae
Nerium odorum	,,
Thevetia nerifolia	,,
Wrightia tomentosa	,,
Agave americana	Agavacea
Aloe vera	Liliaceae
Pittosporum	Pittosporaceae
Madhuca indica	Sapotaceae
Pedalium murex	Pedaliaceae
Argemone mexicana	Papaveraceae

Plants for energy production.

Name of the plant	Family
Memecylon edule	Melastomataceae
Ipomoea carnea	Convolvulaceae
Clerdendron inerme	Verbenaceae
Eucalyptus globulifera	Myrtaceae
Ficus bengalensis	Moraceae
Moringa tinctoria	Moringaceae

(*Contd.*)

Name of the plant	Family
Lannea coromandelica	Anacardiaceae
Grevellia robusta	Proteaceae
Ailanthus excelsa	Simaroubaceae
Dipterocarpus indicus	Dipterocarpaceae
Casuarina equisetifolia	Casuarinaceae
Butea monosperma	Fabaceae

Centre of origin, and centre of production of some crop plants.

Crop	Centre of production	Centre of origin
Groundnut	India	Branzil and Peru
Oilplam	Malysia	Tropical Africa
Rubber	Malysia and Indonasia	Brazil
Wheat	North America	Central Asia
Potato	Eastern Europe	Peru
Monterey pine	Australia	California
Pine apple	Hawaii	Brazil
Cocoa	Africa	Brazil
Coffee	Brazil and Central America	Ethiopia
Maize	Mid-West America	Tropical America

List of plants used as spices.

Name of the spice	Botanical name	Source of spice	Family
Black pepper	*Piper nigrum*	Fruits	Piperaceae
Cardamom	*Elettaria cardomomum*	Seeds	Zingiberaceae
Nutmeg	*Myristica fragrans*	Leaves	Myristicaceae
Anise	*Pimpinella anisum*	Seeds	Apiaceae
Fenugreek	*Trigonella foenum-graceum*	Seeds	Fabaceae
Peppermint	*Mentha piperita*	Leaves	Lamiaceae
Indian mustard	*Brassica juncea*	Seeds	Brassicaceae
Caraway	*Carum carvi*	Seeds	Apiaceae
Coriander	*Coriandum sativum*	Seeds	Apiaceae
Cumin	*Cuminum cyminum*	Seeds	Apiaceae

(*Contd.*)

Name of the spice	Botanical name	Source of spice	Family
Ginger	*Zingiber officinale*	Rhizome	Zingiberaceae
Turmeric	*Curcuma longa*	Rhizome	Zingiberaceae
Clove	*Eugenia aromaticum*	Flower bud	Myrtaceae
Saffron	*Crocus sativus*	Flower	Iridaceae
Cinnamon	*Cinnamomum zeylanicum*	Bark	Lauraceae
Cassia	*C. cassia*	Bark	Lauraceae

Plant used as green manure.

Common name	Botanical name
Sun hemp	*Crotolaria juncea*
Cowpea	*Vigna sinensis*
Cluster bean	*Cyamopsis tetragonoloba*
Lentil	*Lens esculenta*
Sesbania	*Sesbania aculenta*
Horse-gram	*Macrotyloma uniflorum*
Egyptian clover	*Trifolium alexandrium*

Types of coffee with it special characteristics.

Coffee variety	Type	Special characters
Arabican coffee	Plantation	Long and wide beans, grayish or bluish in colour.
	Cherry	Same as above, but colour is golden brown to brown.
Robusta	Plantation	Broad oval beans, smaller in size than Arabica, grayish in colour.
	Cherry	Same as above, but colour golden brown to brown

Composition of Opium latex (25 types of alkaloids).

Name of the alkaloid	Total %
Morphine	9-14%
Codeine	0.5-2%
Nascaphine	3-10%

(Contd.)

Name of the alkaloid	Total %
Papaverine	0.8-1%
Thebaine	0.2-1 %
Narcine	0.2-4%

ECONOMIC BOTANY—A QUICK GLANCE

1. The study of plants useful to human beings and application of botanical knowledge comes under the economic botany.
2. The study of medicinal plants in known as pharmacognosy.
3. The study of action of medicine is known as pharmacology.
4. Hippocrates is known as father of medicine.
5. Dioscorides is known as father of pharmacognosy. He wrote 'Materia Medica'.
6. *Cinchona* also known as Peruvian bark or Countless bark.
7. The maximum quantity of quinine is obtained from *Cinchona ledgeriana*, it is generally grown in Java island.
8. *Cinchona* was discovered by La-Condamine in South America (Peru).
9. The name *Cinchona* was given by C. Linnaeus.
10. Anderson introduced this plant in India from Java Island.
11. Cereals are also called stuff of life because these are most important source of food for human being which is rich in carbohydrates and also contain proteins, fats, vitamins and minerals.
12. The medicine quinine is obtained from the bark of this plant of *Cinchona.*
13. In India, wheat is basically a rabi crop (winter crop), spring wheat have short growing season of about 100 days and winter wheat are of long duration, need relatively low temperatures during early stages of development.
14. Total 25 types of alkaloids are present in the bark of *Cinchona* out of these Quinine, quinidine cinchonine and cinchonidine are most important.
15. Wheat, rice, maize, barley, oat and rye are also called true cereals and all belongs to family Poaceae.
16. Pulses are the rich source of proteins (17-30% on a dry weight basis), they also contain carbohydrates (about 60%) but very low in fat contents (1-2% only).

17. In India the main source of raw material of paper are *Bambusa arundinacea, Dendrocalamus stictus*, *Eulaliopsis binata*, *Pinus roxburghii* and *Boswellia serrata*.
18. Quinine is isolated from the bark in the form of quinine sulphate.
19. Quinine is effective on malaria parasite's asexual Schizont stage.
20. Curcumine and zingiberine are the most important coloured alkaloids in turmeric.
21. *Ferula asafoetida* is also known as Devil's dung, it belongs to family Apiaceae.
22. Legumes are rich source of lysine and threonine amino acids and cereals are rich source of sulphur containing amino acids.
23. Gluten is obtained from wheat grains used for the production of monosodium glutamate, which intensifies the flavour of food.
24. Highly refined wood pulp is used as a basic material in the manufacture of rayon (artificial silk), nitrocellulose membrane and cellophane (transparent film), bioplastic and lacquers.
25. Scutellum in cereal grains is the flat cotyledon, it is in intimate contact with the endosperms and acts as an absorptive organ.
26. Majority of the cultivated fruits belongs to Rosaceae and Rutaceae families.
27. The medicinal product Hing is obtained from the roots, it is an acellular oleo-gum-resin.
28. Hing contains carbon disulphide, pinin, umbelliferone and volatile oil ferulic acid.
29. Hing can cure flatulent colic and can be used as spasm reliever.
30. Opium is also known as Black gold.
31. Maize is also called America's greatest gift to mankind.
32. Citrus is the good source of vitamin C but amla is the richest source of vitamin C.
33. Rice and maize are the dominant cereals of tropical regions but rye and oat dominated in colder regions.
34. Wheat and barley are the most important cereals of temperate zones of the earth.
35. There are total more than 25 alkaloids are present in opium, the most important are: Morphine, codeine, nascaphine, papaverine, thebaine and Narcine.
36. Popcorn (*Zea mays* var. *everta*) contains high percentage of hard starch with a little soft starch in the centre.
37. Most of the cashew nut comes from Kerala followed by Tamil Nadu, Karnataka and AP.
38. Diacetyl morphine (Heroin) is several times stronger than opium.
39. Morphine is a strong analgesic and it is used as a pain killer.
40. The name *Rauwolfia serpentina* was given by Charles Plumier in the honor of German Scientist Leonard rauwolf.

41. More than 50% of world's population whole or partially depends on rice for food.
42. Popping is possible with the kernels of popcorn, when they are exposed to high temperatures, they explode and the grain is turned inside out, thus, snow-white, fluffy palatable mass is obtained which can be eaten as such.
43. Botanically a fibre is a special kind of elongated cell with thick walls and tapering ends.
44. Reserpine and serpentine are the most important alkaloids present in *Rauwolfia serpentina*.
45. Aconite is obtained from the roots of *Aconitum napellus* and it is used in Neuralgia.
46. Cocaine is obtained from the leaves of *Erythroxylum coca* and is used as local anesthesia.
47. Strychnine is isolated from the seeds of *Strychnos nuxvomica* and is used in nervous disorders.
48. India and Indo-China are the centres of origin of rice and then it was introduced into the new world.
49. The kernels of popcorn explode during popping because the central portion of the endosperm contains a fairly large proportion of water, which on heating become vaporize and the steam pressure builts up resulting in explosion.
50. The animal fibres are made up of proteins and vegetable fibres are made up of cellulose.
51. Digitoxin is obtained from the leaves of *Digitalis purpurea* and it is used in heart related problems.
52. Tannins are the derivatives of glycosides and water soluble. These are generally present in unripe fruits.
53. Leaves of tea contain 18% tannin and catha is obtained from the inner-wood of *Acacia catechu* and they also contain tannins in high concentration.
54. Resins are the oxidative products of volatile oils. Conifers contain resins in abundance, e.g. *Abies balsamea*, *Pinus* etc.
55. Due to the presence of glycoside solanine the green potatoes are toxic.
56. Corn oil, corn starch, corn syrup and corn sugar are the useful products of maize.
57. Rice with hull is called paddy, lemma and palea form the hull or husk of the rice.
58. Gum is the water soluble compounds which is obtained from the degradation of cell wall.
59. Gum Arabic is obtained from *Acacia senegal* and *A. arabica*.
60. Rice contains 90% starch and 8-10% proteins, dosa, idli, uppma, uttapam, pulao, kheer etc. are prepared from rice.
61. Zeatin - a cytokinin is obtained from maize grains.
62. The term 'Totaquine' is used for the four alkaloids isolated from *Cinchona*.
63. Guar gum is obtained from the pods of *Cyamopsis tetragonoloba* and Bengal Kino is obtained from the stem of *Butea monosperma*.

64. Plants which produce latex are also called as petrocrops. E.g. *Hevea brasiliensis*, *Papaver* etc.
65. Barley is the principal grain used in malting for brewing beer, whisky, brandy etc.
66. Latex is used for protection, wound healing and for nutrition.
67. *Fagopyrum esculentum*, *Chenopodium quinoa*, *Amaranthus* sp. *Euryale ferox* etc. are also called pseudo-cereals because these are used in manner similar to true cereals and they have a similar grain composition but botanically they are unrelated to cereals.
68. Essential oils are chemically terpenes, they attract insects, repellent animals and are used as antiseptic. E.g. Rose oil, sandal wood oil, *Cinnamomum camphora* oil and *Eucalyptus* oil etc.
69. *Avena sativa* (oats) is the most nutritious of all cereals because it is a rich source of proteins (13.8%), fats (4.3%), vitamin B and minerals.
70. Certain types of acids are also present in plant cells, like, formic acid in *Urtica dioica* (striking nettle), tartaric acid in grapes and Imali, citric acid in citrus and orange, oxalic acid in *Oxalis* and malic acid in *Cicer aritinum.*
71. The biggest starch grains (1/100 mm) are present in *Canna* and smallest in rice cells (1/2000 mm).
72. International Rice Research Institute was established in 1962 at Philippines.
73. Concentric starch grains are present in endosperm of wheat and cotyledons of *Pisum sativum.*
74. Eccentric starch grains are present in potato and rice.
75. Aleurone layer is present in the cereals grains; it is made up of protein. It has two parts: crystalloid and globoid.
76. Wheat is the staple food for one-third of world's population, it is hexaploid (2n = 42) species which is growing in the temperate regions of the world.
77. Crystalloid is basically made up of protein and amides and globoid is made up of protein and minerals, like, enositol phosphate salts.
78. Hemicellulose is present as stored food in Phytelephas (Ivory palm) and used in the preparation of artificial ivory and billiards bolls.
79. *Triticum aestivum* is also called common bread wheat. It is occupy about 86% of acreage in India.
80. Dahalia and Artichoke roots contain inulin (monopolysaccharide) which prepare feather like crystals in the cells.
81. Janus green, neutral red, methylene blue, malachite green are used as vital stains.
82. Fuelgen reaction is used for the staining of DNA and it was developed by Fuelgen and Rosenbeck.
83. *Linum usitatissimum* (flax) fibers is obtained from the pericycle of stem.
84. Cricket bats are manufactured from the wood of *Salix alba* (willow) tree.
85. The wood of *Morus alba* is used for the manufacture of hockey sticks, tennis and badminton rackets and cricket stumps.

86. *Sesbania rostrata* plants have root nodules as well as stem nodules and they fix atmospheric nitrogen with the help of *Rhizobium* in root nodules and *Azorhizobium caulinodans* in stem nodules.

87. Nitrogen fixing bacteria *Frankia* is associated with non-leguminous plants. E.g. *Casuarina*, *Rubus*, *Alnus myrica* etc. contains *Frankia* in their root nodules.

88. *Azotobacter*, *Clostridium* (anaerobic), *Beijerinckia* and *Bacillus polymyxa* are the free living nitrogen fixing bacteria.

89. The loose association of some bacteria with the roots of higher plants was discovered by Dobriner.

90. *Azospirillum lipoferum* is loosely associated with the roots of maize and Brazilian grass and they fix nitrogen.

91. Blue green algae *Anabaena*, *Tolypothrix*, *Aulosira*, *Plectonema* etc. are the examples of nitrogen fixing cyanobacteria.

92. *Aulosira fertilissima* grow and fix nitrogen in rice fields.

93. *Azolla pinnata* is an aquatic fern. It has *Anabaena azollae* in its leaves to fix nitrogen. This fern is used in the rice field.

94. About 97% of world paper is made from wood pulp and out of it, 85% is derived from conifers (e.g. *Picea* sp., *Abies* sp. and *Pinus* sp.).

95. Natural variations are found in the natural homeland of plants.

96. According to Vavilov crop plants have originated from there wild relatives and are found to have maximum variations from them.

97. The new world was discovered by Christopher Columbus in 1442 when he was landed at Bahamas Island (Caribbean Sea).

98. The place where wild relative of a crop plant are present in maximum is called primary home.

99. *Saccharum officinarum* or noble sugarcane is cultivated in most of the states in India and *Pyrilla perpusilla* is a common pest on sugarcane.

100. Secondary home of a crop plant is where the crop produces at its maximum.

101. Out of total 3, 50, 000 known plants species only about 100 plant species are used by the human beings. So many plant species are less known or underutilized crops or new crops which can be used as an alternative when main crop plants are not available.

102. Sorghum is shown as both rabi and kharif crops in India.

103. *Psophocarpus tetragonoloba*, Jojoba, Guayule, Subabul and some oil yielding plants are the examples of less known plants.

104. *Psophocarpus tetragonoloba* is also called winged bean. It is tuberous crop and is full of protein and oil. Plant is also grown for green manure and fodder.

105. *Simmondsia chinensis* or Jojoba is native of Mexican desert. Its seed contains 50% liquid wax

which is similar to sperm whale oil or spermaceti. It is used in cosmetics and in lubricants. This plant can be easily grown in poor, desert soil.

106. *Parthenium argentatum* or Guayule belongs to family Asteraceae and they are native of America. This plant is also known as congress grass in India. It is a shrub and grown as weed. It can be good source of rubber in future. The Guayule rubber is obtained from this plant. It is similar to Para rubber or Hevea rubber.

107. Subabul or *Leucaena leucocephala* is native of Central America. They are small tree which belongs to family Legumenoceae. Leaves of this plant is used as fodder, wood is used in charcoal production, in preparation of paper pulp, rayon and as timber.

108. Subabul is widely used in social forestry. It can be used as wind break, fire breaks and plant cover in desert areas.

109. Margosa or *Azadirachta indica* seeds contain oil which is used in soap making. Other less known oil yielding plants are *Cucurbita foetidissima*, *Citrullus colocynth*is, *Madhuca indica*, *Salvadora persica*, *Pongamia pinnata*, *Schleichera oleosa* etc. The oil obtained from all these plants can be edible after processing.

110. The leaves of *Ilex paraguariensis* (Paraguay tea) is used for tea making.

111. Arabian tea is obtained from the leaves of *Catha edulis* or khat.

112. *Paullinia cupana* or yoco is also is used as non-alcoholic beverages in South America.

113. Tye is also called the grain of poverty and used in the preparation of black bread.

114. Triticale is a man-made cereal which is produced by crossing of *Triticum* ($2n = 28$) with rye (*Secale*, $2n = 14$). The grains of triticale is larger than wheat, which contain more proteins but gluten content is very low.

115. *Acacia nilotica* or babul, *Albizia lebbeck* or siris, *Morus alba* or white mulberry, *Sesbania grandiflora* etc. can be used as fodder plants.

Figures:

Capsicum sp.
Family: Solanaceae
Useful Part-Fruits
Green unripe capsium fruit is rich in vitamin A,C & E. It is used in various food stuffs. The chief colouring matter is the carotenoid capsanthin is present

Brassica sp.
Family: Brassicaceae
Useful Part: Seeds
It is used as a cardinent in pickles, salads etc. oil is used as cooking medium. The oil cake is used as cattle feed.

Figure 3.1 Plants of economic importance

Clove

Syzygium aromaticum
Family: Myrtaceae
Useful Part: Flower Buds

Clove is used to flavour meat fishes taking confeetionaries. It is used in medicinal preparation. Clove oil is used in tooth-ache. It is used as stimulant and carminative.

Saffron

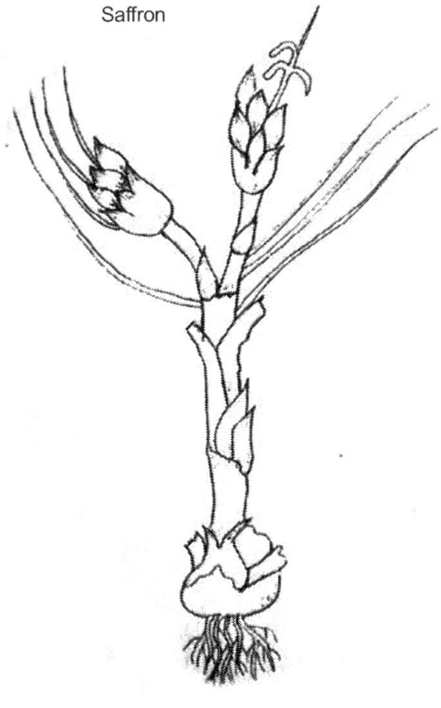

Crocus sativus
Family: Iridaceae
Useful Part Flowers

Saffron has a fine flavour and is used for flavouring puddings, desserts, sweet dishes,and confectionaries. It is used in various religious functions.

Figure 3.2 Plants of economic importance

Drying Oil

Linseed Oil
Linum usitatissimum
Family Linaceae
Useful Part: Seeds

Seeds yield oil used for manufacturing of Paints, vainishes, oil cloth and linoleum. Plant yields strong fibre used in manufacture of ropes. Oil cake used as cattle feed.

Sesame Oil

SemiDrying Oil
Seaamum indicum
Family: Pedaliaceae
Useful Part: Seeds.

Sesamum seeds are elible and used as food items, often made into swear meals. Cake obtained after oil extraction is rich in protein, calcium, vitamins etc.

Non-Drying Oil

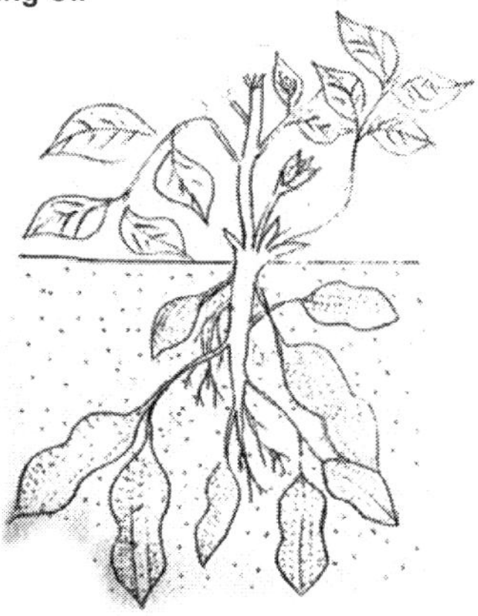

Peanut Oil
Arachis hypogaca
Family Fabaceae
Useful Part Seeds

It is used in preparation of vanaspathi ghee, peanut butter. It is used for cooking, soap making illuminats etc.

Figure 3.3 Plants of economic importance

Rose Oil

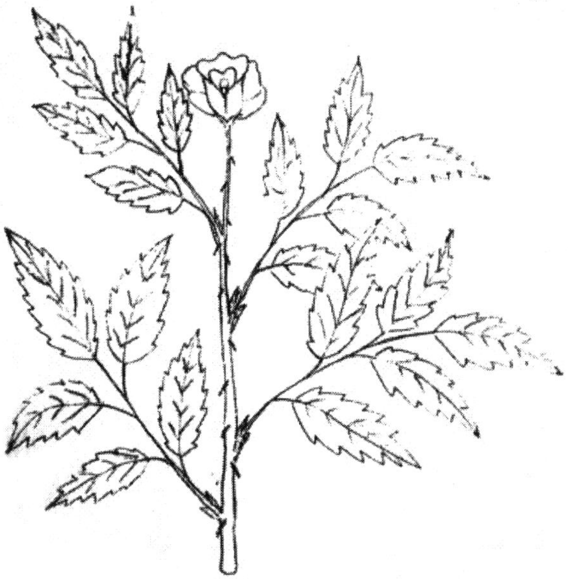

Rosa damascena
Family Rosaceae
Useful Part: Flower

The fragrant oil of Roses are widely used in perfumes other products are Rose-water and Gulkanel. It is an expensive oil used in soaps.

Jasmine Oil

Jasminum sp.
Family: Oleaceae
Useful Part: Flowres

Jasmine oil is used in the preparation of hair oil. It is used in the perfume industry soap industry and medicinal oils.

Figure 3.4 Plants of economic importance

Tobacco

Nicotiana tabacum
Family Solanaceae
Useful part Leaves

It is used for smoking in the form of cigarettes, bidi, cheroot cigar etc. It is also used for chewing and for stuff.

Betel leaf

Piper betle
Family: Piperaceae
Useful Part: Leaves

Betel leaves are widely used as mastigatory. It is commonly chewed smeared with lime (chuna) and careche (katha) used as digestive stimulant and carminative. Betel leaves are applied over boils and wounds.

Betel Nut

Areca catechu
Family: Palmae
Useful Part-Seeds

Betel nut is chiefly used as a mastigatory alone or in combination with betel leaf. The juice of tender nuts in small doses used as laxative.

Figure 3.5 Plants of economic importance

Tea

Camellia sinesis
Family: Theaceae
Useful part: Leaves

The leaves are edible they are dried and powdered. This is used as non-alcoholic beverages. The decoction is used for hair drying. It has a stimulating and refreshing property.

Coffee

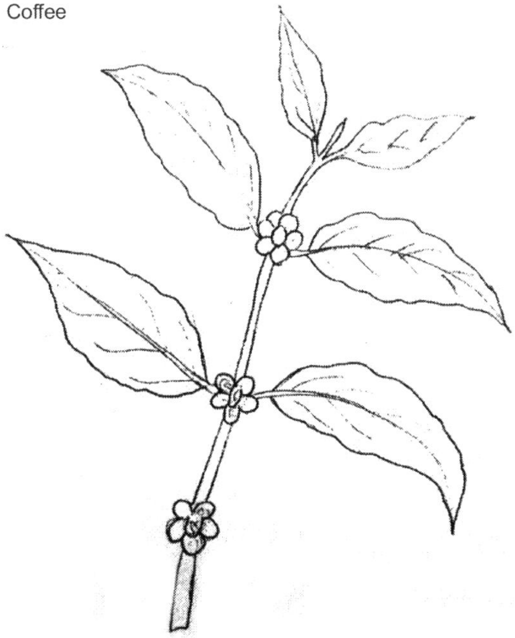

Coffea arabica
Family Rutaceae
Useful Part-Seeds

Seeds are dried and powdered. The powder is used as a refreshing. Non alcoholic beverage.

Figure 3.6 Tae and coffee plant's useful parts

Cotton

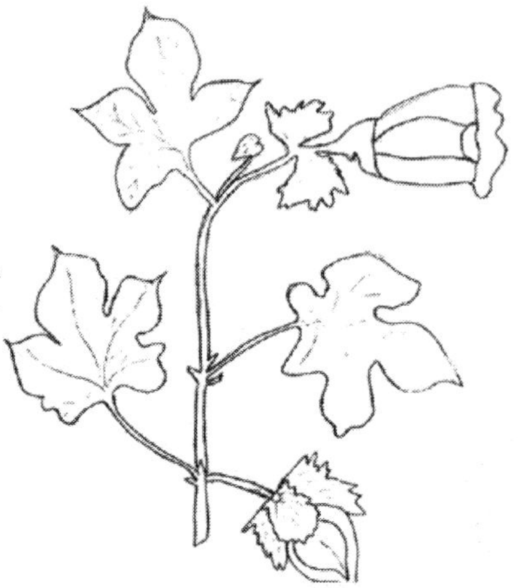

Gossypium herbaceum
Family: Malvaceae
Useful Part Pods

Cotton is a major cash crop. It is used for making clothing househeld articles and industrial articles. The cotton seeds are used for oil extraction.

Jute

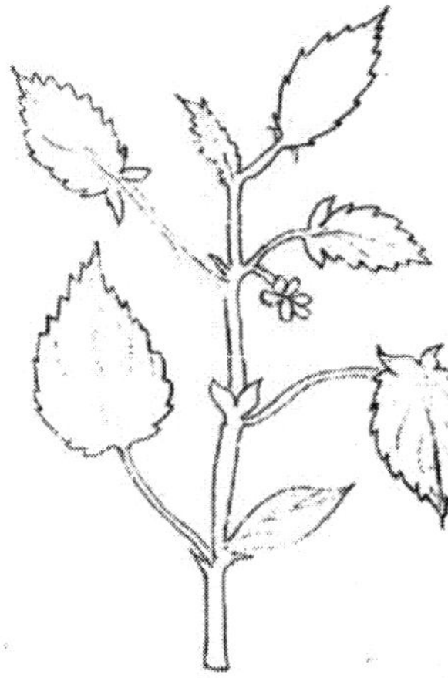

Corchorus olitorius
Family: Tiliaceae
Useful Part: Bark of plant

Fibre of this plant is finer, stronger and lustious jute is the important textile fibre next to cotton. It is used in making ropes, paper, mats and linolium.

Kapok

Cieba pentandra
Family Bombaceae
Useful Parts Pods

Kapok is used in filling and upholstroy They are useful for the manufacture of life boats and belts. It is used for making sound proof covers of walls. Refined kapok seed oil is used for substitue to cotton oil.

Figure 3.7 Fibre yielding plants

Rubber

Para Rubber
Hevea brasiliensis
Family: Euphorbiaceae
Useful part: Latex

It is used in the manufacture of tyres, tubes and other articles used in hospital supplies, shock absorbers, sport goods, eraserbs toys, washers, gaskets, rubber bands, fabrication of battery boxes etc.

Paper

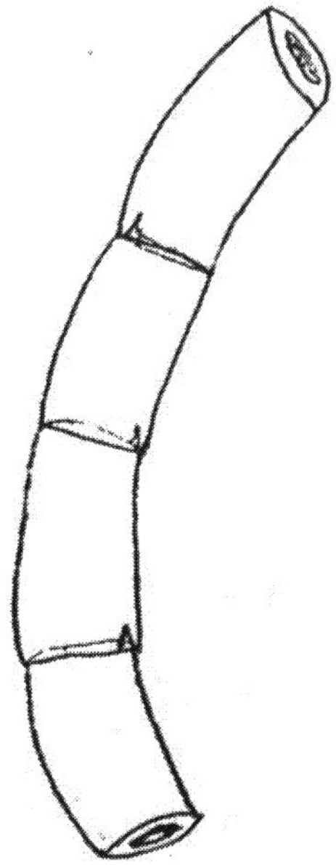

Bamboo
Bambusa sp.
Useful Part Portion of stem

The first step in manufacture of paper is the conversion of the fibrous material into pulp, then the pulp is converted into paper. Paper is used in the manufacture of tissue paper in napkins. Pulking papers writing purpcese, printing etc.

Figure 3.8 Rubber and paper yielding plants

Tannin

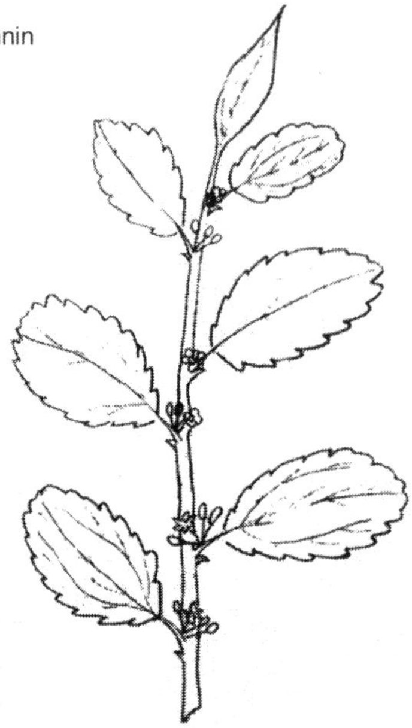

Zizyphus jujuba
Family Rhamnaceae
Useful part Bark

These are small moderate sized deciduous trees. The bark yield some quantity of tannin. Mostly utilized in Northen India.

Dyes Henna

Lawsonia inermis
Family: Lythraceae
Useful Part Leaves

The leaves yield a orange dye.
It is used for dying hair, finger nails and palms etc. The leaves are dried and ground to yield paste of medicinal value.

Figure 3.9 Tannin and dye (henna) yielding plants

ANSWERS

1. (b)	2. (d)	3. (b)	4. (c)	5. (d)	6. (d)	7. (a)	8. (b)	9. (c)	10. (c)
11. (d)	12. (a)	13. (b)	14. (c)	15. (d)	16. (b)	17. (c)	18. (c)	19. (c)	20. (c)
21. (b)	22. (c)	23. (b)	24. (b)	25. (b)	26. (b)	27. (a)	28. (b)	29. (d)	30. (d)
31. (c)	32. (c)	33. (c)	34. (c)	35. (a)	36. (d)	37. (a)	38. (b)	39. (d)	40. (c)
41. (b)	42. (a)	43. (c)	44. (a)	45. (d)	46. (d)	47. (b)	48. (a)	49. (d)	50. (d)
51. (c)	52. (d)	53. (a)	54. (d)	55. (b)	56. (c)	57. (a)	58. (c)	59. (c)	60. (b)
61. (c)	62. (b)	63. (c)	64. (a)	65. (b)	66. (a)	67. (a)	68. (d)	69. (a)	70. (b)
71. (d)	72. (c)	73. (d)	74. (a)	75. (a)	76. (b)	77. (d)	78. (a)	79. (c)	80. (c)
81. (c)	82. (a)	83. (d)	84. (d)	85. (d)	86. (a)	87. (c)	88. (d)	89. (a)	90. (b)
91. (b)	92. (b)	93. (b)	94. (c)	95. (c)	96. (a)	97. (a)	98. (c)		

General Health Related Question Bank Related to Plants

ANSWERS

1. (b)	2. (b)	3. (d)	4. (c)	5. (c)	6. (c)	7. (b)	8. (d)	9. (d)	10. (a)
11. (d)	12. (d)	13. (c)	14. (c)	15. (d)	16. (d)	17. (c)	18. (a)	19. (b)	20. (a)
21. (b)	22. (c)	23. (b)	24. (c)	25. (b)	26. (c)	27. (b)	28. (c)	29. (a)	30. (a)
31. (d)									

4

EMBRYOLOGY

MULTIPLE CHOICE QUESTIONS

1. Which of the following characteristics is a wind pollinated flower likely to have
   (a) Attractive corolla (b) Large exposed stamens
   (c) Feathery protruding stigmas (d) Heavy spiky pollen
2. When embryo develops from some cells of the embryo sac other than the egg, the process is known as
   (a) Apogamy (b) Parthenocarpy
   (c) Parthenogenesis (d) Apospory
3. When pollen tube enters the ovule through micropyle, the process is called
   (a) Aporogamy (b) Mesogamy
   (c) Chalazogamy (d) Porogamy
4. Stamen is equal to
   (a) Megasporophyll (b) Microsporophyll
   (c) Microsporangium (d) Megasporangium
5. The formation of pollen grains from archesporium is known as
   (a) Megasporogenesis (b) Microsporogenesis
   (c) Microsporophyll (d) megagametogenesis
6. Caruncle is a
   (a) Integumentary out growth (b) Embryo sac outgrowth
   (c) Endosperm haustoria (d) Nucellus outgrowth
7. The fertilized egg is called
   (a) Seed (b) Zygote (c) Ovule (d) Embryo

8. When the embryo is formed by some process in which normal meiosis and syngamy have been eliminated, this phenomenon is called
   (a) Agamospermy (b) Parthenogensis
   (c) Double fertilization (d) Amphimixis
9. Layering is used in vegetative propagation of
   (a) Rose (b) Mango (c) Jasmine (d) All of these
10. Simultaneous cytokinesis of microspore mother cell leads to tetrad
   (a) Isobilateral (b) Tetrahedral
   (c) Linear (d) T-shaped
11. Egg apparatus consists of
   (a) Egg and antipodals (b) Egg and synergids
   (c) Polar nuclei (d) None of these
12. If any somatic cell of sporopyte produces gametophyte without reduction division, it is
   (a) Parthenocarpy (b) Apospory
   (c) Apogamy (d) Amphimixis
13. Flowers emit fragrance
   (a) To purify air (b) To attract insects
   (c) To drive away flies (d) To perform all the above functions
14. A microspore is a first cell of
   (a) Male gametophyte (b) Male sporophyte
   (c) Female gametophyte (d) Female sporoph yte
15. Vegetative propagation by bulbils occur in
   (a) *Agave* (b) *Zingiber* (c) *Vallisneria* (d) *Curcuma*
16. The scientist who first time saw the process of fertilization in angiosperm
   (a) M.S. Caves (b) S.G. Nawaschin
   (c) Edward Strasburger (d) Gwenta Davis
17. Comparative embryology is mainly concerned with the
   (a) Taxonomic evaluation of embryological characters
   (b) Critical study of various embryological processes
   (c) Control in the embryological processes
   (d) All of the above
18. When the filament runs along the back of anther, it is called
   (a) Versatile (b) Adnate
   (c) Longitudinal (d) Syngenecious

19. In Cyperaceae, each microspore mother cell forms
    (a) Two microspore (b) Only one microspore
    (c) Four microspore (d) Numerous microspore
20. Pick the odd one out
    (a) Chalazogamy (b) Allogamy (c) Mesogamy (d) Porogamy
21. Pollen grains are able to withstand extremes of temperature and desiccation because there exine is made up of-
    (a) Cutine (b) Callose (c) Sporopollenin (d) Suberin
22. Which plant propagate through leaves
    (a) *Agave* (b) Kalonchoe (c) *Gladiolus* (d) Potato
23. An inflorescence always forms a
    (a) Simple fruit (b) Multiple or composite fruit
    (c) Aggregate fruit (d) Dry dehiscent fruit
24. Embryo sac is also known as
    (a) Microsporangium (b) Megasporangium
    (c) Meggemetophyte (d) None of these
25. Development of megaspore within ovule is known as
    (a) Embryogenesis (b) Megasporogenesis
    (c) Parthenogenesis (d) Megagametogenesis
26. The flower may be described as perigynous if the floral parts
    (a) are inserted below the ovary (b) surround the ovary
    (c) are inserted above the ovary (d) none of the above
27. Axillary buds are used to raise crop of
    (a) Wheat (b) Rice (c) Groundnut (d) Sugarcane
28. The point where funicle joins with ovule is known a
    (a) Chalaza (b) Micropyle (c) Hilum (d) Integument
29. Double fertilization was first discovered by Nawaschin (1898) in
    (a) *Lilium* and *Fritillaria* (b) Mango and sugarcane
    (c) Papaya and pea (d) *Brassica* and mango
30. In a young anther the four rows of cells, which later produce pollen, are called
    (a) Antheridium (b) Archesporium
    (c) Tapetum (d) Zoosporangium
31. Censer mechanism of seed dispersal is found in the members of family
    (a) Liliaceae (b) Rosaceae (c) Papavaraceae (d) Leguminosae

32. Allogamy is best favoured by
    (a) Dichogamy (b) Dicliny
    (c) Cleistogamy (d) Homogamy
33. Cleistogamy occurs in-
    (a) *Pisum sativum* (b) *Oryza sativa*
    (c) *Commelina benghalensis* (d) *Allium cepa*
34. Members of the family Onagraceae is characterized by the universal occurrence of the
    (a) Oenothera type of embryo sac (b) Fritillaria type of embryo sac
    (c) Allium type of embryo sac (d) Adoxa type of embryo sac
35. Development of female gamete into a new individual without fertilization is called
    (a) Androgenesis (b) Parthenogenesis
    (c) Sporogenesis (d) Parthenocarpy
36. Translator apparatus is found in
    (a) Mustard (b) *Calotropis* (c) Mango (d) Pea
37. The edible part of ripe mango is morphologically
    (a) Epicarp (b) Mesocarp (c) Pericarp (d) Endocarp
38. The outer whorl of *Bougainvillaea* flower consists of
    (a) Bracts (c) Epicalyx (c) Sepals (d) Petals
39. *Ficus* is an example of which kind of inflorescence?
    (a) Spike (b) Raceme (c) Hypanthodium (d) Cyathium
40. Anemophilous flowers have
    (a) Sessile stigma (b) Small smooth stigma
    (c) Large feathery stigma (d) Coloured flower
41. Parthenogenesis is defined as development of
    (a) Embryo without fertilization (b) Synergid without fertilization
    (c) Fruit without fertilization (d) Fruit without pollination
42. Parthenogenesis was discovered by
    (a) Coulter and Chamberlain (b) Muller and Melvil
    (c) Theophrastus (d) Juel and Murbeck
43. The position of shoot apex in monocot embryo is
    (a) Terminal (b) Sub-terminal (c) Basal (d) Lateral
44. The fusion product of polars nuclei and male gamete is
    (a) Triple fusion (b) Secondary nucleus
    (c) Primary endosperm nucleus (d) Zygote

45. In which of the following production of seedless fruits in undesirable

    (a) Fruits without pollination (b) Fruits without fertilization
    (c) Seeds without fertilization (d) Only seeds and no fruits

46. When a diploid (male) is crossed with a tetraploid (female), the ploidy of the endosperm cells in the resulting seed is

    (a) Triploid (b) Diploid (c) Pentaploid (d) Tetraploid

47. After penetrating stigmatic and style tissue, the pollen tube usually grows down towards the egg cell because

    (a) it has no other passage to follow
    (b) it grows under the control of egg nucleus
    (c) the egg cell attracts the pollen tubes as they have dissimilar electric charge
    (d) the filiform apparatus of synergid is believed to attract the pollen tube

48. Double fertilization means

    (a) Fusion of two eggs
    (b) Fusion of eggs and pollen nuclei of two pollen grains
    (c) Fusion of one male gamete with egg and the other with the secondary nucleus
    (d) Fusion of one male gamete with the egg and the other with the synergids

49. Caruncle is derived from

    (a) Peduncle (b) Cotyledons
    (c) Outer integument (d) Inner integument

50. Chalazogamy is found in

    (a) *Cucurbita* (b) *Pistacia* (c) *Pisum sativum* (d) *Casuarina*

51. The micropyle in a seed helps in the entry of

    (a) Water (b) Pollen tube
    (c) Male gamete (d) None of these

52. The megasporangium of the angiosperms on maturation gives rise to

    (a) Fruit (b) Seed (c) Embryo (d) Cotyledon

53. The embryo sac of a typical dicot at the time of fertilization is

    (a) 8 celled (b) 7 celled (c) 6 celled (d) 5 celled

54. Dichogamy favouring cross pollination is a type of floral mechanism where

    (a) anthers and stigma are placed at different levels
    (b) stamens and stigma mature at different times
    (c) structure of anther and stigma act as barrier
    (d) pollen is unable to germinate on its own stigma

55. The fruit developed from hypanthodium inflorescence is called
    (a) Drupe (b) Sorosis
    (c) Synconus (d) Siliqua
56. Stigma is always rough and sticky in-
    (a) Entomophilous flower (b) Anemophilous
    (c) Hydrophilous flowers (d) All types of flowers
57. Pollen tube was discovered in 1824 by an Italian mathematician named
    (a) G. B. Amici (b) Robert Brown (c) A. Takhatajan (d) T. S. Eliot
58. Double fertilization is characteristic of
    (a) Bryophytes (b) Pteridophytes
    (c) Gymnosperms (d) Angiosperms
59. The process of double fertilization was discovered by
    (a) Strasburger (b) Rober Koch (c) Nawaschin (d) Drouchet
60. Perisperm is the remains of
    (a) Integument (b) Archegonium (c) Endosperm (d) Nucellus
61. In angiosperms, the functional megaspore of linear tetrad is the
    (a) First nearest to the micropyle
    (b) Second from the micropyle
    (c) Third from the micropyle
    (d) Fourth from the micropyle
62. When a pollen tube enters the embryo sac by piercing through the integuments it is called
    (a) Mesogamy (b) Porogamy (c) Chalazogamy (d) Pseudogamy
63. Antony van Leeuwenhoek, the first to discover bacteria is from
    (a) U.K (b) Sweden (c) France (d) Holland
64. The term xenia denotes the effect of pollen on the
    (a) Endosperm (b) Egg (c) Seed coat (d) Nucellus
65. The best example of polyembryony is
    (a) *Cocos* (b) *Citrus* (c) *Capsicum* (d) *Cycas*
66. Fusion of dissimilar gametes is known as
    (a) Fertilization (b) Dichogamy (c) Autogamy (d) Allogamy
67. An arthotropous ovule is one in which micropyle and chalaza are
    (a) at right angles to funicle (b) parallel to the funicle
    (c) in straight line of funicle (d) parallel along with ovule

68. Entomophily is the pollination by
   (a) Birds (b) Bats (c) Wind (d) Insects
69. Which of the following fruit is not possible without seed?
   (a) Pomegranate (b) Grapes (c) Orange (d) Lemon
70. Scutellum is a
   (a) Endosperm of gymnosperms (b) Shield shaped cotyledon
   (c) Protective covering of radical (d) Protective covering of plumule
71. In angiosperms the female gametophyte is the
   (a) Egg apparatus (b) Embryo (c) Synergids (d) Embryo sac
72. A gymnospermic leaf carries 16 chromosomes. What will be the number of chromosomes in its endosperm
   (a) 16 (b) 24 (c) 12 (d) 8
73. If there are four cells in anther what will be the number of pollen grains
   (a) 8 (b) 4 (c) 12 (d) 16
74. Ubisch bodies found in tapetal cells help in formation of
   (a) Pollen kit and pollinia (b) Exine
   (c) Sporopollenin (d) Exine and pollen kit
75. Formation of gametophyte directly from spore mother cell without meiosis is called
   (a) Apospory (b) Apogamy (c) Apomictic (d) Apomixes
76. The process of growing male gametophyte is known as
   (a) Sporogenesis (b) Androgenesis
   (c) Embryogenesis (d) Parthenogenesis
77. Pollination of flowers below water level is called
   (a) Hypohydrophily (b) Epihydrophily
   (c) Cheiropteriphily (d) Ornithophily
78. Pollination by snails and slugs is
   (a) Malacophily (b) Cheiropteriphily
   (c) Ornithophily (d) Entomophily
79. Bicarpellary gynoecium and oblique ovary occurs in
   (a) Banana (b) Brinjal (c) Cocotut (d) *Pisum*
80. The correct match for edible part of fruit is
   (a) Maize- cotyledon (b) Date palm- epicarp
   (c) Guava- pericarp with thalamus fused (d) Tomato- thalamus

81. In banana, edible part is
    (a) Fleshy epicarp
    (b) Rudimentary mesocarp and fleshy endocarp
    (c) Rudimentary endocarp and fleshy mesocarp
    (d) Pericarp
82. Aril is the edible part in
    (a) Banana (b) Litchi (c) Mango (d) Apple
83. The hairs present in maize corn cob is
    (a) Seed hairs (b) Styles
    (c) Modified hairs of bract (d) Stigma
84. Pollination by ants is
    (a) Myrmecophily (c) Cheiropteriphily
    (c) Ornithophily (d) Entomophily
85. Formation of embryo directly from nucellus and integument is
    (a) Apogamy (b) Apomixes
    (c) Adventive polyembryony (d) Apospory
86. A piece of potato tuber will form a new plant if it possesses
    (a) Roots (b) Eyes
    (c) Branches (d) Stored food
87. Winged fruits are commonly known as
    (a) Samara (b) Siliqua
    (c) Regma (d) Cypsela
88. Which one of the following events in a botanical garden is never directly influenced by light
    (a) Flowering (b) Photosynthesis
    (c) Fertilization (d) Transpiration
89. Food material in exalbuminous seed is mainly stored in
    (a) Nucellus (b) Testa (c) Embryo (d) Endosperm
90. Fruit of *Abelmoschus esculentum* is
    (a) Cypsela (b) Capsule (c) Legume (d) Pome
91. Integumentry tapetum is alternatively called
    (a) Endothecium (b) Endothelium (c) Hypostate (d) Epistate
92. When the entire or most of the internal surface of the ovary wall is covered with ovules, the placentation is described as
    (a) Marginal (b) Basal (c) Superficial (d) Lamellar

93. A monosporic type of embryo sac which does not have any antipodals
    (a) Allium type (b) Peperomia type
    (c) Polygonam type (d) Oenothera type
94. Self pollination which occur between two different flowers present on the same plant is called
    (a) Geitonogamy (b) Homogamy
    (c) Autogamy (d) Herkogamy
95. The condition in which anthers and stigma of a flower matures at the same time is called
    (a) Chasmogamy (b) Dichogamy
    (c) Homogamy (d) Cleistogamy
96. During the process of fertilization the pollen tube of pollen enters through integument is called
    (a) Porogamy (b) Mesogamy
    (c) Chalazogamy (d) None of these
97. The fruit developed from monocarpellary superior ovary with marginal placentation is called
    (a) Legume (b) Nut
    (c) Capsule (d) Siliqua
98. By hormone application it is possible to obtain seedless fruits. In which of the following is it undesirable to obtained seedless fruits
    (a) Tomato (b) Orange
    (c) Pomegranate (d) Watermelon

SMALL ANSWER QUESTIONS

1. What is parthenocarpy?

   Ans. Formation of fruit without fertilization is called parthenocarpy.

2. How many antipodals occur in a polygonum type of embryo sac?

   Ans. Three antipodals.

3. Name the nuclei that fuse to form endosperm?

   Ans. Two polar nuclei (secondary nucleus) and one male gamete.

4. Who wrote the book 'An Introduction to the embryology of Angiosperms'?

   Ans. P. Maheshwari.

5. What are the two phases of life cycle of an angiospermic plant?

   Ans. Diploid sporophytic phase and haploid gametophytic phase.

6. What is the prerequisite for pollination?

   Ans. Dehiscence of anther.

7. What is double fertilization?

   Ans. One male gamete fuses with egg (syngamy) to form zygote, while second male gamete fuses with two polar nuclei to form primary endosperm nucleus (triple fusion), the process is called double fertilization.

8. Which phase is dominant phase in the life cycle of angiospermic plant?

   Ans. Sporophytic phase.

9. Name the pollinator for cheiropteriphilous pollination.

   Ans. Bats.

10. What is polyembryony?

    Ans. When more than one embryo is present in a fertilized ovule, this condition is known as polyembryony.

11. Name the first cells of the gametophytic phase in an angiospermic plant.

    Ans. Microspore and megaspore.

12. Who wrote the book 'Systematic Embryology of Angiosperms'?

    Ans. G. Davis.

13. What is bulbil?

    Ans. When floral buds become fleshy, this can give rise to new plants, e.g. *Agave*.

14. Name three routes by which pollen tube may enter the ovule.

    Ans. Porogamy, chalazogamy, mesogamy.

15. Name the first sporophytic cell in angiosperm.

    Ans. Diploid zygote.

16. What are the constituents of pollenkitt?

    Ans. Carotenoids and lipids.

17. Name the kind of sterile appendages of the flower.

    Ans. Sepals and petals.

18. Name a monocot family in which endosperm does not form.

    Ans. Orchidaceae.

19. Why pollen grains were usually found in preserved forms as fossil?

    Ans. Due to the presence of sporopollenin, which is resistant to physical and biological disintegration?

20. What are sepals?

    Ans. These are the sterile appendages of a typical flower, which together form the calyx.

21. Who used the term 'xenia' first time?

    Ans. Focke in 1881.

22. What is coleorhiza?

    Ans. Protective cap on radicle of monocot plants.

23. What are perianth?

    Ans. Sepals and petals differ in dicots but in some monocot families these are alike, then they collectively called perianth.

24. What are parthenocarpic fruits?

    Ans. The fruits formed without fertilization are called as parthenocarpic fruits. Seedless fruits can be also prepared by this technique.

25. What is the function of sepals?

    Ans. Protect inner floral appendages.

26. Name the plant where ca

32. What is unisexual or imperfect flower?

    Ans. When the flower posses only one fertile appendage (either stamens or carpels)

33. What kind of pollination is present in *Vallisneria*?

    Ans. Hydrophilous type.

34. Name the cells formed by the first division of the microspore nucleus?

    Ans. One generative and one vegetative cell.

35. What is staminate flower?

    Ans. Unisexual flowers which have only stamens (carpels absent).

36. What is law of destination in embryology?

    Ans. In any given species the cells of proembryo give rise to clearly determined parts, and always to the same parts of the embryo.

37. Name the two angiospermic plants where polyembryony observed.

    Ans. *Plantago* and *Argemone.*

38. What is pistillate or carpellate flower?

    Ans. Unisexual flower which has only carpels (stamens absent).

39. What is homogamy?

    Ans. It is a condition in which anthers and stigma of a flower mature at the same time.

40. What is proembryo?

    Ans. Proembryo stage starts from two cell stage and end with the differentiation of the organs.

41. What does the term monoecious mean?

    Ans. When both male and female flowers are born on the same plant.

42. Give the example of two plants where ategmic ovules are observed.

    Ans. *Liriosma* and *Crinum*.

43. What is hypogynous flower?

    Ans. If the floral parts are inserted below the ovary.

44. Give an example of onagrad type of embryogeny.

    Ans. *Capsella bursa-pastoris*.

45. What is epigynous flower?

    Ans. If the floral parts are inserted above the ovary.

46. What is perigynous flower?

    Ans. If the floral parts surround the ovary.

47. What is orthotropus type of ovule?

    Ans. In this type of ovule the body is upright with micropyle, chalaza and funicle falling in a straight line.

48. What type of polyembryony generally found in gymnosperms?

    Ans. Cleavage polyembryony.

49. What is the term used for plant bearing only female flowers?

    Ans. Dioecious.

50. Name the types of ovules present in angiosperms?

    Ans. Orthotropous, anatropous, campylotropous, amphitropous, hemianatropous and circinotropous.

51. Name the condition in which more than four microspores are found in a tetrad.

    Ans. Polyspore.

52. Name the plant where amoeboid tapetum is present.

    Ans. *Actaea, Alisma, Typha,* etc.

53. Name the types of pollen tetrads observed in angiosperms?

    Ans. Tetrahedral, isobilateral, decussate, T-shaped and linear.

54. How many ovules are present in an ovary exhibiting basal placentation?

    Ans. Single.

55. What is the function of ubisch bodies?

    Ans. Help in the formation of exine of pollen grains.

56. Name the type of cytokinesis in microspore mother cell.

    Ans. Successive type and simultaneous type.

57. What type of embryo sac is the most common in flowering plants?

    Ans. Polygonum type.

58. Name an insectivorous plant where compound pollen grains are present.

    Ans. *Drosera*.

59. Give an example of monocarpic monocot.

    Ans. *Agave*.

60. Name the two types of bisporic embryo sac.

    Ans. Allium type and endymion type.

61. Name a bat pollinated flower.

    Ans. Kadamba.

62. Name a tetrasporic embryosac in which the arrangement of the nuclei is the same as in polygonum type?

    Ans. Adoxa type.

63. What is scutellum?

    Ans. Single cotyledon in caryopsis of maize is called scutellum.

64. How many sporangia are present in a typical anther?

    Ans. Four.

65. Name a tetrasporic embryo sac in which egg apparatus consists of a single cell and that represents the egg.

    Ans. Plumbagella type.

66. What type of fruit tomato is?

    Ans. Berry.

67. Give three examples of false fruit.

    Ans. Apple, Pineapple and jackfruit.

68. At which stage tapetum is showing maximum development?

    Ans. At tetrad stage of microsporogenesis.

69. How many types of pollen tetrads observed in angiosperms?

    Ans. Five types.

70. What is funiculus?

    Ans. The stalk of an ovule.

71. Which group of plants usually shows secretory type of tapetum?

    Ans. Angiosperms.

72. How does the mustard inflorescence differ from banana inflorescence in arrangement?

    Ans. In mustard, the inflorescence is raceme; flowers are arranged in acropetal order on floral axis, whereas in banana, inflorescence is spadix, where fleshy axis remains covered by spathe.

73. Who discovered the pollen tube in flowering plants?

    Ans. S. G. Nawaschin.

74. Why are cucurbits referred to as monoecious?

    Ans. Because male and female flowers are borne on same plant.

75. What is the main difference in the endosperm of angiosperm and gymnosperm?

    Ans. In angiosperm endosperm in triploid but in gymnosperm it is haploid.

76. What is geitonogamy?

    Ans. It is type of pollination where pollen is transferred from anther of a flower to stigma of another flower of the same plant.

77. What is the accessory seed covering formed by an outgrowth at the base of the ovule?

    Ans. Caruncle, aril, etc.

78. What is allogamy?

    Ans. It is a type of pollination where the pollens move from the anther of one flower to the stigma of another flower of a different plant, if is referred to as cross pollination, xenogamy or allogamy.

79. How many pollen grains are produced by each microspore mother cell in the members of family Cyperaceae?

    Ans. One only.

80. What is ornithophily?

Ans. The pollination by bird.

81. Name the phenomenon where the embryo is formed from a cell of the embryo sac other than the egg.

Ans. Apogamy.

82. What types of tapetum usually present in angiosperms?

Ans. Amoeboid and secretory.

83. Name the plant which has dimorphic tepetum?

Ans. *Nigella damascena.*

84. Name the families which have pollinia?

Ans. Asclepiadaceae and Orchidaceae.

85. What is agamospermy?

Ans. A process of reproduction, where seeds are formed without fusion of gametes.

86. In which plant endothecium is 2-layered?

Ans. *Coccinia indica.*

87. How many nuclei are present in a typical embryo sac?

Ans. Eight haploid nuclei.

88. Which technique is used for propagating improved varieties of mango and rose?

Ans. Grafting.

89. What is megasporogenesis?

Ans. The development of megaspore in ovule is called megasporogenesis.

90. Name the families of angiosperms which do not form endosperm.

Ans. Trapaceae, Podostemaceae and Orchidaceae.

91. Which flower is pollinated by bat?

Ans. *Kigelia pinnata.*

92. Name the finger like projections present on the wall of the synergids.

Ans. Filiform apparatus.

93. Which structure act as the mother cell of the male gametophyte?

Ans. Microspore.

94. What are the environmental factors which induce male sterility in plants?

    Ans. Temperature and photoperiodic exposers.

95. Why is date palm plant referred to as dioecious?

    Ans. Because date palms are unisexual, where male and female flowers are born on different plant, and therefore, they are referred to as dioecious.

96. What are the various parts of gynoecium?

    Ans. A proximal ovule bearing part- the ovary, distal stigma and in between is long style.

97. Name the plant bearing anemophilous flower?

    Ans. Maize.

98. A technique to produce genetically identical individuals from a single cell is known as-

    Ans. Cloning.

99. A population of genetically identical plants derived from an individual is called-

    Ans. Clone.

100. Study of pollen grains is known as-

     Ans. Palynology.

101. During grafting, the part that becomes the supporting portion is called-

     Ans. Stock.

102. The fruit that develops from the whole of inflorescence is called-

     Ans. Sorosis.

103. Genetically, the megaspore mother cell is haploid or diploid.

     Ans. Diploid.

104. A small part in the ovule through which the pollen tube enters is called-

     Ans. Micropyle.

105. The ovary wall develops into a fruit wall is called-

     Ans. Pericarp.

106. Nutritive layer of cells around pollen sacs in anther is called as-

     Ans. Tapetum.

107. The flowers in catkin are-

     Ans. Unisexual.

108. The tip of the pedicel on which different floral parts are inserted is known as-

     Ans. Thalamus.

109. *Bryophyllum* is the best example of propagation by-

     Ans. Leaf.

110. The process of nuclear fusion to form zygote is called-

     Ans. Syngamy.

111. Inside ovary, ovules develop from a special tissue called-

     Ans. Placenta.

TRUE OR FALSE QUESTIONS

1. The flower is described as hypogynous if the floral parts are inserted above the ovary.

   Ans. False

2. Drusa type of embryo sac is characterized by large number of antipodals.

   Ans. True

3. Nutritive layer of cells around pollen sac in anther is tapetum.

   Ans. True

4. An ovule is called anatropous when its chalaza and micropyle are in an opposite direction on a straight line.

   Ans. True

5. The mature anther wall is made up of four layers.

   Ans. True

6. The development of embryo is of onagrad type in the members of Onagraceae.

   Ans. True

7. A carbohydrate layer present on the outer side of grains of maize, wheat etc. is called aleurone layer.

   Ans. False

8. The embryo sac formed in the nucellus, is the female gametophyte.

   Ans. True

9. The nucleus of the male gametophyte divides to from a small vegetative nucleus and a relatively large generative nucleus.

   Ans. True

10. The embryology of the Orchidaceae is characterized in the absence of an endosperm.

    Ans. True

11. Allium type of embryo sac is bisporic.

    Ans. True

12. The transfer of pollen grains from anther to stigma is known as fertilization.

    Ans. False

13. The flower with superior ovary is called hypogynous.

    Ans. True

14. Wall of fruit having epicarp, mesocarp and endocarp is called periderm.

    Ans. False

15. A characteristic embryological feature of Onagraceae is the universal occurrence of the oenothera type of embryo sac which is not found in any other family except as an abnormality.

    Ans. True

16. The allium type of embryo sac is developed from chalazal diad.

    Ans. True

17. The part of the plant that is grafted on the stalk is called scion.

    Ans. True

18. The functional megaspore functions as mother cell of the male gametophyte.

    Ans. False

19. Oenothera type of embryo sac does not have any antipodals.

    Ans. True

20. Dioecious species are necessarily cross pollinated.

    Ans. True

21. Caruncle is present in the members of family Euphorbiacea.

    Ans. True

22. When both staminate and pistillate flowers are born on the same plant, the taxon is said to be dioecious.

    Ans. False

23. In bisexual flowers, to prevent self-pollination, stamens are removed in bud condition, the technique known as parthenogenesis.

    Ans. False

24. Crassinucellate ovules are usually bitegmic.

    Ans. True

25. The fusion of the one of the male gamete with egg is known as double fertilization.

    Ans. False

26. In oenothera type of embryo sac, the endosperm is diploid in nature.

    Ans. True

27. The triploid nucleus, formed as a result of double fertilization, functions as the primary endosperm nucleus.

    Ans. True

28. In parietal placentation, the gynoecium is multicarpellary and syncarpous.

    Ans. True

29. The cells of the secretory tapetum remain attached to the middle layer till the development of pollen grains.

    Ans. True

30. Cellular endosperm is mostly confined to the dicotyledonous families.

    Ans. True

31. In marginal placentation the gynoecium is monocarpellary and unilocular.

    Ans. True

32. In members of Cyperaceae, each microspore mother cell forms only single microspore.

    Ans. True

33. Helobial endosperm is mostly confined to the monocotyledonous families.

    Ans. True

34. The nucellar tissue present around the embryo sac is known as perisperm.

    Ans. True

35. Pseudo-embryo sac is a characteristic feature of the Podostemonaceae.

    Ans. True

36. Cross pollination requires abiotic or biotic agent.

    Ans. True

37. *Cocos nucifera* is the classical example of nuclear endosperm.

    Ans. True

38. Only meiotic divisions are involved in asexual reproduction.

    Ans. False

39. The phenomenon of substitution of sexual process by asexual methods is known as apomixis.

    Ans. True

40. The effect of pollen grains on the tissue of seed coat is called metaxenia.

    Ans. True

41. The development of embryo from any diploid cell of the ovule lying outside the embryo sac is referred to as adventives embryony.

    Ans. True

42. The development of diploid embryo sacs from the cell of the nucellus is known as generative apospory.

    Ans. False

43. The embryogeny of *Capsella bursa-pastoris* is a classical example of dicotyledonous embryogeny and is widely used as a teaching type.

    Ans. True

44. With reference to higher plants haploids may be defined as sporophytes with gametophytic chromosome constitution.

    Ans. True

45. Parthenocarpy is the development of egg into a new individual without fertilization.

    Ans. False

46. Cleavage polyembryony is of widespread occurrence in the angiosperms.

    Ans. False

47. In the members of Cyperaceae, the microsporogensis is of simultaneous type, whereas in Poaceae, it is of successive type.

    Ans. True

48. Polyembryony can be exhibited *in Citrus*.

    Ans. True

49. Cleavage polyembryony is rare in the angiosperm.

    Ans. True

50. Allium type of embryo sac is tetrasporic.

    Ans. False

51. S. G. Nawaschin was the first person who coined the term' metaxenia'.

    Ans. False

52. The embryos developed from nucellus are called adventive embryos.

    Ans. True

53. The embryo sac formed in the nucellus, is the female gametophyte.

    Ans. True

54. Generative apospory is the development of diploid embryos from diploid cells of the archesporium.

    Ans. True

55. In campylotropous ovules, the funicle is very long and forms a complete circle around the body of the ovule.

    Ans. True

56. Crassinucellate ovules are usually bitegmic and tenuinucellates are unitegmic.

    Ans. True

57. The arrangement of ovules in the ovary is called placentation.

    Ans. True

58. In superficial placentation, most of the internal surface of the ovary wall is covered with ovules.

    Ans. True

59. Successive type of cytokinesis in microspore mother cell is present in dicots.

    Ans. False

60. The fruits which develop without fertilization are called parthenogenic fruits.

    Ans. False

61. Microspores are dipoloid.

    Ans. False

62. In *Vallisneria*, the male flowers are release in air.

    Ans. False

63. The megaspore mother cell is the last cell of the sporophytic generation.

    Ans. True

64. The embryology of the Orchidaceae is characteristic in the absence of an endosperm.

    Ans. True

65. At maturity, drusa type of embryo sac is characterized by large number of antipodals.

    Ans. False

66. A cotyledon of bean seed is called scutellum.

    Ans. False

67. Experimental embryology aims to control and induce changes in the embryological processes.

    Ans. True

68. Production of flowers which never open is known as cleistogamy.

    Ans. True

69. The maintenance of purity of a given variety of plants by vegetative propagation is called cloning.

    Ans. True

70. In fritillaria type of embryo sac the antipodal unclei are triploid.

    Ans. True

71. Antipodals are altogether absent in oenothera type of embryo sac.

    Ans. True

72. Two cells present on either side of egg cell in an embryo sac are termed as antipodals.

    Ans. False

73. Simultaneous type of cytokinesis of microspore mother cell is present in monocots.

    Ans. False

74. According to classical view, the flower is a compressed shoot.

    Ans. True

75. In fritillaria type of embryo sac the polar unclei are tetraploid.

    Ans. True

76. In oenothera type of embryo sac there is only one polar nucleus.

    Ans. True

77. The egg apparatus of peperomia type of embryo sac is characterized by a single synergid.

    Ans. True

78. Cross pollination is also known as allogamy.

    Ans. True

79. Double fertilization or triple fusion is characteristic of angiosperms.

    Ans. True

80. When both staminate and pistillate flowers are born on the same plant, the taxon is called monoecious.

    Ans. True

81. A bean seed is albuminous, while a maize seed is exalbuminous.

    Ans. False

82. The process of entry of pollen tube through integuments for fertilization is called porogamy.

    Ans. False

83. Three cells present at the chalazal end in the embryo sac are known as antipodals.

    Ans. True

84. When staminate and pistillate flowers are born on the different plant, the taxon is called dioecious.

    Ans. True

85. The outer wall of pollen is made up of cutin and suberin like substances.

    Ans. True

86. Gametophytic stage in angiosperms is extremely reduced.

    Ans. True

87. The plumbago type of embryo sac is characterized by the absence of synergids and antipodals.

    Ans. True

88. Monoecious species are necessarily cross pollinated.

    Ans. False

89. The gametophyte is completely depends on the sporophyte in angiosperms.

    Ans. True

90. The functional megaspore functions as mother cell of the female gametophyte.

    Ans. True

91. The condition when the anthers mature before the stigma becomes receptive is called protogyny.

    Ans. False

92. Microsporogenesis is takes place in the ovary of the flower.

    Ans. False

93. The triploid nucleus formed as a result of double fertilization and triple fusion.

    Ans. True

94. The insect visit to flower to help in pollination.

    Ans. False

95. The pollen grains produce more than one male nucleus.

    Ans. True

96. The female gametophyte of the flowering plants itself acts as an endosperm.

    Ans. False

97. The main causes of seed dormancy are hardness of seed coat and presence of inhibitory substances.

    Ans. True

DESCRIPTIVE TYPE QUESTIONS

1. What are the changes take places when ovary converts into fruit?

   The changes take place when ovary converted into fruit are as follows.

	Ovary		**Fruit**
A.	**Ovary wall-**	1.	Pericarp (fruit wall)
	i. Epicarp		
	ii. Mesocarp		
	iii. Endocarp		
B.	**Ovule-**	2.	Seed
	a. Funicle		Stalk
	b. Hilum-		Hilum
	c. Micropyle-		Micropyle
	d. Integuments-		Seed coat
	i. Outer integument-		Testa
	ii. Inner integument-		Tegmen
	e. Nucellus-		Generally degenerates, if persists, called perisperm
	f. Embryo sac-		
	i. Synergids-		Degenerate
	ii. Egg cell-		Zygote- embryo
	iii. Secondary nucleus-		Endosperm
	iv. Antipodals-		Degenerate

2. What is the relationship between fruit, seed and embryo of angiosperms?

   The seed is a complex structure and develops from the ovule only after the process of pollination and fertilization. Fertilization acts as a powerful stimulus with the result of a series of changes followed in the ovary. The fertilized egg gives rise to the embryo, the ovule to the seed, and the ovary as a whole to the fruit.

3. What are the special features of embryology angiospermic plant's ?
   - The presence of stamens and carpels into a structure called the flower, which is typically only of the angiosperm.
   - The ovules are always enclosed in an ovary which is the basal region of the megasporophyll.
   - The production of two kinds of spores, microspores (pollen grains) and megaspores, such type of condition is called heterosporous.
   - Presence of a single functional megaspore which permanently retained within the nucellus.
   - Flowers are adapted to insect pollination.
   - Spore dimorphism having resulted in the production of gametophytes, male and female.
   - Extreme reduction in size, duration of existence and complexity of the structure of the gametophytes which are entirely parasitic.
   - The male gametophyte has reached the limits of reduction.
   - The female gametophyte lacks any extensive development of vegetative tissue.
   - The presence of pollen tube to carry the male cells to the egg apparatus.
   - The seed remains enclosed in the ripened ovary called the fruit.
   - The phenomenon of double fertilization and triple fusion is only present in angiosperms.
   - Endosperm is tripoloid.

4. Define hypogyny?

   In a hypogynous flower, the thalamus is conical, dome-shaped or flat, and the ovary is being situated at the top on the thalamus, whereas the stamens, petals and sepals are separately and successively inserted below the ovary. Here, the ovary is said to be superior and rest of the floral appendages inferior. E.g. *Mustard, Solanum, Magnolia,* etc.

5. What is perigyny? Give examples.

   In a perigynous flower the margin of the thalamus grows upward forming a cup like structure known as the calyx tube, which encloses the ovary but remains free from it, and the sepals, petals and stamens are being situated upon it. Here the ovary is said to be half inferior. E.g. rose, plum, etc.

6. What is epigynous flower condition in guava?

   In an epigynous flower the margin of the thalamus grows further upward, to enclose the ovary completely, of course it fuses with it, and bears the sepals, petals and stamens above the ovary. Here the ovary is said to be inferior and the rest of the floral appendages are superior.

7. What are pollinia? Give examples.

   In some plants the pollen cells of each pollen sac are united into a mass known as the pollinium, instead of reporting into loose grains. Pollinia are present in *Calotropis* and orchids.

8. Define is hypostase?

   At the level of origin of two integuments and just beneath the embryo sac, there is often a well-defined, irregular group of nucellar cells which are generally poor in cytoplasmic contents but possess partially lignified or suberized walls of a highly refractive material. Van Tieghem (1901), for the first time, gave the name hypostase to these cells. *Zostera* is a good example of a well-developed hypostase.

9. What is epistase?

   According to Van Tieghem (1901) the occasional presence of a similar well-marked tissue in the micropylar part of the ovule is called the epistase. Generally the epistase originates from the apical cells of the nucellar epidermis, which show a marked radial elongation and become somewhat lignified or subrerized.

10. What are the forms of ovules present in angiospermic plants?

    On the basis of the relative position of the micropyle, chalaza and funicle the mature ovule can be categorized as follows (Fig. 11.1):

    ✦ Orthotropous or straight: In this type the ovule is erect or straight so that the funicle, chalaza and micropyl lie in one and the same vertical line. E.g. *Polygonum*.

    ✦ Anatropous or inverted: In this type the ovule bends along the funicle so that the micropyle lies close to the hilum. The chalaza lies at the other end. It is most common type of ovule.

    ✦ Amphitropous or transverse: In this type the ovule is placed transversely at a right angle to its stalk or funicle. E.g. *Lemna*.

    ✦ Hemitropous or hemianatropous: In this type the body of the ovule is straight but twisted in such a way that it is placed transversely at right angle, and so the chalazal micropyle line is at right angle to the funicle. E.g. *Ranunculus*.

    ✦ Campylotropous or curved: In this type the transverse ovule is bent round like a horse-shoe so that the micropyle and the chalaza do not lie in the same straight line. E.g. *Capparis*.

    ✦ Circinotropous: In this type the nucellus and the axis remain in the same line in the beginning but due to rapid growth on one side, the ovule gets inverted. This curvature continues and thus the ovule turns completely and once again the micropyle faces upwards. E.g. *Opuntia* and *Plumbago*.

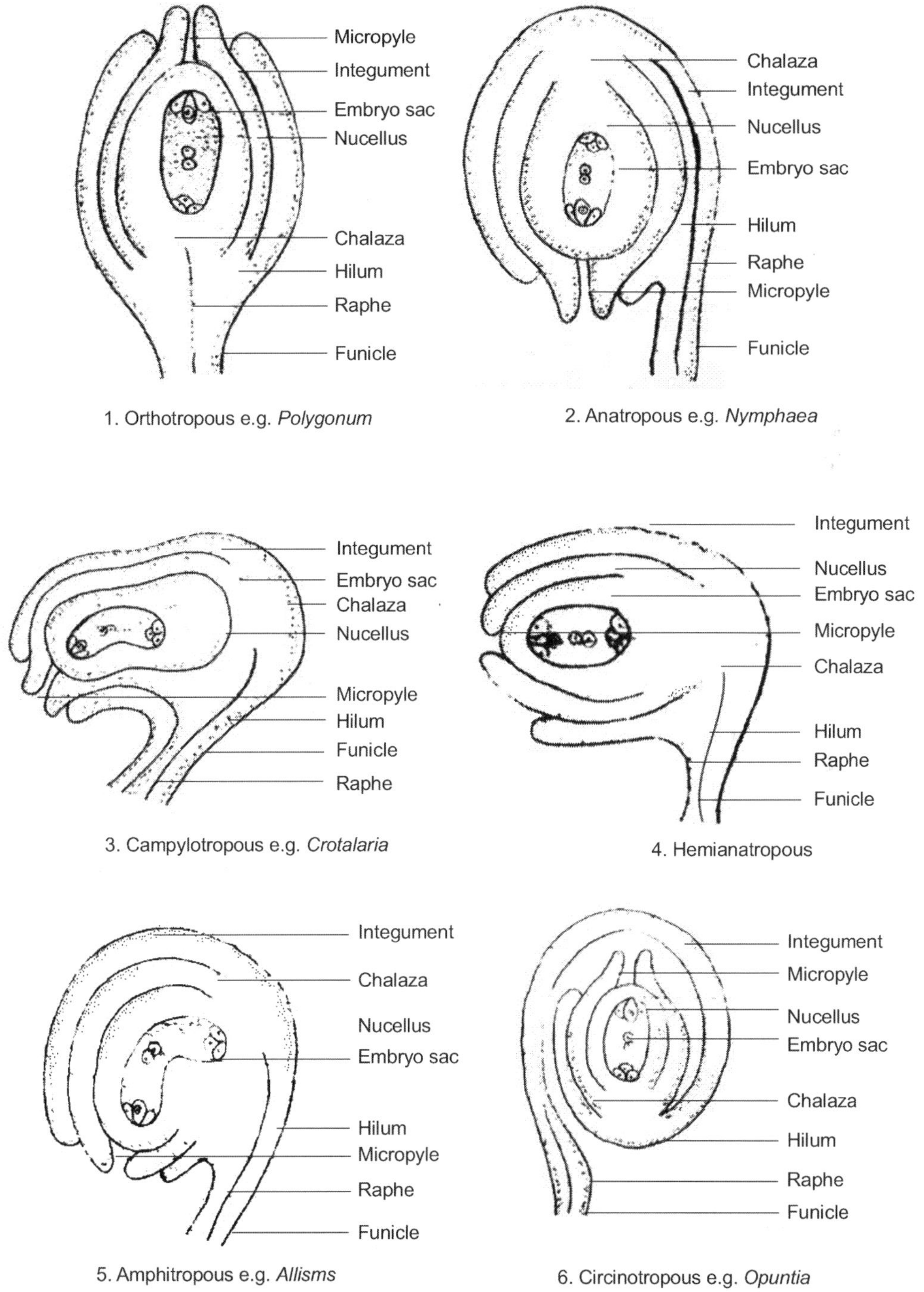

Figure 4.1 Types of Ovule in Angiosperms

11. Briefly describe the sexual process in angiosperms.

    The major events are as follow:

    - Microsporogenesis: In this process pollen grains or microspores are formed from microspore mother cells inside the anther.
    - Megasporogenesis: This process leads to the formation of megaspores from megaspore mother cell, inside the ovule.
    - Microgametogenesis: This involves the formation of male gametes from microspore.
    - Megagametogenesis: The events involving the formation of embryo sac from megaspore are included in this process.
    - Pollination: This is the transfer of pollen grains from the anther to the stigma.
    - Fertilization: In this process, fusion of male and female gametes takes place, resulting in the formation of zygote.
    - Embryogeny: The process involves development of embryo from zygote.
    - Formation of endosperm: This includes the development of endosperm as a result of triple fusion.

12. What is xenia and metaxenia?

    The term xenia was originally used for the direct effect of pollen grains on the characters of seeds and fruits. However, this term is now being used to denote effect of pollen grains on endosperm only. Maize is the classical example of xenia.

    The term metaxenia is used to indicate the effect of pollen grains on the tissues situated outside the endosperm. Date palm is an good example of metaxenia.

Tables

Various types of embryo sacs of angiosperms.

Types	Types	Examples
Monosporic- 8 nucleate	Polygonum	*Polygonum divaricatum*
Monosporic- 4 nucleate	Oenothera	Members of family Onagraceae
Bisporic- 8 nucleate	Allium	
Bisporic- 8 nucleate	Endymion	
Tetrasporic- 16 nucleate	Peperomia	*Peperomia*
Tetrasporic- 16 nucleate	Penaea	Malpighiaceae and Euphorbiaceae families
Tetrasporic-16 nucleate	Drusa	*Drusa*
Tetrasporic- 8 nucleate	Fritillaria	*Fritillaria*
Tetrasporic- 4 nucleate	Plumbagella	
Tetrasporic- 8 nucleate	Plumbago	*Plumbaginaceae*
Tetrasporic- 8 nucleate	Adoxa	*Adoxa*

Different pollinating agents and types of pollination.

Living/ non-living agents	Actual agent	Technical term
Biotic agents (Animals)	Insects	Entomophily
	Birds	Ornithophily
	Bats	Cheiropteriphily
	Snails	Malacophily
Abiotic agents (Non-living)	Wind	Anemophily
	Water	Hydrophily

Homologies in different terms used in embryology.

✦ Carpel-	Megasporophyll
✦ Ovule-	Megasporangium
✦ Embryo sac-	Female gametophyte
✦ Egg-	Female gamete
✦ Stamen-	Microsporophyll
✦ Anther (pollen sac)-	Microsporangium
✦ Pollen grain	Microspore
✦ Germinating pollen grains-	Male gametophyte

EMBRYOLOGY—A QUICK GLANCE

1. The study of flower and flowering plants is called anthology.
2. Cultivation of trees and shrubs is called arboriculture.
3. C.F. Wolf is known as father of embryology.
4. Prof. P. Maheshwari is known as father of Indian embryology, he has written "An Introduction to Embryology of Angiosperms".
5. Angiospermic plants originated in Mesozoic era.
6. Angiospermic plants are originated in the middle of Cretaceous and Jurassic periods.
7. Angiosperms are dominated on the earth in Cenozoic Era; this era is also known as Golden period of angiospermic plants.
8. N. Grew first time told that anthers are the male reproductive parts of plants.
9. The details of sexual reproduction was first time given by Jacob Camerarius, he also told that anthers are male and ovary, style and stigma are the female parts of a flower and interaction in these is essential for the formation of seeds.
10. The importance of insects in pollination is first time discussed by Joseph Kolreuter.
11. Plants which produce flowers and fruits once in life are known as monocarpic plants. E.g. Bamboo, Banana, Century plant etc.

12. Plants which produce flowers and fruits several times in life are known as polycarpic plants. E.g. All perennial plants.
13. Mint and *Chrysanthemum* are the example of suckers, and *Typha, Allocasia*, lotus and turmeric are the example of rhizomes.
14. Onion, garlic and lily are the examples of bulbs, and *Amorphophellus*, gladiolus, *Crocus* and *Colocasia* are corms.
15. *Cynodon* (doob) *Centella* and *Oxalis* are runners, *Fragaria* and *Vellisneria* are stolons, and *Pistia* and *Eichornia* are the examples of offsets.
16. In *Bryophyllum*, *Begonia* and *Santpolia*, the adventitious buds developed on the leaves which detached to the leaves and developed in to a separate plant.
17. In aquatic plants special types of buds developed on the stems which are known as 'Turions'. E.g. *Utricularia.*
18. Monothecous and bisporangiate type of pollen chambers are present in the members of family Malvaceae.
19. Monothecous and bisporangiate type of pollen chambers are present in *Moringa* and *Wolfia.*
20. Monothecous and monosporangiate type of pollen chambers are present in *Arceuthobium.*
21. In Gentianaceae members the tapitum is less developed, some of the sporogenous cells become sterile and provide nutrition to the remaining active sporogenous cells and later to the pollen grains.
22. Pollinium is present in the Asclepiadaceae and Orchidaceae families.
23. Translator apparatus is found in *Calotropis* plant it is helpful in pollination.
24. More than four pollen grains are present in pollen tetrad in *Cuscuta.*
25. Compound pollen grains are found in *Drocera, Typha, Drimys* and *Elodea.*
26. Massullae type of pollen grains is present in *Mimosa pudica* and *Neotia.*
27. The pollen grains which cause allergy are known as aeroallergens. E.g. *Parthenium*, *Chenopodium*, *Prosopis*, *Ameranthus*, *Sorghum vulgare*, *Ricinus communis* etc.
28. Allergy developed by the pollen grains of Ambrosia plant cause Hay fever.
29. In Malvaceae, spiny pollen grains are present.
30. Dumbbell shaped pollen grains are present in *Vinca herbasea.*
31. Smallest pollen grains are found in *Myosotis alpestris.*
32. Largest pollen grains are found in *Mirabilis jalapa.*
33. Pappus is present in the Compositae flowers, the sepals are modified into thread-like structures are called pappus.
34. The longest pollen tube is found in *Zea mays.*
35. The pollen tube was dicovered by G.B. Amici in *Portulaca.*
36. Aril is present in *Myristica* and *Asphodelus.*
37. The presence of perianth is a common feature in monocots and these are made up of tepals.

38. Arillode is found in Lichi and *Inga dulcis*. The edible part in lichi is actually arillode.
39. Gynophore is present in *Gynandropsis*, *Capparis*, *Pterospermum*, it is an inter node which is present between the stamens and the gynoecium.
40. Sometimes, both androphore and gynophore are present in the same flower, this condition is called gynandrophore.
41. Translator is present in the members of family Asclepiadaceae (e.g. *Asclepias*, *Calotropis* etc.), the two adjacent pollinia are united together in pairs by translator. Each translator has two parts 1. The corpusculum- it is a gland like body and 2. Retinaculi or connectives- these are pairs of arms.
42. Sarcotesta is found in the members of family Magnoliaceae.
43. The stem like nature of thalamus can be seen in *Capparis*, *Silene* and *Gynandropsis*.
44. The contents of the pollen tube are discharged in the synergid from where the first sperm is transferred to the egg cell, while the other one moves to the central cell through the cytoplasmic current.
45. Operculum on micropyle is found in the members of family Lemnaceae.
46. Caruncle or strophiole found in castor, it absorb water at the time of germination and help in seed germination.
47. Comose type of seeds found in *Calotropis* and *Gossypium*.
48. The point of attachment of funicle with chalazal end is called hilum.
49. The flower of *Citrus* has a swollen disc or gynobase which is present at the base of ovary.
50. Androphore or gonophores is an inter node present between petals and stamens. E.g. *Passiflora* and *Gynandropsis*.
51. Endothelium or integumentary tapitum is present in the members of family Compositae.
52. Syconus fruit develops from the hypanthodium type of inflorescence.
53. Pseudo-embryo sac found in the plants of family Podostemaceae.
54. Endosperm is absent in the families of Orchidaceae, Podostemaceae and Trapaceae.
55. Jack fruit is an example of composite fruit.
56. Coronary corona is found in *Calotropis* and stamina corona in *Nerium*.
57. 2n endosperm is found in *Oenothera*.
58. 5n endosperm present in *Plumbago* and *Plubegilla*.
59. Anthophore is an inter node which is present between sepals and petals in *Silene*.
60. The fruit developed from a multicarpellary, apocarpous ovary is called aggregate fruit.
61. Branched stamens are found in *Ricinus*, half-fertile in *Salvia* and arrow-head like anthers are present in *Thevetia* and *Vinca*.
62. 9n endosperm is found in *Pepromea* plant.
63. Single seeded, winged dry fruit is called samara.
64. Mosaic endosperm present in maize and tomato.
65. Thalamus is suppressed end of floral axis on which floral parts are inserted.

66. Meroblastic embryo development is present in angiospermic plants.
67. Nutritive aleurone layer is found in the seed of cereal plants.
68. The *Rafflesia* flowers are pollinated by carrion flies.
69. Obdiplostemonous condition is found in *Citrus*, gynostegium (formed due to the union of stigma and stamens) is found in *Calotropis* and Orchids.
70. Persistant nucellus or perisperm is found in *Arica* nut and black pepper.
71. Most of the seeds of diocot plants are non-endospermic but castor seeds are endospermic.
72. The dispersal of seeds in *Ruellia* is by Jaculator method.
73. The *Lodoicea sechellarum* (double coconut) is the largest fruit.
74. The swollen base of the style is called stylopodium.
75. Achlamydous flowers are naked flowers (without sepals and petals). E.g. members of family Piperaceae.
76. Longest leaf is present in *Raphia vinifera* (10-15 meters) and widest leaf is found in *Victoria resia* (1-1.5 meter).
77. Cotyledons are absent in *Cuscuta* seeds.
78. The spines present on outer thick leathery covering of jack fruit are hardened stigmas.
79. Persistent wheel-like stigma is present in Poppy and funnel-like stigma in *Crocus*.
80. The smallest leafless parasitic angiosperm is *Arceuthobium*.
81. The smallest seeds found in orchids and these are also known as Dust seeds.
82. A flower is a modified shoot.
83. In Iris plant seeds the growth inhibitor is present in the endosperm.
84. Spathe of a branch is called spethella and a sterile pistil is called pitillode.
85. The term 'Xenia effect' was given by Focke and it can be observed in maize.
86. The manner in which placentae are distributed in the cavity of ovary is called placentation.
87. Metaxenia can be observed in Date palm. The effect of pollen on the outer tissue of ovule is known as metaxenia.
88. Perianths present in grass flowers is called lodicules.
89. The seed dormancy in the seeds of *Gingko* and *Eranthis hyemelis* is due the under developed embryo.
90. 2-chloroethanol, gibberellins and thio-urea can be used to break bud dormancy.
91. The production of flowers on old stem from dormant buds is called cauliflory. E.g. *Ficus*, *Artocarpus*.
92. Chalaza is the term used for basal region of angiosperm ovule, where funicle is attached.
93. *Commelina* and *Viola* are the example of cleistogamy.
94. Grafting includes a shoot or part of the plant (also called scion) is inserted into another plant (also called stock) so as to be nourished by it and united with together. Grafting is successful in dicotyledonous plants where stems have a cambium layer.

95. The type of inflorescence in Compositae family is also called head, it is most advance type of inflorescence because aggregation of florets ensures enhancement of attraction, maximum protection of flowers and a single insect can pollinate many flowers.
96. The stamens are modified into cladodes in the genus *Asparagus*.
97. Keel, the special structure, only present in the flowers of legume plants.
98. The opening of flowers at the time of maturation of reproductive parts is known as anthesis or chasmogamy.
99. A transversely dehiscing capsule is called pyxidium.
100. Herkogamy (morphological barrier in between anther and stigma) is found in *Gloriosa* and *Calotropis*.
101. Anthotaxy is the arrangement of flowers on the peduncle.
102. Heterostyly is observed in the primrose flowers.
103. Self-sterility is found in flowers of petunia, tea, orchids etc.
104. Advertisement flag is found in *Musenda*.
105. The direction of maturation of flowers in racemose inflorescence is acropetal and centripetal.
106. The recemose and cymose types of inflorescence are also called pure inflorescence.
107. Layering includes the induction and development of roots on a stem before detaching it from the parent plant for propagation.
108. Lever mechanism of pollination is present in the *Salvia*.
109. Trapdoor mechanism is found in the *Ficus* flowers due to the presence of hypanthodium type of inflorescence.
110. In Eusporangiate type the development of an anther is the anther it develops from a group of cells.
111. Caruncle facilitates seed dispersal (it is sugary and eaten by ants leading to dispersal of seeds) and being hygroscopic it absorbs water and passes it to embryo during germination.
112. The regeneration of new plants from vegetative organs is called vegetative or somatic reproduction.
113. Pitfall flowers are present in *Aristolochia* and flytrap type of pollination is found in it.
114. The anther wall is composed of epidermis, endothecium and innermost is the tapetum.
115. The pollination in *Ophrys speculum* is by wasp with the pseudo-masturbation method because the morphology and smell of the flower is just like female wasp (Mimicry).
116. Hanstein was the first to follow the sequence of early development of embryo.
117. *Vallisneria* and *Ruppia* are the hydrophytic plants and apihydrophilly type of pollination is found in these plants.
118. The cells of endothecium are radially elongated and hygroscopic in nature which helps in the dehiscence of the anther.
119. *Jostera* and *Ceratophyllum* hydrophytes, and hypohydrophilly type of pollination is found in these plants..

120. A pome fruit is said to be false because the fruit is present in fleshy edible thalamus.
121. In Bignonia the pollination is performed by humming bird and honey bird (*Nectrina alpha*) perform pollination in *Strelitzia*.
122. Pollenkitt is an oily layer found outside of the mature pollen grains and consists of lipids and carotenoids.
123. When the pollination performed by bats, it is known as chiropterophilly. E.g. *Enthocephallus cadamba*, *Bauhinea* and *Kigalia*.
124. Censer mechanism of seed dispersal is found in the members of family Papaveraceae.
125. The main function of tapetum is the production and transport of enzymes, hormones and nutrients for the proper development of pollen grains.
126. Endosperm is a triploid tissue, because it develops after fertilization and from secondary nucleus, it is nutritive tissue for the developing embryo in angiosperms.
127. Adventives embryony in citrus is due to nucellus.
128. The pollination by snails is known as malacophilly. E.g. *Chrysanthemum*.
129. In *Erythrina* plant, pollination is performed by Crow and squirrels.
130. Sexual reproduction of flowering plants was discovered by Camerarius.
131. Cytokinesis in the developing pollen grains is two types, simultaneous and successive. The successive is advanced type of cytokinesis in pollen grains.
132. In *Centaurea* plant pollination performed by piston mechanism.

ANSWERS

1. (c)	2. (a)	3. (d)	4. (b)	5. (b)	6. (a)
7. (b)	8. (a)	9. (c)	10. (b)	11. (b)	12. (b)
13. (b)	14. (a)	15. (a)	16. (c)	17. (a)	18. (b)
19. (b)	20. (b)	21. (c)	22. (b)	23. (b)	24. (c)
25. (b)	26. (b)	27. (d)	28. (c)	29. (a)	30. (b)
31. (c)	32. (a)	33. (c)	34. (a)	35. (b)	36. (b)
37. (b)	38. (a)	39. (c)	40. (c)	41. (a)	42. (d)
43. (d)	44. (c)	45. (b)	46. (c)	47. (d)	48. (c)
49. (c)	50. (c)	51. (a)	52. (b)	53. (a)	54. (b)
55. (c)	56. (a)	57. (a)	58. (d)	59. (c)	60. (d)
61. (d)	62. (a)	63. (d)	64. (a)	65. (b)	66. (a)
67. (c)	68. (d)	69. (a)	70. (b)	71. (d)	72. (d)
73. (d)	74. (d)	75. (a)	76. (b)	77. (a)	78. (a)
79. (b)	80. (c)	81. (c)	82. (b)	83. (b)	84. (a)
85. (c)	86. (b)	87. (a)	88. (c)	89. (c)	90. (b)
91. (b)	92. (c)	93. (d)	94. (a)	95. (c)	96. (b)
97. (a)	98. (d)				

5

PLANT BREEDING

MULTIPLE CHOICE QUESTIONS

1. The reason why vegetatively reproducing crop plants are best suited for maintaining hybrid vigour is that
   (a) These are more resistant to diseases
   (b) These can be easily propagated
   (c) These have longer life span
   (d) Once a desired hybrid has been produced there few chances of losing it
2. Which of the following genes can be used for production of male sterile lines for hybrid seed production in a crop
   (a) gus (b) barnase (c) cat (d) npt II
3. Name the cotton hybrid variety among these
   (a) 170- CO (b) Moti (c) Sona (d) H4
4. National Botanical Research Institute is located at
   (a) New Delhi (b) Shimla (c) Lucknow (d) Jaipur
5. It is possible now to breed plants and animals with desired characters through
   (a) Tissue culture (b) Chromosome engineering
   (c) Genetic engineering (d) Hybridization
6. In backcross method for transfer of a dominant trait, in which generation, would you start the selfing experiments
   (a) Back cross 2 (b) Back cross 4
   (c) Back cross 6 (d) Back cross 8

7. Which of the following is not a part of T-DNA, which is resides in Ti-plasmid
   (a) Shooty loci (b) os region
   (c) vir region (d) Border sequences
8. Desired improved varieties of economically useful crops are raised by
   (a) Mutation (b) Biofertilizer
   (c) Hybridization (d) Natural selection
9. Which one of the following institute engaged in potato research
   (a) CDFD (b) DTRI (c) CPRI (d) CRRI
10. Which method will be useful to transfer of resistance against a gene, controlled by a single dominant gene
   (a) Modified bulk method (b) Backcross method
   (c) Pedigree method (d) Pure line selection method
11. In order to obtained triploid plant which of the following plants should be crossed
   (a) Diploid x diploid (b) Diploid x haploid
   (c) Tetraploid x haploid (d) Tetraploid x diploid
12. Which of the following genes is widely used for imparting herbicide resistance in crop plants
   (a) bar gene (b) CpTi gene (c) Bt cry genes (d) Coat protein gene
13. Which one of the following is a cross pollinated plant
   (a) Rice (b) Wheat (c) Maize (d) Groundnut
14. The performance of a double cross will be predicted as average of how many single crosses-
   (a) 3 (b) 4 (c) 5 (d) 6
15. Which of the following varieties was a direct release
   (a) Lerma Rojo (b) Kalyan Sona (c) Sonalika (d) Moti-5
16. Which of the following method is most time consuming
   (a) Pedigree method (b) Back cross method
   (c) Bulk method (d) Direct selection
17. The condition, where the hermaphrodite flowers never open to avoid cross pollination is known as
   (a) Herkogamy (b) Cleistogamy (c) Dichogamy (d) Agamospermy
18. Mutation was induced into a plant. It did not reappear when the plant was grown by seeds. The reason being
   (a) The mutation was recessive (b) That it was not useful
   (c) That it was a somatic mutation (d) That mutations are not sexually inherited

19. Homozygosity of a plant raised from androgenic haploid can be maintained by using
    (a) PEG (b) Chloramphenicol (c) IAA (d) Colchicine
20. Which of the following is not a reporter gene for production of transgenic crops
    (a) CaMV 35 S (b) cat (c) gus (d) npt II
21. Suggest a condition, where the MMC directly gives rise to the megaspore without undergoing meiosis
    (a) Dioecism (b) Apospory (c) Dichogamy (d) Diplospory
22. Which one of the following variety developed by pedigree method
    (a) K 68 (b) CSH 1
    (c) Tift 23 A (d) Composite cross II
23. In an economically important plant species, embryo mortality is very high. What method do you suggest to check the mortality?
    (a) Meristem culture (b) Micropropagation
    (c) Hybridization method (d) Embryo culture
24. Haploids are useful because
    (a) They are easy to maintain
    (b) These have only a few chromosomes
    (c) Homozygous diploids can be produced in very short period
    (d) New and useful recombination can be introduced
25. The selection can be started at which generation in bulk method
    (a) F3 (b) F 6 (c) F 9 (d) F 4
26. The performance of future crosses is predicted on the basis of
    (a) Single cross (b) Double cross (c) Top cross (d) Basic cross
27. Which of the following is an anther tapetum specific promoter, used for the expression of barnase and barstar genes on the anther
    (a) HSE (b) TA29 (c) LRE (d) CaMV 35S
28. Which one of the following is self incompatible crop
    (a) Sorghum (b) Cotton (c) Tobacco (d) Maize
29. Which Institute in our country looks after the plant introduction work
    (a) CRRI (b) CDFD (c) NBPGR (d) ICRISAT
30. Haploids produced from microspores or pollen grains are known as
    (a) Androgenic haploids (b) Homozygous mutants
    (c) Parental haploids (d) Heterozygotes

31. Which one of the following legume is an often cross pollinate
    (a) Pea (b) Lentil (c) Chickpea (d) Pigeon pea
32. One of the consequences of inbreeding is increase in the relative frequency of
    (a) Dominant homozygotes only (b) Recessive homozygotes only
    (c) Both homozygotes (d) None of these
33. Which is the New World spices that has become an essential part of Indian cuisine
    (a) Ginger (b) Cardamom (c) Red pepper (d) Black pepper
34. Induced mutation can be maintained through
    (a) Hybridization (b) Sexual reproduction
    (c) Vegetative propagation (d) All of these
35. Which of the following genes is widely used for imparting insect resistance in crop plants
    (a) Bt cry genes (b) PsbA gene (c) bar gene (d) bxn gene
36. Why the over dominance hypothesis for heterosis could not be accepted
    (a) Epistatic interaction at different dominant loci is known
    (b) Partial dominance between alleles is known at many loci
    (c) Over dominance at individual loci could never be demonstrated
    (d) It assumes presence of multiple alleles
37. In which crop, dwarf varieties were introduced in India from Mexico
    (a) Bajra (b) Rice (c) Wheat (d) None of these
38. Sarbati Sonora variety of wheat was obtained by
    (a) X-ray treatment
    (b) Hybridization between wild grasses
    (c) Irradiation of Sonora 64 with gamma rays
    (d) Crossing with wild varieties of wheat
39. General combining ability of inbreeds is tested on the basis of
    (a) Top cross (b) Double crosses (c) Single crosses (d) Basic cross
40. Which international agency has the major responsibility of looking after the wheat germplasm and its introduction
    (a) IRRI (b) IPGRI (c) CIMMYT (d) ICRISAT
41. *Triticum durum* was possibly fertilized by wild grass which is thought to have given rise to
    (a) *Aegilops speltoides* (b) *Aegilops squarrosa*
    (c) *Triticum aestivum* (d) *Triticum turgidum*
42. Why the decrease in size and vigour in early generation of inbreeding is
    (a) Loss of hybrid vigour (b) Loss of superior genes

(c) Reduction in the frequency of superior genes
(d) Expression of many lethal and sub-vital recessive alleles

43. Banana is generally seedless because it is
(a) A sterile triploid
(b) Store of reserve food material
(c) Artificially treated with auxins
(d) Cultivated by vegetative methods

44. Which of the following is not a physical delivery method for gene transfer in higher plants
(a) PEG stimulated DNA uptake
(b) Microinjection
(c) T-DNA mediated DNA uptake
(d) Electroporation

45. One of the major developments resulting from human efforts is
(a) Increased human population
(b) Elimination of forest
(c) Increase in food production
(d) Management of energy crisis

46. The hand emasculation is generally recommended in crops with
(a) Dioecious condition
(b) Very small flowers
(c) Large flower
(d) Flower without calyx

47. What is green revolution
(a) Maintaining soil fertility
(b) Use of more green plants for covering the earth so that soil erosion can be controlled
(c) Development of new crop verities which helped to overcome hunger
(d) Growing green plants to establish balance of nature

48. Who was awarded Nobel Peace Prize in the area of plant breeding
(a) B.P. Pal
(b) M.S. Swaminathan
(c) R.K. Pachouri
(d) N.E. Borlaug

49. The process of bringing wild species of plants under cultivation and training them to suit human needs is called
(a) Farming (b) Selection (c) Domestication (d) Selection

50. Which of the first crop to be widely used for imparting insect resistance in crop plants
(a) Rice (b) Wheat (c) Brinjal (d) Cotton

51. Cytoplasmic male sterility is commonly used in
(a) All breeding methods
(b) Pedigree method
(c) Hybrid seed production
(d) Back cross methods

52. The weediness and presence of large amount of food reserves in the seeds were the characters of annuls which were most suitable for
(a) Cultivation
(b) Crop improvement
(c) Hybridization
(d) Domestication

53. Who recommended for the first time the use of recurrent selection as a breeding scheme
    (a) Jenkins (b) Hull (c) Lonnquist (d) Jones
54. Triploids are said to more useful than other polyploids because these are
    (a) Sterile and show increased vigour
    (b) Fertile and more healthier
    (c) Fruitless and are always propagated by stem cuttings
    (d) Produced when diploid definitive nucleus fuses with a male gamete
55. In what types of situation, emasculation will not be needed, although hermaphrodite flowers are present
    (a) Homogamy (b) Self incompatibility
    (c) Genetic male sterility (d) Cytoplasmic male sterility
56. How many years would be needed for the production of a variety involving hybridization followed by clonal selection in a vegetatively propagated plant
    (a) 6 (b) 7 (c) 8 (d) 4
57. In an effort to improve the plant species, two varieties of a species were artificially pollinated with the desire to obtain a new variety that combines useful characters of both these individuals. The process involved is known as
    (a) Selection (b) Pollination (c) Hybridization (d) Selfing
58. Which of the following selection method will be suitable for maintaining purity of a variety in a self pollinated crop
    (a) Progeny selection (b) Pure line selection
    (c) Mass selection (d) Clonal selection
59. Heterosis is referred to
    (a) Difference in meiosis
    (b) Two dissimilar stage occurring in meiosis
    (c) Hybrid vigour
    (d) Mating of two genetically dissimilar plants
60. Who recommended the use of early testing for improvement of inbreds
    (a) Jenkins (b) Hull (c) Lonnquist (d) Jones
61. Progeny rows are evaluated and superior progenies are identified in which of the selection
    (a) Mass selection (b) Progeny selection
    (c) Clonal selection (d) Pure selection
62. The hybridization programme for sugarcane improvement is undertaken at Sugarcane Research Institute located at
    (a) Jaipur (b) Hyderabad (c) Chennai (d) Coimbatore

63. Polyploidy is often the result of
    (a) Fertilization of egg by more than one male gamete
    (b) Failure of meiosis during gamete formation
    (c) Abnormal mitosis causing increase in chromosome number in sex cells
    (d) All of these
64. No single plant progenies are grown during
    (a) Mass selection (b) Progeny selection
    (c) Clonal selection (d) Pure selection
65. Heterosis is best maintained by
    (a) Continuous inbreeding (b) Maintaining parental lines
    (c) Vegetative reproduction (d) All of these
66. A method of selection that is particularly useful for self-pollinated plants is
    (a) Mass selection (b) Pure line selection
    (c) At random selection (d) None of these
67. In which generation after hybridization, clonal selection will be initiated
    (a) F1 (b) F2 (c) F3 (d) F4
68. Cytoplasmic male sterile line, Tift 23A that was introduced from Tifton, Georgia, belongs to the which of the following crop
    (a) Onion (b) Cotton (c) Rice (d) Bajra
69. Heritable variation due to recombination can also arise in pure lines, due to
    (a) Mechanical mixture (b) Chromosomal aberrations
    (c) Natural crossing (d) Mutations
70. The plants preferred for cultivation during earlier days included those
    (a) With maximum food value
    (b) Which could be grown easily
    (c) Having multiple use
    (d) Competing the life cycles in short periods of time
71. ICRISAT look after the genetic resources and plant introduction for the following crops
    (a) Pigeon pea, chickpea, groundnut, sorghum and pearl millet
    (b) Pigeon pea, chickpea, barley, sorghum and pearl millet
    (c) Pigeon pea, chickpea, groundnut, cotton and pearl millet
    (d) Pigeon pea, rice, wheat, sorghum and pearl millet
72. Which one of the following crops was subjected the most to continuous selection and domestication by man
    (a) Wheat (b) Rice (c) Potato (d) Corn

73. Which of the following species of *Solanum* was utilized for the improvement of potato
    (a) *S. nigrum* (b) *S. melongena* (c) *S. khasianum* (d) *S. andigena*
74. A plant that may provide oil, fibres, jiggery, etc is
    (a) Coconut palm (b) Hemp (c) Agave (d) Rubber
75. In cross pollinated crops, mass selection can be used for improvement of
    (a) Inbreds (b) Hybrids (c) Clones (d) Synthetics
76. Who recommended the production of double cross hybrids in maize
    (a) Jenkins (b) Hull (c) Lonnquist (d) Jones
77. Which one of the following factors will not affect the result of mass selection
    (a) Heritability (b) Available variability
    (c) Selection intensity (d) The extent of mutations
78. The methods which bring together various useful characters from different plants into one are collectively known as
    (a) Hybridization (b) Crop improvement
    (c) Pure line (d) Micropropagation

Small Answers Questions

1. Define polyallele crossing.

   Ans. The crossing of several inbred lines with each other to determine their respective combing ability is known as polyallele crossing.

2. What are inbreds?

   Ans. The production of pure lines in cross pollinated plants is known as inbreds.

3. What is bud mutation?

   Ans. The mutation in the bud which causes variation in it and gives rise to a branch, flower or fruit unlike of the plant, is termed as bud mutation.

4. What is law of homologous series in variation?

   Ans. This law was formulated by Vavilov, stated that 'characters found in one species also occur in other related species'.

5. Name two common dioecious crops grown in India.

   Ans. Cucurbits and papaya.

6. Who develop the progeny test?

   Ans. Louis de Vilmorin.

7. Define protoplasm fusion?

   Ans. Wide crosses are generally sterile and unsuccessful, but this difficulty can be overcome by the technique of protoplasm fusion.

8. Write the major disadvantage of all selection methods of plant breeding?

   Ans. No variability can be introduced.

9. Name any one important monoecious crop.

   Ans. Maize.

10. What is protandry?

    Ans. Male organ matures earlier than female in bisexual flower is known as protandry.

11. What is gene fund?

    Ans. It indicates the genetic resource or gene pool for breeding.

12. Who told that 'hybrid vigour denotes the manifest of heterosis'?

    Ans. Whaley (1944).

13. Who introduced the concept of recurrent selection?

    Ans. Stadler (1845).

14. Which is the primary center of origin for cereals?

    Ans. Abyssinian.

15. Describe recurrent parent?

    Ans. In back cross method, the variety that needs to be improved is described as the recurrent parent.

16. Name the plasmids, which facilitated *Agrobacterium* mediated gene transfer in crop plants.

    Ans. Ti and Ri plasmids.

17. Give the example of some cleistogamous flower plants.

    Ans. Groundnut.

18. Who introduced the concept of early testing?

    Ans. Jenkins (1935).

19. Write the full form of CRRI.

    Ans. Central Rice Research Institute it is in Kottak..

20. Name the two wheat varieties developed through pure line selection.

    Ans. NP 4 and NP 552.

21. What are the chemicals, for which foreign genes have been introduced in crops demonstrating the potential of transgenic crops for molecular farming.

    Ans. Cyclodextrin and Polyhydroxybutyrate (PHB).

22. Define homogamy.

    Ans. The condition, where anther and stigma mature at the same time.

23. What are barnase and barstar?

    Ans. These are two bacterial genes for male sterility and restorer.

24. What is recurrent selection?

    Ans. It involves several generations of regular selection and inter-mating.

25. Which character was improved in golden rice?

    Ans. Vitamin A.

26. What is the full form of CSH variety in Jowar?

    Ans. Coordinated Sorghum Hybrid.

27. Who gave the term gene fund?

    Ans. Zhukovsky (1970).

28. Who found that embryo size was found to be little importance in expression of heterosis?

    Ans. Shull (1940).

29. Name one important vegetatively grown vegetable crop.

    Ans. Potato.

30. Describe donor parent?

    Ans. In back cross method, the variety, which carries the describable gene to be transferred, is described as donor parent.

31. Name the three most important characters that were initially improved in transgenic corps.

    Ans. Herbicide resistance, insect resistance and virus resistance.

32. Name a recently released hexaploid wheat variety.

    Ans. Arjun.

33. What is dichogamy?

    Ans. The condition, where pollen and stigma mature at different times.

34. Name recently released tetraploid wheat variety.

    Ans. Meghdoot.

35. Define noblization.

    Ans. The technique involving repeated crosses with the wild species, *Saccharum officinarum*, used for improvement of sugarcane.

36. What are cat, npt II and gus?

    Ans. These are reporter genes that have been employed in the production of transgenic crops.

37. What temperatures are generally used in cold water treatment for emasculation?

    Ans. $<6^{o}C$.

38. What is Dee-geo-woo-gen?

    Ans. It is dwarf genes in semi-dwarf rice varieties derived from Japonica dwarf variety.

39. Name one important vegetatively grown plant species, which is dioecious.

    Ans. *Asparagus*.

40. Define self incompatibility.

    Ans. When mature pollen fails to bring about self fertilization.

41. What is the major advantage of progeny selection?

    Ans. It allows the separation of environmental effects from the genetic effect.

42. What is germplasm collection?

    Ans. The sum total of hereditary material or genes present in a species is known as the germplasm of that species.

43. Give names of two wheat varieties, which were primary introduction.

    Ans. Lerma Rojo, Sonora 64.

44. What is Norin?

    Ans. The dwarf wheat that was introduced from Mexico carried this gene for dwarfness.

45. Name two legume crops, which are often, cross pollinated.

    Ans. Broad bean and Pigeon pea.

46. N.I. Vavilov belonged to which country?

    Ans. Russia.

47. What are the chemical gametocides used for emasculation of flowers?

    Ans. Ethrel and FW 450.

48. While transferring a dominant trait using back cross method, would you self the hybrid plants after each back cross, before making the next back cross.

    Ans. No.

49. Who explained the genetic basis of pure line selection?

    Ans. Johannsen (1903).

50. How Kufri sindhuri developed by the Indian researchers?

    Ans. It is a selection from a cross between Kufri red and Kufri Kundan.

51. Name a crop plant where self incompatibility is common.

    Ans. Tobacco.

52. Who proposed the eight centres of origin of crop plants?

    Ans. N.I. Vavilov.

53. What temperatures are generally used in hot water treatment for emasculation?

    Ans. > 40°C.

54. Write the full form of ICRISAT and the its location.

    Ans. International Crop Research Institute for Semi-Arid Tropics, Hyderabad.

55. Name any hybrid variety of wheat.

    Ans. N.P. 165.

56. Names of two wheat varieties, which were secondary introduction.

    Ans. Kalyan Sona, Sonalika.

57. Name two hybrid crops other than potato and sugarcane, which have been improved through hybridization followed by clonal selection.

    Ans. Malus, Strawberries.

58. Name two vegetatively grown crops where interspecific hybrids are used for clonal selection.

    Ans. Potato and sugar cane.

59. Who proposed multiple factor hypothesis?

    Ans. Nilsson Ehle.

TRUE OR FALSE QUESTIONS

1. The species of *Saccharum* are complex polyploids with high chromosomal numbers.

   Ans. True

2. Domestication is the process of bringing wild species under human management.

   Ans. True

3. Rice is cultivated very extensively in the tropical and sub-tropical regions of the world.

   Ans. True

4. In crops having small flowers, hand emasculation is generally not possible.

   Ans. True

5. Plant introduction means of taking a plant species of variety into a new environment where it has not been grown before.

   Ans. True

6. Papaya is a dioecious plant.

   Ans. False

7. Mass selection cannot be practices in cross pollinated crops.

   Ans. False

8. Puskarnath is related to Central Potato Research Institute, Shimla.

   Ans. True

9. NBPGR is situated in New Delhi.

   Ans. True

10. In self pollinated crops, mass selection is mainly used for purification.

    Ans. True

11. The revolution in the production of wheat and rice in India was brought about due to the introduction of dwarf varieties.

    Ans. True

12. In monoecious and dioecious plants, hybridization does not require emasculation of individual flowers.

    Ans. True

13. Introduction may be primary and secondary.

    Ans. True

14. In pedigree method of plant breeding, individual plant selection for quantitative traits is initiated in F2 generation.

    Ans. False

15. Mass selection is not suitable for improvement of characters low heritability.

    Ans. True

16. Top cross is generally used for early testing of inbreds.

    Ans. True

17. Semi dwarf wheat variety is a directly released variety after introduction.

    Ans. True

18. Dioecious crops bear male and female flowers on separate plants.

    Ans. True

19. In mass pedigree method, single plant progenies are grown in early generations, but seed from selected plants is bulked in later generations.

    Ans. False

20. The general effect of inbreeding in out-crossing population is a reduction in size and vigour.

    Ans. True

21. Royal Botanical Garden is situated in Kew, U.K.

    Ans. True

22. Give the name of two wild species, of genus *Solanum*, which have been used for interspecific hybridization aimed at the improvement of potato.

    Ans. *S. curtilobum* and *S. andigena.*

23. Pureline is a progeny of a single, homozygous, self-pollinated plant.

    Ans. True

24. For transfer of a simple inherited trait, backcross method is normally used.

    Ans. True

25. Hybrid corn is showing heterosis.

    Ans. True

26. Exploration is the primary source of all the germplasm maintained in germplasm collections.

    Ans. True

27. Mild inbreeding involves open pollination but no intercrossing.

    Ans. False

28. Inbreds in maize are always obtained by selfing an open pollinated variety, and never involve any hybridization step.

    Ans. False

29. Pedigree record is maintained in pedigree method of plant breeding so that the plant breeder may be able to repeat the experiment, whenever it needed.

    Ans. False

30. In self-pollinated crops, selection permits reproduction in plants with the desirable characteristics only.

    Ans. True

31. Mass selection cannot be effective if the base population has a low level of variability.

    Ans. True

32. In mass selection a large number of similar phenotype are selected and their seeds are mixed together to constitute a new variety.

    Ans. True

33. Give the name of two wild species, of genus *Saccharum*, which have been used for interspecific hybridization aimed at the improvement of sugarcane.

    Ans. *S. barberi* and *S. spontaneum.*

34. During hybridization, emasculated buds need to be bagged, but no such bagging is required for selfing.

    Ans. False

35. Pure selection is very time consuming and more expensive than mass selection.

    Ans. True

36. Hybrid verities are never produced in self pollinated crops.

    Ans. False

37. Pedigree method allows improvement in both, the simply inherited traits and the complex traits.

    Ans. True

38. A number of hybrid varieties have been successful in several pulse crops.

    Ans. False

39. Hybrid varieties in India were more successful in sorghum and bajra than maize.

    Ans. True

40. The condition, where flowers never open, thus ensuring self pollination is called as cleistogamy.

    Ans. True

41. Wheat is a monoecous crop.

    Ans. False

42. While transferring a recessive trait using backcross method, would you allow one generation of selfing of the backcross progeny, before making every alternate backcross.

    Ans. True

43. The variation observed within a clone will be only partly environmental.

    Ans. False

44. N. E. Borlaugh is related to wheat breeding.

    Ans. True

DESCRIPTIVE TYPE QUESTIONS

1. What is heterosis?

   Heterosis is also considered as hybrid vigour may be defined as the increased vigour, size, growth or other functions of the hybrid (F1 plants) over their parents resulting from cross of genetically unlike organisms. Heterosis has been clearly demonstrated in corn for its marked effect on yield improvement. It is also successfully employed in sorghum, bajra, wheat, potato etc.

2. Define geitonogamy?

   When the pollens of a flower fall on the stigma of another flower on the same plant, it is termed as geitonogamy and genetically is equal to self pollination. In maize, pollen from male flower may fall on the stigma of the female on the same plant resulting in geitonogamy.

3. What is forceps and scissors method?

   It is a process where the opening of flowers and removal of anther are done with the help of forceps and scissors.

4. What is the use of hot water emasculation technique?

   In plants like rice, sorghum, bajra etc., where the flowers are very small, the whole panicle is dipped in a bottle containing hot water having a desired temperature for a definite period. Thus the temperature kills all the anthers of the flowers sterile.

5. Define male sterility method.

   In case of plants like sorghum, barley, onion, bajra, etc., the whole operation of emasculation is eliminated by the use of male-sterile plants. Male sterility may be due to cytoplasmic or genetic cause. It can be induced by the application of some chemicals like 2,4-D, NAA, maleic hydrazide etc.

6. What is Gamma garden?

   Gamma rays have ionizing effects on plants like that of X-rays. Thus materials can be treated with Gamma rays for induction of mutation. Restricted sites where facility for gamma irradiation of plants exists have been developed in several research institutes. These sites are called Gamma Gardens. The first Gamma Garden of the world was established in Long Island near New York. In India the first Gamma Garden was established at the Bose Research Institute, Kolkata in 1959 and the second in IARI, New Delhi in 1960.

7. What is mutation breeding?

   After the discovery of x-rays induced mutation by Miller in 1927, L.J. Stadler, a pioneer American plant breeder induced mutations in plants in 1928 and exploited the possibilities of crop improvement by mutation breeding for the first time.

8. Define plant breeding.

   It is the applied branch of botany that deals with the improvement of agricultural crops and also production of improved new crop varieties.

9. What are the objectives of plant breeding?

   The main objectives of plant breeding are as follow:

   - Higher yield of grain, fodder, fiber, oil and other plant products.
   - Better quality with regard to shape, size, colour, nutrition, taste, malting, milling, baking, keeping, cocking etc., of food grains, vegetables, fruits and flowers.
   - Resistance to diseases, insects and pests, frost, flood, drought, wind storms etc.
   - Change in duration, earliness or lateness in maturity as needed.
   - Change in growth habits such as induction of dwarfness, restricted branching and less tillering so as to prevent the crop plants from lodging; or tallness, profuse branching and more tillering so as to increase the straw for fodder.
   - Winter hardiness.
   - Response to heavy manuring.

- Easier thrashability.
- Awned or awnless ears.
- Adaptability to different climatic regions.
- Suitability to machine harvesting in mechanized area.

10. Discuss the contribution of B.P Pal in plant breeding.

He is a pioneer Indian agricultural scientist particularly as plant breeder, who took leading part in the improvement of Indian agriculture and production of famous N.P. wheat varieties.

11. Write the name of some Indian plant breeders.

- M.S. Swaminathan is famous wheat breeder and also known as father of Indian green revolution.
- K. Ramaiah is a renowned Indian Rice breeder of International repute.
- Puskarnath is famous Indian potato breeder.
- Athwal is famous breeder, worked on hybrid bajra.
- Boshi Sen is famous Indian maize breeder.
- T.V.S Venkataraman is a pioneer Indian plant breeder, famous for surgarcane improvement.

12. Name the different plant breeding methods.

A number of plant breeding methods are available. Choice of breeding methods depends upon several factors, such as the nature of plant, objective of plant breeding, mode of reproduction in the concerned plant, degree of fertility and incompatibility etc. Some of these methods are as follows:

- Selection (mass, pure line, clonal, pedigree, bulk selection).
- Hybridization.
- Plant introduction and acclimatization.
- Mutation breeding.

13. What is selection?

Selection is the oldest breeding method and is the basis of all crop improvement. By selection process, individual plants or groups of plants are sorted out from mixed population. Selection methods are generally three types:

- Mass selection
- Pure line selection
- Clonal selection

14. What is pure line selection?

It is a collection of plants descended as a result of self-pollination from a single homozygous individual. Consequently, a variety developed by pure line selection is genetically more homogeneous than a variety developed by mass selection. Pure line is progeny of a single homozygous self-pollinated plant.

15. What is clonal selection? Write the uses of clonal selection.

    It is practiced in vegetative propagated crops (e.g. sugarcane, banana, potato, sweet potato etc.) where a group of plants obtained vegetatively from a single plant termed as clone, and the method of developing varieties from the clones is known as clonal selection. Clonal selection is just like pure line selection in vegetatively propagated crops. It is a method of selection based on clones. It does not involve sexual reproduction.

16. What is hybridization?

    Hybridization is the method of producing new crop varieties in which two or more plants of unlike general constituents are crossed together. Depending upon the parents involved, hybridization is of the following types:

    ✦ Intravarietal hybridization
    ✦ Intraspecific hybridization
    ✦ Interspecific hybridization
    ✦ Intergeneric hybridization

17. Briefly describe the process of hybridization.

    The various steps involved in the process of hybridization. These are as follows:

    ✦ Selection of parents: The plants with desired characters are selected as parents.
    ✦ Emasculation: Removal of stamens from parents before they burst and have shed their pollens.
    ✦ Bagging: The bagging of male and female flowers separately before pollination.
    ✦ Crossing: Desired viable pollens are transferred to the stigma of desired emasculated female parent.
    ✦ Labeling: The crossed flowers are tagged with all necessary particulars.
    ✦ Harvesting of seeds and raising of F1 generation to find out the results.

18. Explain the hybridization methods for self-pollinated crops.

    ✦ Pedigree method: Here individual plants selected in F2 and subsequent generations (up to F6) and a pedigree record is maintained. The pedigree record may be used as basis of selection in later generation.
    ✦ Bulk method: In this method, F2 and subsequent generations are grown in bulk, usually without artificial selection. In the end, pure line varieties are developed through individual plant selection.
    ✦ Back cross method: Back cross breeding is based on repeated back-crossing of the F1 and the subsequent generations to the recurrent parent.
    ✦ Multiple cross method: It consists of crossing several pure lines together.

19. Explain the hybridization methods for cross pollinated crops.

    ✦ Single cross (A x B): Crossing of two inbred lines (A and B).

- Three way cross: Cross between a single cross (A x B) as a female and an inbred line (C) as a male, i.e. (A x B) x C.
- Double cross (A x B) x (C x D): Crossing two single crosses involving four inbred lines (A, B, C and D).
- Multiple cross: Crossing of several inbred lines.
- Synthetic cross: Combination of a number of inbreeds from different sources.

20. What is plant introduction and acclimatization?

Plant introduction consists of taking of a genotype or a group of genotypes of plant into new environment where they have not been grown before. It may involve new varieties of crop already grown in some other area, a wild relative of the crop species or a totally new crop species for the area, e.g. Grape cultivation in Haryana and Punjab, of wheat in West Bengal, and of rice in Punjab. Often the materials are introduced from other countries or continent.

Acclimatization on the other hand is the adaptation or adjustment of an individual or a group of plants under changed climates for a number of generations.

21. Name some introduced plants in India.

- Introduction of rice strains Taichung Native-1 (TN-1) from Formosa and IR-8 from Philippines.
- Ridley, a wheat variety introduced from Australia.
- Sonara-63, Sonara-64 and Lerma Rojo are the Mexican wheat varieties introduced in India.
- Almost all N.P. wheat and hybrid maize varieties released in recent years have some of the superior characters incorporated from the introduced varieties.

22. What is Breeder's seed?

The improved seed produced from the nucleus seed under the guidance of original plant breeder. It is genetically and phenotypically cent percent pure and used as a source for foundation seed. (Syn. Breeder's Stock Seed).

23. Define Nucleus seed.

The original seed produced for the first time by the plant breeder. It is 100 percent pure in all genetical and physical qualities and used as a parent for the multiplication of Breeder's stock seed.

24. Define hybrid sterility.

The sterility found in hybrids and caused by hybridization due to which the hybrids do not produce viable seeds. This basis of hybrid sterility may be genic, chromosomal or cytoplasmic.

25. What is amphimixis?

It is the normal sexual reproduction in which the morphologically dissimilar male and female gametes unite together to form zygote.

26. Define apomixes.

An abnormal sexual reproduction in which the embryo develops from the egg cell or cell associated with it without fertilization and with or without meiosis. It may be of three types:

- Parthenogenesis
- Apospory
- Apogamy

27. What is the contribution of M.S. Swaminathan?

He is a leading Indian Agriculture scientist and pioneer radiation breeder under whose guidance mutation breeding work was introduced at IARI, New Delhi.

- His early research (1947-60) included the elucidation of the origin and differentiation of cultivated potato; cytogenetic interrelationship among *Triticum* species; induction of mutations for qualitative and phylogenic traits in wheat.
- His later researches (1960-80) had their main focus on the conservation of biological diversity with particular reference to rice and wheat, modification of plant architecture and growth rhythm for raising yield in wheat and rice.
- Recent researches (1980-2010) relate to rice breeding and biotechnology, conservation of coastal biodiversity with particular reference to mangrove ecosystem.

He was the director of IARI, Director General of ICAR, Director General of International Rice Research Institute, Philipines and Principal Secretary of Ministry of Agriculture.

Tables

Differences between mass and pure line selection.

Mass selection	Pure-line selection
1. In mass selection, plants are selected on the basis of their phenotype and their seeds are composited without progeny testing.	1. It is collection of plants descended as a result of self-pollination from a single homozygous individual.
2. Mass selection is generally adopted in cross-pollinated plants.	2. Pure-line selections are commonly used to improve self-pollinated crops.
3. It actually takes about eight years to produce a new variety.	3. Pure-line plants are homozygous and are uniform in characters, so it takes only a few years for isolation of stable variety.
4. Mass selection has some limitations for improvement of crops, yet it is the most common method of crop improvement among farmers.	4. Pure-line plants may be used as new improved varieties, because it is the only method to improve local varieties of self-pollinated crops, and the varieties produced by pure-line selection are highly uniform in appearance.

(*Contd.*)

Mass selection Pure-line selection	Pure-line selection
6. The variety developed is suited to several areas under varying climatic conditions.	6. The variety developed is homozygous and therefore, is not suited under varying conditions but is adapted to a limited region.
7. Examples: TMV-1 and 2, Bahra: T-5, T-55, Potato: K-122.	7. Examples: wheat: N.P. 4 and 6, K- 54, Rice: Co-4 and 6, cotton: A-10 etc.

Some important improved cereal crops by plant breeding methods.

Name of the crop	Improved varieties
Zea mays	Decan Hybrid, Ganga Safed-2, Hi Starch, Ganga-5, Deccan -101, Deccan -103, Kanchan, Tarun etc.
Triticum aestivum	Kalyan Sona, Sonalika, Arjun, C-306, UP-262, Rohini, Lok-1, HD-2285 etc.
Oryza sativa	Jaya, Jangannath, RI-8, Pusa-2-21, Pusa-33, Pant Dhan-4, Cauvery, Jaishree etc.
Sorghum bicolor	CSH-1, CSH-9, CSH-6 etc.
Pennisetum typhoides	RHR-1, RCB-2, BK-560 etc.

Some important improved fruit crops by plant breeding methods.

Name of the crop	Improved varieties
Mangifera indica	Hybrid-10, Hybrid-17, Dusheri, Amarapali, Mallika, Lal Sundari, Sensation etc.
Carica papaya	Washington, CO-1 CO-2, Pant Papita-1 etc.
Vitis vinifera	Thompson Seedless, Red Prince, Arko Hans, Pusa Seedlless, Angur Kalan etc.
Citrus sp.	Coorg, Troyer, Citrange, Rangpur etc.
Prunus amygdalus	Marcona, Nikitisky, HS-6, HS-9 etc.

Some important improved pulses crops by plant breeding methods.

Name of the crop	Improved varieties
Cicer arietinum	Pusa-209, Pusa-408, Pusa-413, Pant G-114, Radhey, C-214, K-468 etc.
Cajanus cajan	Pusa Ageti, Prabhat, Vishakha, Pant A-1, Pusa-85 etc.
Vigna mungo	Pant-U19, Type-9S1, LBG-17 etc.
Vigna unguiculata	C -152 etc.
Pisum sativum	T -163, Rachna etc.

Some important improved vegetable crops by plant breeding methods.

Name of the crop	Improved vatieties
Raphanus sativus	Pusa Rashmi, Pusa Chetki, Pusa Himani, Japanese White etc.
Daucus carota	Pusa Kesar, Nantes etc.
Abelmoschus esculenta	Pusa Sawani, Sel-2, Sel-10, Sel-12 etc.
Solanum melongena	Pusa Purple, Pusa Makhmali, Azad Kranti, Smrat Rituraj etc.
Brassica oleracea	Pant Shubhra, Pusa Himjyoti, Kunwari, Deepali etc.
Solanum tuberosum	Kufri Chandrimukhi, Kufri Jyoti, Kufri Sindhuri etc.
Lycopersicon esculentum	Pusa Sheetal, Pusa Ruby, Pusa Red, Hybrid 1 and 2 etc.

Sugarcane improved varieties.

Saccharum officinarum	Co – 1148, Co-313, Co J-64, Co- 1158, Co-802 etc.

Outline of the procedure for pure-line selection in self-pollinated crop.

Years	Practices
1st year (a mixture of pure-lines)	200-3000 plants are selected from old local variety or a mixture of pure-line on the basis of their phenotype.
2nd year (individual plant progenies)	Individual plant progenies are grown. Undesirable progenies are rejected (can be repeated).
3rd year (preliminary yield trials)	Remaining progenies are planted in a preliminary yield trial. Inferior progenies are rejected.
4th year (multilocation yield trials)	Replicated yield trials conducted at several locations. Inferior progenies are rejected. Disease resistance and quality tests are conducted. Best progeny released as a new variety.
8th year (seed multiplication)	Seeds are multiplied for distribution.

PLANT BREEDING—A QUICK GLANCE

1. Plant breeding is an art and a science, which together deals with genetic improvement of plants in relation to their economic use.
2. In other words it is the branch of biology concerned with changing the genotype of plants so that they become more useful to man.
3. The methods which bring about useful characters together into one individual plant and then multiply such individual to the extent that it is easily available for common use of farmers, gardeners etc. are collectively called as crop improvement.
4. Agricultural practices first time started around 7000-13000 years back, on the banks of Indus, Tigris and Nile rivers.
5. Crop plants were originated at a time both in old (Asia, Africa, Europe) and new world (America).
6. Cereal plants domesticated first.
7. Plant breeding involves the domestication and systematic improvement of crop species, has been basis of past and present agriculture.
8. Crop improvement most commonly involves plant breeding which has now become highly specialized and the aim of plant breeding is to synthesize a genetically superior quality plant which is adaptable to a specific set of environment.
9. The first hybrid in plants was artificially made by Thomas Fairchild in 1717, when he made a cross between Sweet William and Carnation.
10. Joseph Kolreuter made first crosses in tobacco for getting artificial hybrids.
11. Vilmorin developed better sugar beet successfully using his method of progeny testing in 1856.
12. The individual plant selection was developed by Nilsson and his associates in Sweden around 1900, when the rediscovery of Mendel's principles of inheritance also initiated a new era in the history of plant breeding, so that the breeding methods could be based on science of genetics.
13. Pure line theory was proposed by Johannsen in 1903.
14. G.H. Shull and Edward East started their experiments on inbreeding of maize in 1904 and proposed a 'single cross hybrid' in maize.
15. Earliest work on plant breeding in India was done by British scientists like, Barber (who worked on sugarcane) and Howard (worked on wheat).
16. Some of the pioneer plant breeders of India include T.S. Venkataraman (worked on sugarcane), K. Ramaiah (worked on rice), K.C. Mehta, Ramdhan Singh, B.P. Singh and M.S. Swaminathan, Pushkarnath, Dharampal Singh, Boshi Sen, R.L. Paliwal and C.L. Dhawan, N.G. P. Rao and D.S. Athwal etc.

17. Repeated crossing of the original sugarcane, *Saccharum barberi* (having poor yields and low sugar content) with noble canes of *S. officinarum*, led to significant increase in their yield and sugar content, this technique is called noblization of canes, and all the present day sugarcanes of India are noblized canes.

18. Development of transgenic or genetically modified (GM) crops is the latest method in plant breeding, which is being effectively utilized for the improvement of crop plants.

19. Selection is the oldest method of plant breeding and is the basis of all crop improvement programme. The selection may be either 1. Natural (which takes a longer time for bringing about improvement) and 2. Artificial (which can bring about improvement within few years).

20. Polycross is open pollination (assumed to be random mating) in isolation among a number of selected genotypes.

21. A number of selection methods are available for improvement in self-pollinated, cross-pollinated and asexually propagated crops. These selection methods include the following: 1. Mass selection, 2. Pure line selection, 3. Progeny selection and 4. Clonal selection.

22. Mass selection method is a very easy, quick and simple method, since crossing, selfing and progeny testing are not involved.

23. Through mass selection, synthetic and composite varieties of cross-pollinated crops may be improved for few characters like duration, plant type, grain characters etc.

24. Mass selection is not very suitable for improvement in characters having low heritability.

25. Prepotency is the ability of an individual to produce progeny, which are similar to each other and to itself.

26. A method of breeding in which individual plants are selected in F_2 and the subsequent generations and a pedigree record is maintained is called pedigree method. The pedigree record may be used as basis for the selection in later generations.

27. Almost all the present day varieties of self-pollinated crops are pure lines.

28. Pure line varieties are uniform, and therefore to the farmers and consumers these are more attractive than the varieties developed through mass selection.

29. The varieties developed through pure line selection are less adaptable and less stable to the environmental fluctuations.

30. The time required for variety development in pure line selection is more than the time required in mass selection method.

31. P_1, P_2, P_3 etc. are the first, second and third parents respectively, used for producing a hybrid of a series of hybrids.

32. Mutation is a sudden heritable change in a characteristic of an organism. It may be macromutation (produce sufficiently large change to be easily identifiable) and micromutation (produce small changes) that are studied in terms of mean and variance.

33. Domestication is the process of bringing wild species of plants under cultivation and trains them to suit human needs.
34. The hybridization between plants belonging to two different species of the same genus is called inter-specific hybridization.
35. Adaptation is the process by which organisms become more suited to survive and function in a given environment.
36. Progeny test is the evaluation of the genotype value of an individual on the basis of performance of its progeny.
37. Irradiation is exposure of biological materials to radiation, more particularly to mutagenic radiation.
38. The early civilizations of Incas in South America and Maya in Mexico developed around maize cultivation.
39. The Babylonian and Egyptian civilizations flourished around wheat and barley cultivations.
40. In mass pedigree method, the segregating generations are maintained as a bulk till such time when conditions suitable for selection occur; at this stage pedigree method of breeding is applied.
41. Quadriallel analysis is a type of analysis which is based on double crosses obtained by crossing homozygous lines in a diallel fashion, and then crossing F_1's so generated according to a diallel scheme with the restriction that in any double cross a homozygous line must not occur more than once as a parent.
42. An addition line has one pair of chromosomes from another variety or species in addition to the normal somatic chromosome complement (2n) of the species.
43. The most common methods of crop improvement are: 1. Selection, 2. Hybridization, 3. Polyploid breeding, 4. Induced mutations, 5. Tissue culture and 6. Genetic engineering.
44. Selection is the process of choosing the plants with desirable characters, the plants with useful and desirable characters were often chosen and maintained.
45. Genetic erosion is the gradual disappearance of various forms of a cultivated species and of its wild relatives.
46. The culture of anther or pollen grains on a suitable medium for the production of callus and/or haploid plants is called anther culture or androgenesis.
47. The genes or alleles present in the population of a species are called gene pool.
48. There are two most important methods of genetic conservation: 2. *In situ* conservation and 2. *Ex-situ* conservation.
49. Heterosis is the superiority of F_1 over the parents (or even inferiority to both the parents, e.g. earliness).

50. Cells or plants produced through somatic hybridization, and containing the full somatic complement of one fusion partner but only a part of the somatic complement of the other is called asymmetric hybrid.

51. The conservation of a species in its natural habitat is called *in situ* conservation. For example national parks, wild life centuries and biosphere reserves.

52. The enhancing the activity or changing the substrate specificity of enzymes by changing their amino acid sequences, e.g. through site-directed mutagenesis is called enzyme engineering.

53. Pure line is a progeny of a single self pollinated or self fertilized homozygous individual.

54. Hybridization is crossing of two genetically dissimilar plants having desired genes or genotypes and bringing them together into one individual called hybrid and the process is hybridization.

55. Escaping of a susceptible host plant from damage due to an insect pest, usually, as a result of the host plants being at a much less susceptible development stage when the pest population is at its peak is called avoidance.

56. Evaluation is the process of assessing the performance (yield, quality, disease and insect resistance etc.) of newly developed strains of a crop through appropriate multiplication trails.

57. Removal of stamens or anthers or killing the pollen grains of a flower without affecting the female reproductive organs is known as emasculation.

58. An organism having more than two sets of chromosomes is called polypolid. Polyploidy is often the result of failure of; 1. Meiosis during gamete formation, 2. Abnormal mitosis causing increase in chromosome number in the sex cells, or 3. Fertilization of egg by more than one male gamete.

59. F_1 is the progeny obtained by crossing two different genotypes or it is the first generation from a cross.

60. F_2 progeny obtained by self-fertilization of or intermating among F_1 individuals.

61. A breeding method based on repeated backcrossing of the F_1 and the subsequent generations to the recurrent parent is called backcross breeding.

62. If a selected rare species conserved away from its natural habitat is called *ex-situ* conservation. E.g. *Ginkgo biloba* conserved and growing in special botanical gardens, laboratories, gene banks, cryobanks, zoo, arboreta etc.

63. The haploid chromosome number of the ancestral diploid species of a polyploidy is called basic number and it is represented by x.

64. To make the resistant potato (*Solanum tuberosum*) variety, the resistant genes against Potato leaf roll virus and Potato virus-X is obtained from the wild species *Solanum acaule*.

65. Detasseling is removal of the tassel (the male inflorescence) before it sheds pollen in maize.

66. The resistance genes against wilt fungus *Fusarium* and cyst nematode, *Globodera* is obtained from wild potato species *S. spegazzini*.

67. The resistant gene against Potato virus-Y was obtained from the wild species *S. stoloniferum.*

68. *S. demissum* was the source of resistant gene against the disease causing fungus *Phytophthora infestans.*

69. The conversion of a less useful compound into a more valuable and useful one through the action of an organism by an enzyme produced by an organism is known as biotransformation.

70. Cybrids are the somatic hybrids obtained by fusion of the complete protoplast of one species with such protoplasts of another species, which either have no nucleus or have an inactive nucleus.

71. Polyploidy can also be produced artificially by colchicines treatment, heat and cold treatments, x-rays, gamma rays etc.

72. Protoplast fusion is brought about by suitable agents called fusogens. The most commonly used fusogen is polyethylene glycol (PEG), it gives high frequency of protoplast fusion.

73. Virus cross-protection is a phenomenon of reduced susceptibility of a host plant, which was earlier infected by a mild strain of a virus, to a highly virulent strain of the same virus or even of a related virus.

74. *Raphanobrassica* obtained by the fusion of *Raphanus sativus* and *Brassica oleracea.*

75. *Erucobrassica* produced by the fusion of *Eruca sativa* and *Brassica napus* and *Nicotiopetunia* from *Nicotiana tabacum* and *Petunia inflorata.*

76. *Solanopersicon* produced by the fusion of *Solanum tuberosum* and *Lycopersicon esculentum* and *Daturotropa* from *Datura innoxia* and *Atropa belladonna.*

77. The presence of variations in the plants derived from the callus is called somaclonal variation.

78. Cytoplast is a protoplast containing either no nucleus or an inactive nucleus.

79. Red rot resistant gene in sugarcane was isolated from its wild species *Saccharum spontaneum.*

80. Variety is a strain released for commercial cultivation by a Variety Release Committee.

81. Certified seeds are produced from foundation, its purity is certified by a Seed Certification Agency, and is usually used for commercial crop production.

82. Seeds which cannot destroy or can survive at low temperature (–10 to –20°C) and low moisture are called orthodox seeds. Such types of seeds can be viable at –196°C. E.g. wheat, rice, maize, oat, legume seeds etc.

83. The manipulation of gene or DNA, instead of complete chromosome set, appears to be a better, though difficult possibility for crop improvement. This has however, become possible because of the following two major achievements: 1. Discovery of restriction enzymes which cut DNA sequences only at specific sites and 2. Development of methodology for the production of large amount of DNA containing desirable genes by using bacterial plasmids, bacteriophages, etc.

84. Triple test cross analysis is a genetic analysis based on the progenies derived by crossing a sample of randomly selected F_2 plants with the parental F_1 and the two homozygous parents of the cross.

85. Combination breeding involves the transfer of one or a few characters, usually governed by oligogenes, from one variety into another.

86. Restriction enzymes are very useful in genetic engineering and are called genetic scalpels or molecular scissors.

87. The role of restriction enzymes in the cell is to check bacteriophage infection by degrading the DNA of the infecting bacteriophage without affecting bacterial DNA.

88. Some of the restriction enzymes are *Eco RI* isolated from *Escherichia coli*, *Sal I* from *Streptomyces albus*, *Hind II* from *Haemophilus influenzae*, etc.

89. Embryo rescue is the use of embryo culture to recover such inter-specific hybrids that would otherwise be lost due to endosperm abortion during early stages of embryo development.

90. Sharbati sonara variety of wheat was obtained by irradiation of Sonara 64 with gamma rays and Pusa Lerma variety of wheat obtained from Lerma Rojo 64 by the same method.

91. The seeds which cannot tolerate low temperature (-10 to -20°C) and low moisture are called recalcitrant seeds.

92. 'De-Geo-woo-gen' is a dwarf variety of rice from Taiwan, was used by the International Rice Research Institute (IRRI) to develop higher producing and early maturing varieties of rice, e.g. IR-8 and IR-24 varieties of rice.

93. The use of genetic factors to make the male gamete nonfunctional in self and/or cross pollination is called genetic emasculation.

94. IR-36 variety of rice was developed by Dr. Gurudev Singh at IRRI, Manila by crossing different varieties of rice with wild rice variety *Oryza nivara*. IR-36 rice variety is resistant against grassy stunt virus.

95. In Indonesia, rice form 'Pelita I' gave a high yielding rice cultivar 'Atomita 2' by irradiation with gamma rays and this cultivar is resistant to brown plant hopper, tolerates salty soils and has good palatability.

96. The resistance governed by polygenes and is pathotype-nonspecific is called horizontal resistance.

97. IR-36 rice variety is a higher producing variety and solved major food problem in Asia.

98. Mexican plant breeder N.E. Borlaug developed several dwarf wheat varieties, e.g. Sonora-64, Larma Rojo-64. He is also called as father of green revolution and got Nobel Prize for peace in 1970.

99. Haberlandt, a German botanist is given the credit for beginning of plant tissue culture technique.

100. The first successful anther culture was reported by Maheshwari and Guha (1964) in *Datura innoxia* and haploid plants were obtained.

101. Triticale is also called man made crop. It is produced by the intergeneric hybridization between *Triticum aestivum* and *Secale cereale*. It is two types, hexaploid and octaploid. It is not suitable for bread making because gluten content is very low. But it is very good forage crop.

102. Inbreeding is the mating of individuals more closely related by ancestry than would be expected under random mating and the loss in vigour due to inbreeding is called inbreeding depression.

103. I_1, $I_{2,}$ I_3, etc. are the populations obtained after one, two, three etc. generations of inbreeding.

104. Chromosome doubling by colchicine treatment produces homozygous diploids in a very short period of time but in normal procedure self-fertilization is required to be repeated several times.

105. In mass selection, several plants are selected on the basis of their phenotype and their seed is composted to raise the next generation.

106. Micropropagation is the method of rapid vegetative multiplication of valuable plant material for agriculture, horticulture and forestry.

107. Somatic embryogenesis is a process where embryos are produced from somatic cells in tissue culture. These embryos are known as embryoids and these embryoids show the same developmental stages as normal sexually produced embryos.

108. The improvement of crop plants is required for: 1. To increase seeds, oil, fibres, 2. Getting resistant against insects, diseases and frost, 3. To change maturity period etc.

109. T.S. Venkataraman developed improved sugar cane varieties.

110. Ramdhan Choudhary developed C-591 wheat variety.

111. The NP- series of wheat varieties were developed by Dr. B.P. Pal.

112. Dr. Puskar Nath is well known potato breeder.

113. Dr. Boshi Sen is the famous maize breeder.

114. Dr. M.S. Swaminathan is famous Indian plant breeder; he developed Sharbati Sonora by mutational breeding. He is also called father of Indian green revolution.

115. Seed test is a series of tests on purity, moisture content and germination of seed sample to determine its quality.

116. Reporter gene is a gene which produces a product or phenotype that can be easily and unequivocally detected through simple tests.

117. Selection is the most primitive and simplest method of crop improvement.

118. There are three methods of artificial selection: 1. Pure line selection, 2. Mass selection, 3. Clonal selection.

119. The progeny of self pollinated homozygous plants is called pure line. This method is useful only in self crossing crop plants. E.g. wheat.

120. Seed certification consists of field inspections and seed tests to ensure genetic and physical purity and good germination of seed lots, and issuance of a certificate to that effect.

121. Mass selection method is useful in self pollinated as well as cross pollinated crops.

122. Clonal propagation is used in the vegetative propagated plants. E.g. Sugarcane, Banana, potato etc.

123. Single seed descent method is a method of advancing the generations; one seed from each F_2 plant of a cross and the subsequent generations is bulked to raise the next generation.

124. Variations cannot be produced in clonal selection method.

125. Crossing in between two different genus is called intergeneric cross. Rabbage was produced by crossing of radish and cabbage. Triticale produced by crossing in between *Triticum* and *Secale*.

126. Sibs are the individuals having both the parents common, but derived from different gametes but they are not identical twins.

127. Fairchild mule is the first natural hybrid plant obtained by Thomas Fairchild by crossing Sweet William and carnation.

128. A line in which a pair of chromosomes has been replaced by a pair from another variety of the same species is called substitution line.

129. Kalreuter used hybridization technique practically to improve crop plants.

130. The word heterosis or hybrid vigour were used by Shull.

131. Polyploids are gigantic and this can be used in crop improvement because these plants are bigger than normal plants.

132. Triploidy is naturally present in some crop plants and such type of plants are seedless. E.g. Banana.

133. Synthetic variety obtained by mating in all possible combinations a number of lines that combines well each other.

134. Polyploidy can be produced by colchicine treatment. Colchicine is an alkaloid which is obtained from *Colchicum autumnale*.

135. Remei and atomita-2 varieties of rice are also produced by mutation breeding.

136. Thermotherapy is the exposer of plants/explants to a relatively high temperature (around 40° C) for variable periods with a view to eliminate or inactivate virus particles in them without killing these plant or plant parts.

137. The world famous Aruna variety of castor also developed by mutation breeding. This is highly resistant to diseases and contains more oil.

138. Leaf rust of coffee, fire blight of potato, bunchy top of banana etc. diseases came to India due to uncontrolled introduction of plants.

139. Weeds *Argemone maxicana*, *Eichornia crassipes*, *Parthenium argentatum* are also came in India due to uncontrolled introduction of plants from other parts of the world.

140. A plant in which a gene or genes from another organism has been introduced through genetic engineering and has become integrated in its genome is called transgenic plant.

141. The Ridley variety of wheat is introduced from Australia.

142. Pomato = Potato x Tomato and Bromato = Brinjal x Tomato were developed by protoplast hybridization.

ANSWERS

1. (d)	2. (b)	3. (d)	4. (c)	5. (c)	6. (c)
7. (c)	8. (c)	9. (c)	10. (b)	11. (d)	12. (a)
13. (c)	14. (b)	15. (b)	16. (c)	17. (b)	18. (c)
19. (d)	20. (a)	21. (d)	22. (a)	23. (d)	24. (c)
25. (b)	26. (a)	27. (d)	28. (c)	29. (c)	30. (a)
31. (d)	32. (c)	33. (c)	34. (c)	35. (a)	36. (c)
37. (c)	38. (c)	39. (a)	40. (c)	41. (c)	42. (d)
43. (a)	44. (c)	45. (c)	46. (c)	47. (c)	48. (d)
49. (c)	50. (d)	51. (c)	52. (c)	53. (b)	54. (a)
55. (b)	56. (c)	57. (c)	58. (c)	59. (c)	60. (a)
61. (b)	62. (d)	63. (d)	64. (a)	65. (d)	66. (b)
67. (a)	68. (d)	69. (c)	70. (c)	71. (a)	72. (d)
73. (d)	74. (a)	75. (b)	76. (d)	77. (d)	78. (b)

6

PLANT PATHOLOGY

MULTIPLE CHOICE QUESTIONS

1. Red rot of sugarcane is caused by
   (a) *Alternaria solani* (b) *Colletotrichum fulcatum*
   (c) *Helminthosporium oryzae* (d) *Cercospora arachidicola*
2. Biological control component is central to advance agricultural production. Which of the following is used as a third generation pesticide?
   (a) Pheromones (b) Insect repellant
   (c) Pathogen (d) Insect hormone analogues
3. Body disorders due to harmful physiological changes are known as
   (a) Pathology (b) Infection (c) Disease (d) Susceptibility
4. The pesticides are very useful because these
   (a) Establish the balance of ecosystem by killing harmful insects
   (b) Kill both useful and harmful pests
   (c) Protect high yielding and useful varieties of plants
   (d) Control the vector born diseases and thus population
5. The branch of agricultural, botanical or biological science which deals with the cause, etiology, resulting losses and control of plant diseases is called
   (a) Agronomy (b) Phycology (c) Phytology (d) Phytopathology
6. Which of the following is not a viral disease
   (a) Red rot of sugarcane (b) Tobacco mosaic
   (c) Leaf curl of papaya (d) Tristeza disease of Citrus

7. Irradiation has been used in pest control in
   (a) Purification of chemicals uses as pesticides
   (b) Sterilization of insect pests
   (c) Producing disease resistant plants
   (d) Secretion of insect hormones
8. The pathogen that causes covered smut of Sorghum is
   (a) *Colletotrichum* (b) *Sphacelotheca*
   (c) *Helminthosporium* (d) *Cercospora*
9. The pesticides need to be replaced because these
   (a) are mostly toxic and nondegradable
   (b) are very costly
   (c) Cause abnormalities in the target population
   (d) Cannot be stored for a longer time
10. A diseased leaf shows some areas with dead tissue, such types of symptoms is referred as
    (a) Necrotic spot (b) Downy mildew (c) Chlorosis (d) Blight
11. Column A consists of name of disease and column B consists of causal organisms. Match column A with column B and select the correct answer

Column A	Column B
A. Red rot of sugarcane	1. *Xanthomonas citri*
B. Citrus canker	2. *Colletotrichum falcatum*
C. Tikka disease of ground nut	3. *Alternaria solani*
D. Early blight of potato	4. *Cercospora personata*

    (a) A B C D
        1 2 4 3
    (b) A B C D
        1 4 3 2
    (c) A B C D
        2 4 3 1
    (d) A B C D
        2 1 4 3
12. Pesticides are substances used for killing
    (a) Fungi (b) Bacteria (c) Insects (d) All of these
13. *Bacillus thuringiensis* is commercially useful as
    (a) A fermenting agent (b) A vector for gene transfer
    (c) A bioinsecticide (d) None of these
14. Banana bunchy top is caused by
    (a) Mycoplasma (b) Deuteromycetes (c) *Xanthomonas* (b) Virus
15. Which of the following is not a fungal disease
    (a) Rust disease of crucifers (b) Red rust of tea
    (c) Rust of coffee (d) Rust of wheat

16. A part of pathogen that is transmitted to the host is known as
   (a) Infection (b) Inoculum (c) Incubation (d) Pathogenesis
17. The period of infection the causal organism lands on the host, resumes activity or germinates and gains entry into it directly or indirectly. This invasion of the host tissue by the microorganism is called
   (a) Induction period (b) Incubation period
   (c) Infection (d) Penetration
18. Which of the following pesticide is an acetyl choline esterase inhibitor?
   (a) Aldrin (b) Endosulfan (c) Malathion (d) BHC
19. Smut disease of maize is caused by
   (a) *Ustilago avenae* (b) *U. maydis* (c) *U. hordie* (d) *U. nuda*
20. The scientist who is famous for his research on rust fungi in India is
   (a) Butler (b) K.C. Mehta (c) Sivadasan (d) Swaminathan
21. Which of the following does not help in penetration of pathogen into the host
   (a) Ligninase (b) Pectinase (c) Cellulase (d) Cutinase
22. Cochineal insects have proved very useful for
   (a) Weed control (b) *Eichhornia* prevention
   (c) *Parthenium* eradication (d) Cactus prevention
23. The foundation of modern plant pathology is laid by
   (a) Berkely (b) Morren (c) Anton de Barry (d) Blakslee
24. Aflotoxin is produced as a product in
   (a) *Penicillium* (b) *Aspergillus*
   (c) Pathogenic bacteria (d) *Streptomyces*
25. Leaf spot of ground nut is caused by
   (a) *Sclerospora* (b) *Pyricularia*
   (c) *Helminthosporium* (d) *Cercospora*
26. Which amongst the following is a fungicide?
   (a) 2,4-D (b) BHC (c) DDT (d) Bordeaux mixture
27. Some dead and corky areas with definite margins appeared on the stem, branches and fruits of lemon is known as
   (a) Curl (b) Scab (c) Rot (d) Canker of citrus
28. Pests include those organisms which cause
   (a) Fungal infection and reduce crop yield
   (b) Weeds to grow and increase the intraspecific competition

(c) Epidemic diseases with high mortality
(d) None of these

29. The time lapsing between inoculation and appearance of symptoms is known as
(a) Syndrome (b) Invasion
(c) Incubation period (d) Infection

30. Citrus canker is caused by
(a) *Pseudomonas rubrilineans* (b) *Xanthomonas oryzae*
(c) *Xanthomonas citri* (d) *Pseudomonas solanacearum*

31. *Helminthosporium oryzae* was the cause of disaster
(a) Irish famine in 1845
(b) Collapse of French industry in 1870
(c) Displacement of planters in Ceylon in 1867
(d) Bengal famine in 1943

32. Crown gall disease is caused by
(a) Bacteria (b) Virus (c) Fungi (d) Mycoplasma

33. Late blight of potato is caused by
(a) *Albugo candida* (b) *Fusarium moniliforme*
(c) *Phytophthora infestans* (d) *Alternaria solani*

34. DDT is
(a) A non-biodegradable pollutant (b) A biodegradable pollutant
(c) Not a pollutant (d) An antibiotic

35. Organism which regardless of the case with which they can be cultivated on artificial media, always obtain their food in nature from living tissues on which they complete their life cycle are called
(a) Pathogen (b) Biotrophs (c) Parasite (d) Saprophyte

36. Downy mildew disease is caused by
(a) *Peronospora* (b) *Puccinia* (c) *Albugo* (d) *Phytophthora*

37. When a diseased plant is legally stopped from export, this is done under
(a) Quarantine measures (b) Crop improvement
(c) Biological control (d) Curative check

38. 2,4-D is a
(a) Fungicide (b) Herbicide (c) Insecticide (d) Bactericide

39. Ergot is obtained from
(a) *Claviceps* (b) *Penicillium* (c) *Aspergillus* (d) *Rhizobium*

40. The rust disease is caused by the members of
    (a) Ustilaginales (b) Uredinales (c) Peronosporales (d) None of these
41. Azadirachtin, a common constituent of mosquito repellant coils comes from the plant
    (a) Tobacco (b) *Derris* (c) Pyrethrum (d) Margosa
42. Streptomycin is used to treat the diseases caused by the bacteria
    (a) Gram negative (b) Gram positive
    (c) Both (a) and (b) (d) All types of bacteria
43. The prodigious spread of *Opuntia* in south India was controlled by
    (a) 2,4-D (b) Colchicine
    (c) Hormones (d) Cochineal insect
44. Downy mildews are caused by the members of
    (a) Ustilaginales (b) Uredinales
    (c) Peronosporales (d) None of these
45. Some viruses can cure bacterial diseases due to
    (a) Phagocytosis (b) Pinocytosis
    (c) Transformation (d) Phage action
46. Blast disease of paddy is caused by
    (a) *Rhizoctonia* (b) *Pyricularia*
    (c) *Colletotrichum* (d) *Fusarium*
47. Wilt of arhar is caused by
    (a) *Rhizoctonia* (b) *Pyricularia*
    (c) *Colletotrichum* (d) *Fusarium*
48. Find out the organochlorines among the following
    (a) Malathion, Parathion (b) DDT, BHC
    (c) Carbofuran, aldicarb (d) Pyrethrin, simazine
49. Early blight of potato is caused by
    (a) *Phytophthora infestans* (b) *Alternaria solani*
    (c) *Albugo candida* (d) None of these
50. The phenomenon of heterothallism was first observed by
    (a) Alexopoulos (b) K.C. Mehta
    (c) Blakeslee (d) None of these
51. Vertical transmission of mycoplasmal diseases may be through
    (a) Tubers (b) Corms (c) Bulbs (d) All of these

52. Alkaloids from which of the following plants are used as insecticide?
    (a) Tobacco (b) Rose (c) Neem (d) Tulsi
53. The whip smut of sugarcane is caused by
    (a) *Ustilago hordei* (b) *U. scitaminea*
    (c) *U. maydis* (d) None of these
54. The phenomenon of heterothallism was observed for the first time in the order
    (a) Ustilaginales (b) Mucorales
    (c) Erysiphales (d) None of these
55. A pheromone cylinder used to trap
    (a) Only harmful insects
    (b) All types of insects
    (c) Only male insects are attracted and thus insect population is controlled
    (d) None of these
56. The name smut diseases are given to those produced by *Ustilago* because
    (a) Its mycelium is black in colour
    (b) The host becomes completely black
    (c) It parasitizes cereals
    (d) The fungus produces black sooty spore masses
57. Which one of the following is effective against mycoplasmal diseases?
    (a) Penicillin (b) Chloramphenicol
    (c) Vancomycine (d) All of these
58. Pheromones are
    (a) Enzymes (b) Hormones (c) Proteins (d) Volatile oils
59. Damping off disease is caused by
    (a) *Aspergillus* (b) *Ustilago* (c) *Cercospora* (d) *Pythium*
60. *Erysiphe* causes the disease
    (a) Powdery mildews (b) Downy mildew
    (c) Covered smut (d) Late blight of potato
61. Bacteria were first discovered by
    (a) Robert Koch (b) A.V. Leeuwenhoek
    (c) N.D. Zinder (d) Robert Hooke
62. Bacteria having a tuft of flagella at one end are called
    (a) Amphitrichous (b) Lophotrichous
    (c) Monotrichous (d) Peritrichous

63. The Irish potato famine occurred in North Europe during 1845, was caused by
    (a) *Phytophthora infestans* (b) *Alternaria solani*
    (c) *Synchytrium endobioticum* (d) *Erwinia carotovora*
64. Bordeaux mixture consists of
    (a) Lime and calcium sulphate (b) Sulphur and lime
    (c) Copper sulphate and lime (d) Copper sulphate and sulphur
65. White rust of crucifers is a pseudo-rust because
    (a) The disease is not seen on wheat
    (b) The disease in seen on crucifers
    (c) The disease is not caused by basidiomycetous members
    (d) The colour of the pustules not red
66. A plant disease in which the pathogen is seen as a cottony growth on the surface of the host is called
    (a) Smut (b) Rust
    (c) Powdery mildew (d) Downy mildew
67. White rust of crucifers is caused by
    (a) *Ustilago* (b) *Peziza* (c) *Cystopus* (d) *Puccinia*
68. Tikka disease is caused by
    (a) *Fusarium* (b) *Cercospora* (c) *Puccinia* (d) *Colletotrichum*
69. 'Target board effect' is caused by
    (a) *Alternaria* (b) *Colletotrichum*
    (c) *Pyricularia* (d) *Helminthosporium*
70. Black rust of wheat is caused by
    (a) Yeast (b) *Penicillium* (c) *Puccinia* (d) *Rhizopus*
71. *Pythium debaryanum* is the causative agent of
    (a) Powdery mildew (b) Black wart disease of potato
    (c) Damping off of seedling (d) Soft rot of sweet potato
72. *Claviceps purpurea* is causal organism of
    (a) Smut of barley (b) Ergotism of rye
    (c) Powdery mildew of pea (d) Rust of wheat
73. Disease resistant plant varieties can be produced by
    (a) Heat treatment (b) Colchicines treatment
    (c) Hormonal treatment (d) Crossing with wild varieties of plants

74. The major difference between parasite and pathogen is that
    (a) One causes disease and other fails to do so
    (b) One obtains his food only while other causes disease to the host
    (c) One lives inside while other lives outside the body
    (d) Former is weaker whereas later is stronger in power
75. 'Little leaf disease' of brinjal is caused by
    (a) Bacteria (b) Virus (c) Fungi (d) Algae
76. Leaf curl of potato is caused by
    (a) Bacteria (b) Virus (c) Fungi (d) Mycoplasma
77. A horticultural plant infected with a viral disease is not allowed to be imported to our country but its seeds can be imported because
    (a) Spread of disease is far less from seed than from the plant
    (b) Viral particles can be easily eliminated from seeds
    (c) Viruses generally do not enter seeds beyond the seed coat
    (d) Viruses infect only the endosperm in the seed
78. Mycoplasma is the causative agent for one of the following diseases
    (a) Tobacco mosaic (b) Aster yellow
    (c) Leaf curl (d) Vein yellowing

SMALL ANSWER QUESTIONS

1. What is phytopathology?

   Ans. It is the study of the nature, development, and control of plant diseases. Phytopathology, (phyton-plants; pathos- suffering; logos- knowledge).

2. Define disease.

   Ans. Several definitions of diseases in plants have been proposed. Some of these are as follow:
   - A series of harmful physiological processes caused by continuous irritation of the plant by a primary agent.
   - A harmful deviation from the normal functioning of physiological processes.
   - A continuous impairment of metabolism.

3. What is pathogen?

   Ans. A pathogen is an organism or virus capable of causing disease in a particular host or range of hosts.

4. What is Witches' broom?

   Ans. Broom like growth or massed proliferation caused by the dense clustering of branches of woody plants.

5. What is phylloplane?

   Ans. The surface of a leaf or the other parts of plant body.

6. What is primary inoculum?

   Ans. When the source of primary inoculums serves as infective agene, it becomes primary inoculums.

7. What is secondary inoculum?

   Ans. When the host becomes diseased, the organism subsequently produces spores or infective bodies, which are collectively known as secondary inoculum.

8. What is synergism?

   Ans. The concurrent parasitism of a host by two pathogens in which the symptoms or other effects produced are greater magnitude than the sum of the effects of each pathogen acting alone.

9. What are virusoid?

   Ans. The extra small circular RNA component of some isometric RNA viruses.

10. Define pathotype.

    Ans. This is a subdivision of a species distinguished by the common characters of pahtogenicity, particularly in relation to the range of hosts.

11. What is fumigation?

    Ans. Application of a fumigant to the foliage or roots through the irrigation system.

12. What is pathogenicity?

    Ans. It is the quality or characteristic of a pathogen capable of causing disease.

13. Define sterilization.

    Ans. The elimination of pathogen from soil by means of heat or chemicals is called sterilization.

14. What is horizontal resistance?

    Ans. Partial resistance equally effective against all races of a pathogen.

15. What is pathogenesis?

    Ans. It is the sequence of progress in disease development from the initial contact between a pathogen and its host to the completion of the syndrome.

16. Define epiphytotic.

    Ans. A widespread and destructive outbreak of a disease of plants is called epiphytotic.

17. What is disease cycle?

    Ans. The chain of events in the disease development is known as the disease cycle.

18. What is host?

    Ans. An organism that supports the activities of a parasite.

19. Define vivotoxin.

    Ans. A substance produced by the pathogen, or its host, or both, which operates in the causation of disease but not itself the primary agent.

20. Define ectoparasite.

    Ans. A parasite that lives on the exterior of an organism is called ectoparasite (opposite to endoparasite). E.g. *Erysiphe graminis.*

21. Define viroids.

    Ans. Small, low molecular weight RNAs that can infect plant cells, replicate themselves and cause disease.

22. What is inoculum?

    Ans. It is the infectious material that can cause disease, and it is that portion of individual pathogens that is brought into contact with the host.

23. What is inoculum potential?

    Ans. The degree of infectivity of infectious agents is called inoculum potential.

24. What is disinfectant?

    Ans. An agent that kills or inactivates pathogens in the environment or on the surface of the plant, prior to infection is called disinfectant.

25. Define epiphyte.

    Ans. An epiphyte is a plant growing upon another plant, where no parasitic or symbiotic relation is involved.

26. What is virescent?

    Ans. A normal white or coloured tissue that develops chloroplast and becomes green.

27. What are endophytes?

    Ans. The organisms are contained entirely within the substrate plant. Such organisms may be either parasitic or symbiotic.

28. Define disinfectant.

    Ans. A physical or chemical agent that frees a plant or organ from infection.

29. What is penetration in plant pathology?

    Ans. The initial invasion of the host by an organism.

30. Define infection.

    Ans. The establishment of the pathogen inside the host following penetration in which a parasitic relationship between the two organisms is established.

31. Define vertical resistance.

    Ans. Complete resistance to some races of a pathogen but not to others.

32. What is incubation period?

    Ans. The time between the penetration of the host by a pathogen and the first appearance of symptoms on the host is known as the incubation period.

33. What are dieback symptoms?

    Ans. Progressive death of shoots and roots generally starting at the tip.

34. What is susceptibility?

    Ans. It refers to a condition of a plant in which it is normally subject to attack by a given pathogen.

35. What is virion?

    Ans. A single virus particle.

36. What are symptoms?

    Ans. The external or internal reaction or alterations as a result of a disease are referred to as symptoms.

37. What is cross protection?

    Ans. The phenomenon in which plant tissues infected with one strain of a virus are protected from infection by other strains of the same virus.

38. What is vector in plant pathology?

    Ans. An animal or any other medium which is able to transmit a pathogen.

39. Define epidemiology.

    Ans. The study of the factors affecting the outbreak of an infectious disease is called epidemiology.

40. What is resistant?

Ans. Possessing qualities that hinder the development of a given pathogen.

41. What is canker?

Ans. These are dead and corky areas with definite margins occurring on the stems, twigs or fruits of woody trees. E.g. Citrus canker.

42. What are chronic symptoms?

Ans. Symptoms that appear for a long period of time.

43. Define toxicity.

Ans. The capacity of a compound to produce injury to the living tissues.

44. What is syndrome?

Ans. The diseased plant show several valuable symptoms by which a disease is recognized. These symptoms are collectively known as a syndrome.

45. What is scab?

Ans. Scabs are rough, crust like lesions developed on the infected surface. These are more or less circular in outline. E.g. Apple scab.

46. What is chemotherapy?

Ans. Control of a plant disease with chemicals that are absorbed and are translocated internally.

47. Define infectious organism.

Ans. It is one that may be transmitted from a diseased host plant to a healthy plant and which is capable of inciting the disease in a favourable environment.

48. What is the pathogen of Cedar rust?

Ans. *Gymnosporangium juniperivirginiana.*

49. What is tolerance?

Ans. The ability of a plant to sustain the effects of a disease without dying or suffering serious injury.

50. Define blotch.

Ans. A disease characterized by large, and irregular in shape, spots or lesions on leaves, shoots and stems.

51. What is blight?

Ans. This is characterized by rapid killing of leaves, stem and flowers. It gives a burnt appearance. E.g. Leaf blight of potato.

52. What is resting spore?

Ans. The sexual or other thick walled spore of a fungus that is resistant to extremes of temperature and moisture and which often germinates only after a period of rest from its formation.

53. What is infectious disease?

Ans. It is caused by a transmissible causal organism and on infection the disease can spread from a diseased to a healthy plant.

54. What is bacteriostatic?

Ans. A chemical or physical agent that prevents multiplication of bacteria without killing them.

55. What is virulence?

Ans. The aggressiveness of a parasite to invade and grow in its host plant and to reproduce on or in, it is called virulence of that organism.

56. What are rickettsiae?

Ans. Microorganisms similar to bacteria in most respects but generally capable of multiplying only inside living host cells.

57. Define predisposition.

Ans. The effect of one or more environmental factors which makes a plant vulnerable to attack by pathogen.

58. What is russet?

Ans. Brownish roughened areas on skin of fruit as result of cork formation.

59. What is autoeciuos fungus?

Ans. A parasitic fungus which completes its entire life cycle on the same host.

60. What are rots?

Ans. Softening, discolouration and often disintegration of the succulent plant tissues as a result of fungal or bacterial infection is known as rot. E.g. Stem rot of paddy and fruit rot of cucurbits.

61. What is resistance?

Ans. The inherent ability of a plant to prevent or restrict the establishment and subsequent activities of a potential pathogen.

62. What are symptomless carriers?

Ans. Those plants in which the agent inciting disease is present but shows no symptoms.

63. What is saltation?

Ans. A mutation within an isolated species known to be a pure genotype is called saltation.

64. Define active and passive resistance.

    Ans. Active resistance is due to reaction incited by the attack, while passive resistance is due to qualities innate in the host prior to attack by the pathogen.

65. Define serology.

    Ans. A method using the specificity of the antigen antibody reaction for the detection and identification of antigenic substances and the organisms that carry them.

66. What is pathodeme?

    Ans. Population of a host in which all individuals have a particular character of resistance in common is called pathodeme.

67. Define shock symptoms.

    Ans. The sever and often necrotic symptoms produced on the first new growth following infection with some viruses.

68. What is appressorium?

    Ans. The swollen tip of a hypha or germ tube that facilitates attachment and penetration of the host by the fungus.

69. What is chlorosis?

    Ans. The discolouration or yellowing of leaves is called chlorosis. It is caused by destruction of chlorophyll in the cells or failure of infected cell to form chlorophyll.

70. Define anthracnose.

    Ans. A leaf spot or fruit spot type of disease caused by fungi that produce their asexual spores in an acervulus. E.g. Disease symptoms produced by *Colletotrichum* sp.

71. What is latent infection?

    Ans. The stage in which a host is infected with a pathogen but does not show any symptoms.

72. Differentiate fungitoxic, fungicidal and fungistatic.

    Ans. Fungitoxic are substances which are harmful to fungi and it includes fungicide. Fungicide is any agency which has the ability to kill the fungus. Fungistatic are some chemicals which can not kill fungi but inhibit fungus growth temporarily.

73. What is soil solarization?

    Ans. Attempt to reduce or eliminate pathogen populations in the soil by covering the soil with clear plastic so that the sun rays will raise the soil temperature to levels that kill the pathogen.

74. What is endemic disease?

    Ans. When a disease is constantly present in a moderate or severe form and is confined to a particular, area, country or district is called as endemic disease. E g. Wart disease of potato caused by *Synchytrium endobioticum* is endemic in Darjeeling.

75. What is mold?

Ans. Any profuse or woolly growth of fungus on damp or decaying matter or on surfaces of plant tissues.

76. What is alternate host?

Ans. One of the two kinds of plants on which a parasitic fungus must develop to complete its life cycle. E.g. Black rust of wheat caused by *Puccinia graminis tritici*.

77. Define malignancy.

Ans. A cell or tissue that divides and enlarges autonomously, and its growth cannot be controlled by the organism on which it is growing.

78. What are mosaic symptoms?

Ans. Symptom of certain viral diseases of plants characterized by intermingled patches of normal and light green colour.

79. What are curls?

Ans. Abnormal increase in the size of some infected cells of leaf, it becomes curled, arched or distorted. This symptom is known as curl disease.

80. What is epidemic?

Ans. This type of disease usually occurs widely, but periodically, in a destructive form. E.g. Cereal rust and powdery mildews.

81. Define non-host resistance.

Ans. Inability of a pathogen to infect a plant because the plant is not a host of the pathogen due to lack of something in plant that the pathogen of substances incompatible with the pathogen.

82. What is imperfect stage?

Ans. The part of life cycle of a fungus in which no sexual spores are produced.

83. What are phytoalexins?

Ans. These are the substances that inhibit the development of a fungus on hypersensitive tissue, formed only when host plant cells come in contact with the parasite.

84. What are sporadic diseases?

Ans. The diseases which occur at very irregular intervals and locations and it relatively fewer instances. E.g. Blotch disease of cucumber.

85. What are galls?

Ans. These are globose or irregular outgrowths produced on the infected parts of the host. Ans. Smaller galls are called warts whereas larger ones are called as excrescence. E.g. wart disease of potato.

86. Define pathovar.

    Ans. In bacteria, a subspecies or group of strains that can infect only plants within a certain genus or species.

87. What is pustule?

    Ans. Small blister like elevation of epidermis as spores emerge from the inner tissues is known as pustule.

88. What is fumigation?

    Ans. The spray of fumigant in a chamber for disinfestations.

89. Define scorch.

    Ans. Burning of leaf margins as a result of infection or unfavourable environmental conditions is called scorch.

90. What are pandemic diseases?

    Ans. These are epidemic types of disease which occurs over vast and extensive areas causing severe losses. E.g. Late blight of potato.

91. What is quarantine?

    Ans. Control of export and import of plant to prevent spread of diseases and pests.

92. What is rust?

    Ans. The rusts appear as small pustules of red, brown, black or yellow colour on stem of leaf epidermis of the host. E.g. Black rust of wheat.

93. What is parasite?

    Ans. When a living organism subsists wholly or partially upon living tissue.

94. Define hypersensitivity.

    Ans. Excessive sensitivity of plant tissue to certain pathogens. Affected cells are killed quickly, blocking the advancement of obligate parasites.

95. Who discovered iatrogenic diseases?

    Ans. James G. Horsfall in 1956.

96. Define integrated control mechanism.

    Ans. An approach that attempts to use all available methods of control of a disease or of all the diseases and pests of a crop plant for best control results but the least cost and the least damage to the environment.

97. What are haustoria?

    Ans. The parasite sucks its food from the host cells with the help of specialized absorbing organs called haustoria.

98. What is hypertrophy?

    Ans. Excessive growth due to increase in size of cells.

99. What are smuts?

    Ans. It refers to sooty charcoal like mass of spores of fungal pathogens. Smut spores usually develop on the floral organs of the host. E.g. Loose smut of wheat.

100. What are iatrogenic diseases?

    Ans. The diseases caused by the side effects of the use of chemicals for control one disease.

101. What is vein-clearing?

    Ans. Destruction of chlorophyll in the vein tissue, as a result of infection by a pathogen.

102. Give the classification of diseases based on host plant.

    Ans. According to the types of host plants- (a) Field crop plants, (b) vegetable diseases, (c) fruit diseases, (d) forest disease and (e) ornamental plants diseases.

103. What is hypoplasia?

    Ans. Under development of a tissue or plant due to decreased cell division.

104. Define saprophyte.

    Ans. A saprophyte is an organism which subsists wholly or partly upon dead organic matter and inorganic materials.

TRUE OR FALSE QUESTIONS

1. The removal and burning of infected plant parts is called as sanitation.

   Ans. True

2. The wilting of plant is due to a toxic substance- toxin (a polysaccharide) produced by the pathogen in wilt of arhar.

   Ans. True

3. An organism which lives on dead and decaying organic matter is called parasite.

   Ans. False

4. A physical or chemical agent that frees a plant, organ or tissue from infection is called sterilant.

   Ans. False

5. A group of fungi which do not produce any kind of spores is called sterile fungi.

   Ans. True

6. Bactericidal substances produced by certain strains of bacteria and active against some other strains of the same or closely related species are called bacteriocins.

   Ans. True

7. Rust diseases are systematic diseases.

   Ans. False

8. It is possible to convert rust resistant wheat to susceptible by long term use of DDT.

   Ans. True

9. Nematocide is a chemical which kills fungi.

   Ans. False

10. A disease caused by an environmental factor, and not by a pathogen is called non-infectious disease.

    Ans. True

11. Follidal is a bactericide.

    Ans. False

12. An agent that kills or inactivates pathogens in the environment or on the surface of a plant or plant organ before infection takes place is called disinfectant.

    Ans. False

13. Short filamentous tube put forth by a germinating spore is called germ tube.

    Ans. True

14. The swollen tip of a hypha or germ tube that facilitates attachment and penetration of the host by the fungus is called appressorium.

    Ans. True

15. Smut disease is the examples of seed born disease.

    Ans. True

16. Pusa swamy is a resistance variety of Bhindi.

    Ans. True

17. The event happening between take-off of the inolculum and its deposition on the host surfacc is designated as dissemination.

    Ans. True

18. A disease of wheat caused by the fungus *Tilletia* in which the contents of the cereal grains are replaced by odorous smut spores is called rust.

    Ans. False

19. Bacterial brown rot is caused by *Pseudomonas solanacearum.*

    Ans. True

20. The phenomenon in which plant tissues infected with one strain of a virus are protected from infection by another more severe strain of the same virus is called cross protection.

    Ans. True

21. Partial or complete loss of virulence of a pathogen is called attenuation.

    Ans. True

22. The attempt to prevent pathogens, insects, and weeds from causing economic crop losses by using a variety of management methods that are cost effective and cause the least damage to the environment is known as integrated pest management.

    Ans. True

23. A symptom in which small diseased fragments of leaves fall off and leave small holes in their place is called shot hole.

    Ans. True

24. A mass of thin walled undifferentiated cells, developed as the result of wounding is called canker.

    Ans. False

25. The arrival or transfer of a pathogen onto a host is called inoculation.

    Ans. True

26. Enation is malformation induced by certain virus infections.

    Ans. True

27. Eradication is the control of plant disease by eliminating the plant that carries the pathogen.

    Ans. True

28. Elicitors are the molecules produced by the pathogen that induce a response by the host.

    Ans. True

29. Progressive death of shoots, branches, and roots generally starting at the tip is known as chlorosis.

    Ans. False

30. Latent infection is the state in which host is infected with a pathogen but does not show any symptoms.

    Ans. True

31. Excessive growth due to increase in cell number is known as hyperplasia.

    Ans. True

32. A necrotic, often sunken lesion on a stem, branch, or twig of plant is called callus.

    Ans. False

33. Masked symptoms are the virus infected plant symptoms that are absent under certain environmental conditions but appear when the host is exposed to certain conditions of light and temperature.

    Ans. True

34. A plant outgrowth due to abnormal cell enlargement is hyperplasia.

    Ans. False

35. The imperfect or asexual stage of a fungus is also known as anamorph.

    Ans. True

36. Some pathogens infect the host plant through air and bring primary as well as secondary infection.

    Ans. True

37. Suppressive soil is a type of soil in which certain diseases are suppressed because of the presence of microorganisms antagonistic to the pathogen in the soil.

    Ans. True

38. To bring a pathogen into contact with a host plant is called incubation.

    Ans. False

39. Phytoncide is a chemical substance produced by higher plants which can inhibit the growth of microorganisms.

    Ans. True

40. Epidemic rate is the amount of increase of disease per unit of time in a plant population.

    Ans. True

41. A toxic gas that is used to disinfest certain areas from various pests is called as fumigant.

    Ans. True

42. Control of a plant disease with chemicals that are absorbed and transmitted internally is known as chemotherapy.

    Ans. True

43. Plantibodies produced in transgenic plants expressing the antibody producing genes of a mouse that had been previously injected with a pathogen that infect the plant.

    Ans. True

44. Eradicant is a chemical that destroys a pathogen at its source.

    Ans. True

45. The union of a hypha with another resulting in intercommunication of their genetic material is called anastomosis.

    Ans. True

46. A parasite parasitic on another parasite is hyperparasite.

    Ans. True

47. Black, sunken leaf, stem, or fruit lesions caused by fungi which produce their sexual spores in an acervulus, symptoms are called as necrosis.

    Ans. False

48. Propagative virus can multiply in its insect vector.

    Ans. True

49. A group of races and biotypes of a pathogen species that can infect only plants within a certain host genus or species is known as forma specialis (f. sp.).

    Ans. True

50. Protectant is a substrate that protects an organism against infection by a pathogen.

    Ans. True

51. Short, bunchy habit of plant growth is rosette condition.

    Ans. True

DESCRIPTIVE TYPE QUESTIONS

1. What is plant pathology? Explain in details.

   Plant pathology is the study of-

   - The living entities and the environmental conditions that cause disease in plants.
   - The mechanisms by which these factors produce disease in plants.
   - The interactions between the causal organisms and the diseased plant.
   - The methods of controlling a disease either before or after it develop in a plant.

2. How pathogen can cause a disease?

   The pathogen may cause the disease in plants by-

   - Consuming the contents of the host cells upon contact.
   - Killing or disturbing the metabolism of host cells through toxins, enzymes and growth regulators they secrete.

- ✦ Weakening the host by continuous absorption of food from the host cells for their own use.
- ✦ Blocking the translocation of food, mineral nutrients, and water through the conductive elements by means of secreting toxins and plugging the vessels.

3. Classify the diseases on the basis of pathogens or causal organisms.

They may be grouped as follow according to the pathogens:

- ✦ Animate pathogens

  Diseases caused by fungi

  Diseases caused by bacteria

  Diseases caused by mycoplasma

  Diseases caused by nematodes

  Diseases caused by algae

  Diseases caused by phanerogams
- ✦ Virus pathogens
- ✦ Inanimate pathogens:

Low/high temperature effects, light effects, nutritional disorders etc.

4. What are phanerogamic parasites?

These are heterotrophic plants which are unable to synthesize there food from sunlight. Such plants therefore, must live at the expense of autotrophs or upon decaying organic matter. Depending upon the source from which they obtained their nutrition, heterotrophs may be grouped into two categories- Parasites and saprophytes.

- ✦ **Parasites**: A heterotrophic plant which obtained its food from other living organism without killing, it is called parasite. The parasite sucks its food with the help of haustoria. Again phanerogamic parasites can be classified into two groups: (a) total parasites and (b) partial parasites.

Again total parasite can be classified in two branches: (i) stem parasites (e.g. *Cuscuta*) and (ii) root parasites (e.g. *Orobanche*).

5. Write note on total stem phanerogamic parasites.

Cuscuta or dodder plant is the most common example of the total stem parasites. The stem is thin, slender, pale yellow and has only scaly leaves. It is, therefore, unable to manufacture its own food. It turns around the stem of the host plant. The parasite develops haustoria which penetrate into the host tissue and finally get connected with the conducting tissue of the host. Thus parasite derives its organic food, water and minerals from the host tissue. Eventually, the host plant may die due to exhaustion.

6. What are total root parasites? Explain with example.

Members of families like Orobanchaceae, Balanophoraceae, Rafflesiaceae etc. are common root parasites. *Orobanche* (broomrape) is a large leafless plant. This plant inhabits the roots of several crop plants such as tobacco, turnip, cabbage, potato, brinjal etc.

Rafflesia (stinking corpse lily), is also a total root parasite. It is commonly found in the forests of Jawa, Sumatra and South East Asia. It has very reduced stem and roots. It grows directly on the roots of *Cissus* and figs. The flowers of *Rafflesia arnoldi* are the largest in the world and bright red in colour.

7. Write note on partial stem phanerogamic parasites.

Such parasitic plants have chlorophyll and, therefore, synthesize their organic food themselves. But for mineral and water requirements they depends on other plants. Like total parasites, they grow on the stem and roots of the host plants and can be grouped into two categories: (a) partial stem parasite (e.g. *Cassytha*) and (b) partial root parasite (e.g. *Santalum album*).

The seeds of *Loranthus* grow on the stems of many trees like mango, fig, *Dalbergia* etc. the fruit of this plant is a berry. They adhere to the beaks of the birds which eat the fruits. When bird rubs its beak to get rid of the seeds, they adhere to the branch of the host where they germinate. The parasite is attached to the host with the help of modified adventitious roots or haustoria which penetrate the bark and sap wood. *Viscum album* is also a good example of partial stem parasite on apple, pears, pinus etc.

8. What are partial root parasites? Explain with example.

Santalum album is most common example of partial root parasite. It grows on the roots of *Dalbergia sissoo*. It has green leaves but absorbs minerals and water only from the host plants.

9. What are the effects of infection on host physiology?

Infection induced some important changes in the metabolism of the cell. Some of the visible or invisible physiological changes are as follow:

- Effect on membrane permeability: All types of infections bringing about changes in the permeability of the cell membranes. It appears that toxins secreted by these pathogens react with the receptor sites of the cell membrane. As a result, the membrane becomes leaky and the ionic balance of the infected cell is disturbed.
- Effect on photosynthesis: Most of the pathogen secretes an enzyme chlorophyllase which disintegrates the plastids in the host cells. Therefore, infected plants usually have lower rate of photosynthesis.
- Effect on respiration: Infected plants usually have higher rate of respiration due to two reasons- (a) the toxins secreted by the pathogen uncouples phosphorylation from electron transfer in the host cell and (b) to fight infection, the metabolism of the host cell increases in the early stages, so the rate of respiration also increase.

- Effect on nitrogen metabolism: In diseased cells the nucleic acids and protein contents are usually high.

10. What is necrosis?

The pathogen, after entering the host, secretes some enzymes and toxins which kill the host tissue. This is known as necrosis and the dead tissues as necrotic areas. The pathogens which derive their nutrition from such dead tissue are called necrotrophs. Necrotic areas may appear as spot (e.g. leaf spot of mango) or stripes (e.g. stripe disease of barley).

11. What is damping off?

This symptom is caused by soil inhabiting fungi. The pathogen attacks the base of the stem near the soil surface and due to disintegration of tissue or the basal portion of the stem or seedling, the plant topples down. E.g. damping off of seedlings caused by the species of *Pythium* **(Fig. 6.1)**, *Fusarium* etc.

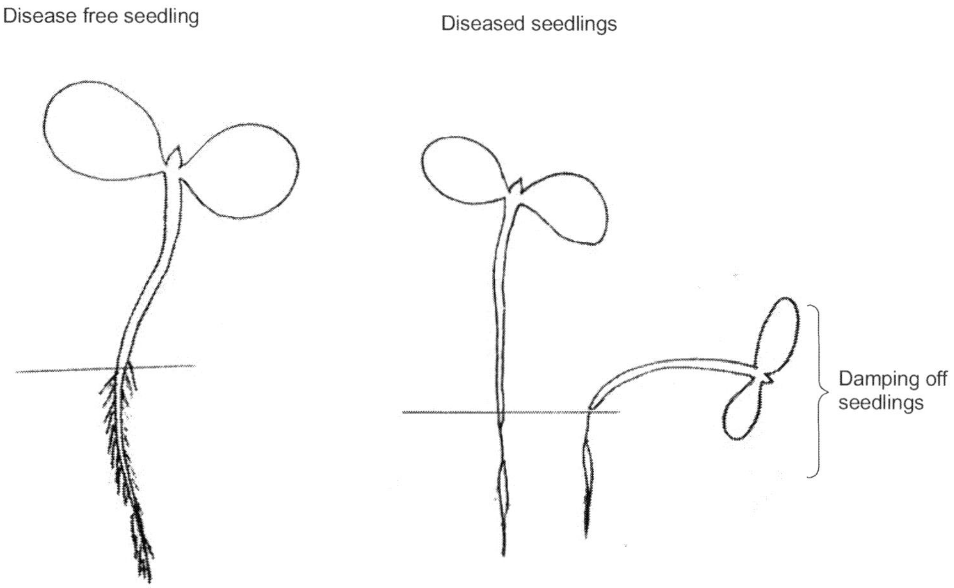

Figure 6.1 Damping off disease symptoms of an seedlings

12. Write a note on mildews symptoms.

In mildew the mycelium and spores of pathogen (fungus) are seen as a whitish growth on the host surface. When the tangled mycelium, sporangiophore and sporangia appear like a downy growth, it is called downy mildew. E.g. Downy mildew of bajra **(Fig. 6.2)**, downy mildew of grapes **(Fig. 6.3)**.

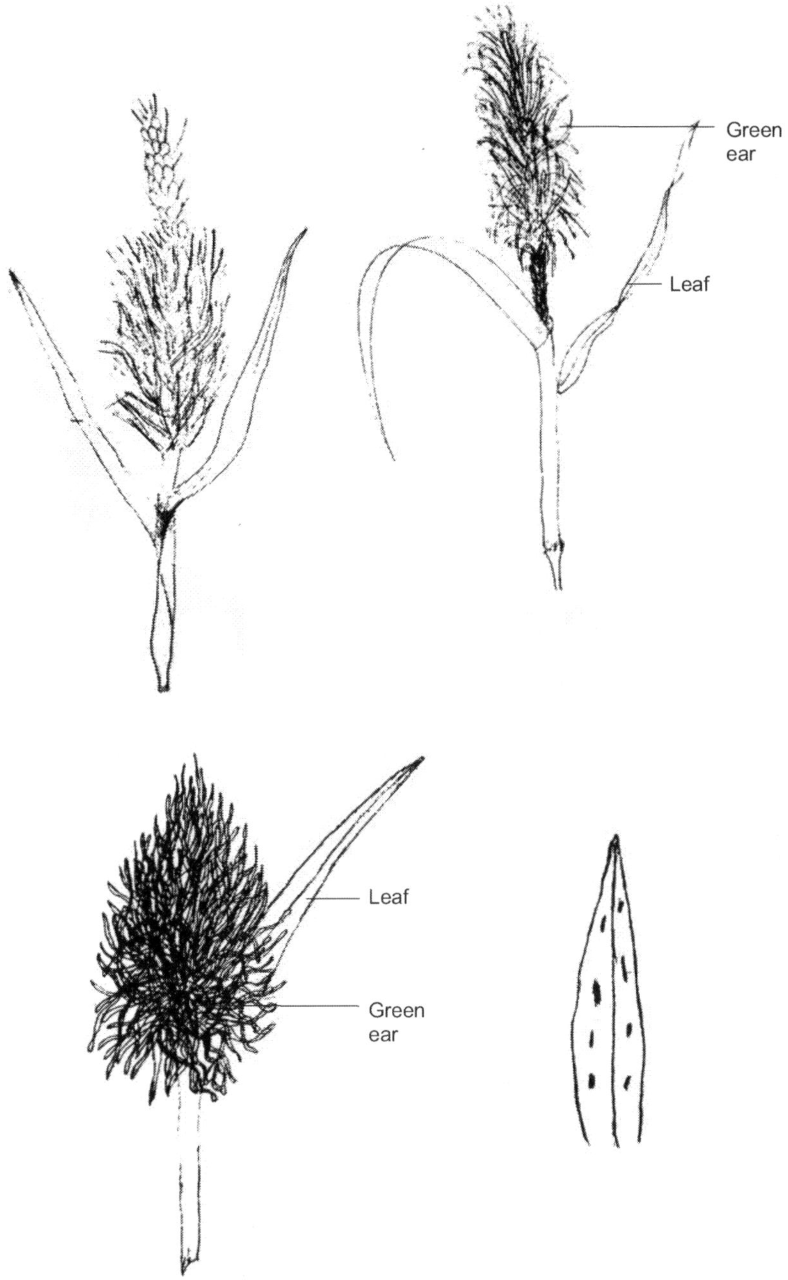

Figure 6.2 Different typtes of symptoms of Downy Mildew of Bajra

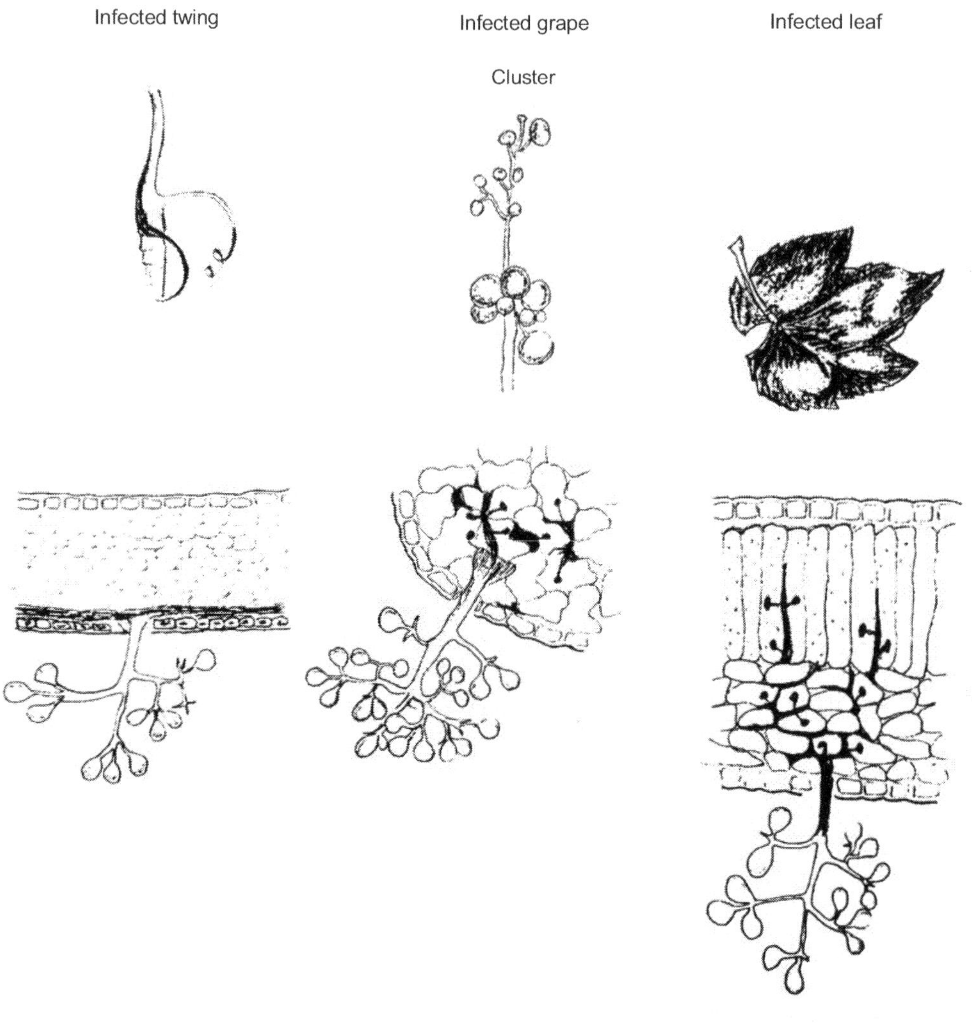

Figure 6.3 Downy Mildew of Grapes : Symptoms and formation of sporangia an host leaft

On the other hand when fungal mycelium along with conidia and conidiophores appear as dull or white powder, it is known as powdery mildew. E.g. Powdery mildew of rose **(Fig. 6.4)**, powdery mildew of cucurbits.

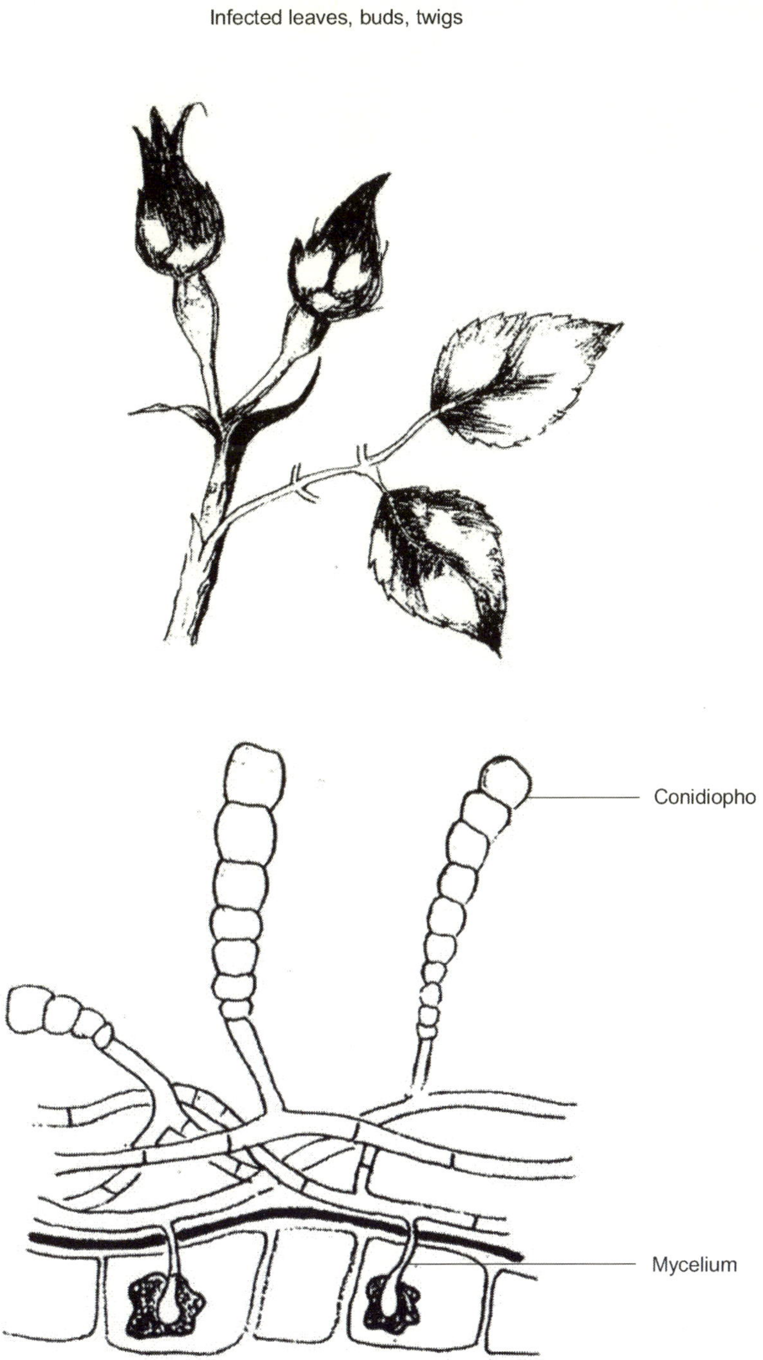

Figure 6.4 Powdery mildew of Rose : Symptoms and Conidial Chain

13. Write a note on fungal plant pathogens.

The majority of phytopathogenic fungi belong to the Ascomycetes and the Basidiomycetes.

Fungal diseases can be controlled through the use of fungicides in agriculture, however new races of fungi often evolve that are resistant to various fungicides.

Significant fungal plant pathogens:

Ascomycetes-

- *Fusarium* sp. (causal agents of Fusarium wilt disease)
- *Thielaviopsis* sp. (causal agents of canker rot, black root rot, Thielaviopsis root rot etc.)
- *Verticillium* sp.
- *Magnaporthe grisea* causes blast of rice and gray leaf spot in turf-grasses.

Basidiomycetes-

- *Rhizoctonia* sp.
- *Phakospora pachyrhizi*, causes soybean rust.
- *Puccinia* sp., causal agents of severe rusts of virtually all cereal grains and cultivated grasses

Oomycetes-

They include some of the most destructive plant pathogens including the genus *Phytophthora* which includes the causal agents of potato late blight **(Fig. 20.5)** and sudden oak death.

Significant oomycete plant pathogens

- *Pythium* sp.
- *Phytophthora* sp., including the causal agent of the Great Irish Famine (1845-1849).

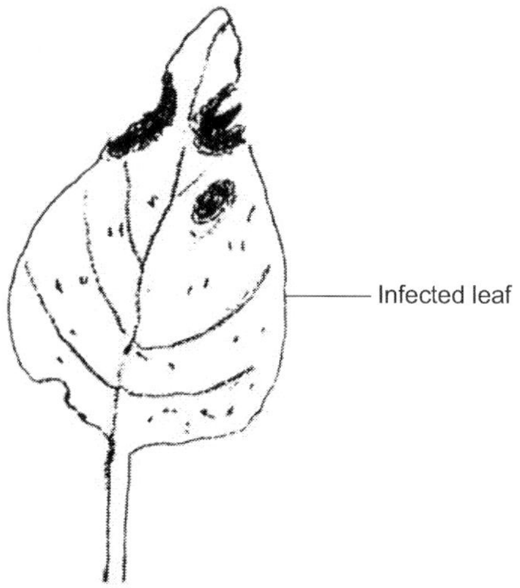

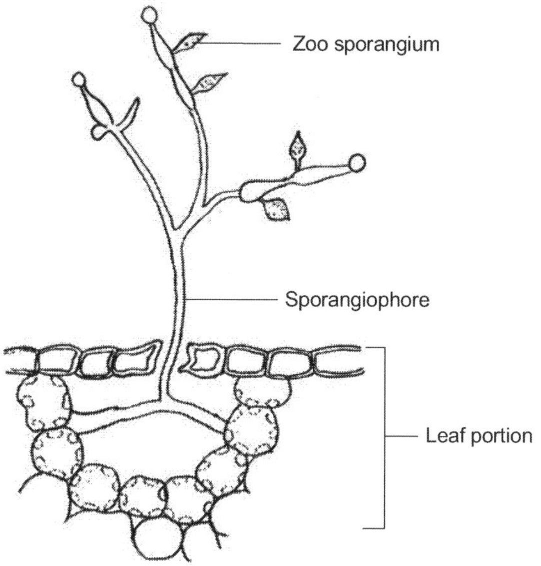

Figure 6.5 Late blight of Potato - Symptoms on leaf and formation of zoosporangium on the surface of leaf.

14. Write notes on bacterial plant pathogens.

Most bacteria that are associated with plants are actually saprotrophic, and do no harm to the plant itself. However, a small number, around 100 species, are able to cause disease. Bacterial diseases are much more prevalent in sub-tropical and tropical regions of the world.

Most plant pathogenic bacteria are rod shaped (bacilli). In order to be able to colonise the plant they have specific pathogenicity factors. Five main types of bacterial pathogenicity factors are as follow:

- Cell wall degrading enzymes: Used to break down the plant cell wall in order to release the nutrients inside. Used by pathogens such as *Erwinia* to cause soft rot.
- Toxins: These can be non-host specific, and damage all plants, or host specific and only cause damage on a host plant.
- Effector proteins: These can be secreted into the extracellular environment or directly into the host cell. Some effectors are known to suppress host defense processes.
- Phytohormones: For example *Agrobacterium* changes the level of auxins to cause tumours.
- Exopolysaccharides: These are produced by bacteria and block xylem vessels, often leading to the death of the plant.

15. What is the mode of transmission of plant viruses?

There are several mode of transmission of plant viruses from one plant to other plant, these are as follow:

✦ **Through sap:**

Viruses can be spread by direct transfer of sap by contact of a wounded plant with a healthy one. Such contact may occur during agricultural practices, as by damage caused by tools or hands, or naturally, as by an animal feeding on the plant. Generally TMV, potato viruses and cucumber mosaic viruses are transmitted via sap.

✦ **Insects:**

Plant viruses need to be transmitted by a vector, most often insects such as leafhoppers. The chosen insect vector of a plant virus will often be the determining factor in that virus's host range. It can only infect plants that the insect vector feeds upon. Depending on the way they are transmitted, plant viruses are classified as non-persistent, semi-persistent and persistent. In non-persistent transmission, viruses become attached to the distal tip of the stylet of the insect and on the next plant it feeds on; it inoculates it with the virus. Semi-persistent viral transmission involves the virus entering the foregut of the insect. Those viruses that manage to pass through the gut into the haemolymph and then to the salivary glands are known as persistent.

✦ **Nematodes:**

Soil born nematodes also have been shown to transmit viruses. They acquire and transmit them by feeding on infected roots. Viruses can be transmitted both non-persistently and persistently. The virions attach to the stylet or to the gut when they feed on an infected plant and can then unattach during later feeding to infect other plants. Examples of viruses that can be transmitted by nematodes include tobacco ring spot virus and tobacco rattle virus.

✦ **Plasmodiophorids**

A number of virus genera are transmitted, both persistently and non-persistently, by soil born zoosporic protozoa. These protozoa are not phytopathogenic themselves, but prarsitic. Transmission of the virus takes place when they become associated with the plant roots. Examples include *Polymyxa graminis*, which has been shown to transmit plant viral diseases in cereal crops. Plasmodiophorids also create wounds in the plant's root through which other viruses can enter.

✦ **Seed and pollen borne viruses:**

Plant virus transmission from generation to generation occurs in about 20% of plant viruses. When viruses are transmitted by seeds, the seed is infected in the generative cells and the virus is maintained in the germ cells and sometimes, but less often, in the seed coat. When the growth and development of plants is delayed because of situations like unfavourable weather, there is an increase in the amount of virus infections in seeds. There does not seem to be a correlation between the location of the seed on the plant and its chances of being infected.

16. What are the symptoms produced by viruses in plants after infections?

The different kinds of symptoms caused by viruses in plants are given below:

- Mosaics: These are characterized by light green, yellow or white areas intermingled with the normal green areas of the leaves or fruits.
- Mottles: It is a kind of mosaic, where an irregular pattern of indistinct light and dark areas develops on the leaves.
- Chlorosis: In extreme cases of mosaics and mottles, the leaf may become almost completely yellow due to chlorosis.
- Leaf curling: The leaves become curled and rolled to varying extents of infected plant. These are common in papaya, tomato, potato etc.
- Enation: These are small outgrowths on leaf, stem etc. This is usually associated with mosaics. E.g. Tobacco enation.
- Vein clearing: There is chlorosis of leaf tissue in close proximity to the veins. The tissue close to veins turns yellow, remaining areas appears green. E.g. Bhindi.
- Vein banding: The parenchyma close to the veins is green and rest of the lamina surface shows chlorosis.
- Ring spots: These are characterized by the appearance of chlorotic or necrotic rings on leaves and sometimes also on fruit and stem.

17. What are the control methods of plant viruses?

Methods to control viral diseases are generally similar to those used for other pathogens. Some important methods are described below:

- Exclusion of virus from the healthy plants: The best way to control a virus disease is to keep it out of area through quarantine, inspection and certification system. Eradication of all sources of inoculums can be effective in annual crops. Diseased plants should be destructed to eliminate the inoculums from the field.
- Control of virus vectors: Controlling the insect vectors and removing the weeds which serve as hosts for vectors may help in controlling viral diseases.
- Use of tested material for plantation: The use of virus free seeds, tubers, buds etc. is the most important method for control of viruses.
- Resistant varieties: This the most important and satisfactory method of virus elimination. Many plant varieties resistant to viruses have been developed by the researchers in some crop plants.
- Tissue culture or micropropagation: This is most cost effective and 100% viable method of virus elimination from the plants. Shoot tips can be cultured in this method to produce virus free stock plant material.
- Physical methods: Heat can be used to inactivate viruses from the planting material. Dormant parts are dipped in hot water (35-55°C) for few minutes to hour where as non-dormant plant parts are kept in growth chambers for several days, weeks or months at 35-40°C.

- Chemical methods: Viricides are so far not available in the market but foliar application of some growth regulators such as gibberellic acid has been effective in stimulating growth of virus suppressed axillary buds in sour cherry yellows.

Modification in agricultural and horticultural practices may be used to reduce viral attack. Practices like reduction of close spacing, mixed cropping pattern etc. are also effective.

Tables

Important diseases of Rice in India.

Name of the disease	Causal organism
Bactirial blight/wilt	*Xanthomonas oryzae* pv. *Oryzae*
Blast	*Pyricularia oryzae*
Brown leaf spot	*Helminthosporium oryzae*
False smut	*Ustilaginoidea virens*
Sheath blight	*Rhizoctonia solani*
Tungro	Viral disease

Important diseases of wheat in India.

Name of the disease	Causal organism
Ear cockle	*Corynebacterium michiganese var. tritici*
Foliar blights	*Helminthosporium sativum* and *Alternaria triticum*
Karnal bunt	*Neovossia indica*
Molya	*Heterodera avenae*
Rust	*Puccinia graminis f. sp. tritici, P. striiformis f. sp. tritici, P. recondita*

Some important pearl millet diseases in India.

Name of the disease	Causal organism
Downy mildew	*Sclerospora graminicola*
Ergot	*Claviceps fusiformis*
Smut	*Tolyposporium penicillariae*

Some important potato diseases in India.

Name of the disease	Causal organism
Bacterial wilt	*Pseudomonas solanacearum*
Late blight	*Phytophthora infestans*
Nematode pests	*Globoderma rostochiensis* and *Meloidogyne sp.*

List of non-parasitic diseases.

Enviromental agent	Effects on plants
Low temperature	Freezing injury
High temperature	Heat canker of flax
Mineral deficiency	Red leaf of cotton or Khaira disease of rice
Mineral excess	Internal bank necrosis of apple
Unfavourable oxygen	Black heart of potato
Unfavourable soil moisture	Flooding injury
Lightning injury	Damage to potato plants due to lightning
Injurious atmosphere	Black tip of mango

Examples of some iatrogenic diseases.

Host	Iatrogenic pathogen	Chemical
Tobacco	TMV	2,4-D
Maize	*Ustilago*	2,4,5-T
Wheat	*Helminthosporium*	2,4-D
Rice	*Rhizoctonia*	2,4-D
Wheat	*Puccinia*	DDT
Maize	*Pythium*	Picloram
Cherry	*Agrobacterium*	Dichlone
Tomato	*Botrytis*	Zineb
Grape	*Erysiphe*	Zineb
Cotton	*Rhizoctonia*	Trifluralin
Cotton	TMV	IAA
Tomato	*Verticillium*	GA3

PLANT PATHOLOGY—A QUICK GLANCE

1. The word pathology has been derived from two Greek words, pathos = suffering and logos = discourse. Therefore the study of diseased plants is called pathology.
2. According to J.C. Horsfall and A.E. Dimond (1959) disease is a malfunctioning process that is caused by continuous irritation. Of course, this process must result in some suffering. And hence, disease is a pathological process.
3. Disease is a complex phenomenon and, is an interaction among the host, the parasite and the environment.
4. A condition of a plant in which it is normally subject to attack by a given pathogen is called susceptibility and lacking the inherent ability to resist disease or attack by a pathogen is called susceptible.
5. The ability of an organism to overcome completely or in some degree, the effect of a pathogen or other damaging factor is called resistance.
6. The study of the symptoms of diseases is called etiology.
7. Pathogen or causal organism is an agency that incites disease; these are also called incitants because they incite the disease under the influences of the other factors.
8. Pathogenesity is the property of a microorganism whereby it may become a part of the causal complex, or the relative capability of a pathogen to cause disease.
9. The broom like overgrowth or massed proliferation caused by dense clustering of branches of a plant is called witches' broom.
10. Quiescence of an organism is the stage when under the influence of some biological, physical or chemical stress the organism or its propagules undergo an inactive state, this is called quiescent stage.
11. The process of events whereby disease development takes place is called pathogenesis.
12. Acervulus is a subepidermal, saucer shaped, sexual fruiting body producing short conidiophores and conidia. E.g. *Colletotrichum.*
13. *Mucor stololifer* causes soft rot disease of vegetables, *Rhizopus arrhizus* causes fruit rot of apple and *R. artocarpi* causes fruit drop of Jack fruit.
14. Pests are the organisms which cause physical and economic loss to the human beings.
15. Parasites are the organisms which subsist wholly or partially upon living tissues.
16. Viruliferous is said of a vector which contain the virus and capable of transmitting it.

17. Elicitors are signals, presumably chemical signals, generated by an organism to bring about metabolic shifts in another organism.

18. Organisms which, regardless of ease with which they can be cultivated on artificial media, always obtain their food in nature from living tissue on which they complete their life cycle are called biotrophs or obligate parasite.

19. Saprophytes are the organisms which subsist wholly or partly upon dead organic matter and inorganic materials.

20. Fungi, bacteria, mycoplasma, algae, phanerogams, nematodes etc. are called animate pathogens and low or high temperature effects, light effects, nutritional disorders, effects of atmospheric impurities are called inanimate pathogens.

21. The pathogen which completes many life cycles and disease cycles in one season is called polycyclic.

22. Tolerance is a type of defense that minimizes crop loss without restricting disease development.

23. Virulence is a measure or degree of pathogenicity of an isolate or race of the pathogen.

24. The green algae *Cephaleuros* is an intercellular parasite in the leaves of a number of angiosperms, such as *Magnolia*, *Rhododendron*, *Camellia sinensis* and *Piper nigrum*.

25. Disease causing algae and algal blooms can be destroyed by copper sulphate, it is known as algicide.

26. Quorum sensing is a regulatory mechanism that connects gene expression to cell density in bacteria. Numerous functions, including pathogenicity are regulated in this way.

27. Fungicides can kill fungi. Bordeaux mixture, Burgundy mixtures and mercuric chloride are the examples of fungicides.

28. Aggressiveness is to describe the capacity of a pathogen to invade and grow in its host plant and to reproduce on or in it.

29. Bordeaux is the first fungicide used at commercial scale. It is discovered by Maillardet and it is the first fungicide discovered first time.

30. Sporodochium is a fruiting structure consisting of a cluster of conidiophores woven together on a mass of hyphae.

31. The name of Bordeaux mixture is based on 'Bordeaux University, France.

32. Immunity of a plant against a given disease is an absolute quality and it denotes that the pathogen cannot establish parasitic relationship with the host.

33. *Cuscuta* (dodder) is a total stem parasite and belongs to the Angiospermic family Convolvulaceae, *C. reflexa*, *C. hyaline* and *C. chinensis* are the most common species. The *Cuscuta* stem twines around the host plant and develop haustoria which penetrate into the tissues of the host to get nutrients.

34. Stem-pitting is a symptom of some viral disease of trees characterized by depressions on the stem under the bark.

35. The pathogen grows from the point of entry to varying extents without showing adverse effect on tissue through which it passes is called systemic infection.

36. When the source of primary inoculums serves as infective agent, it is called primary inoculum and when the host becomes diseased, the organism subsequently produces spores or infective bodies, are called secondary infection.

37. Bordeaux mixture is prepared from $CuSO_4$ + Lime + Water and Burgundy mixture is prepared from $CuSO_4 + Na_2CO_3$ + Water.

38. The complete elimination or eradication of living organisms or pathogens from the object is called sterilization (by means of heat, chemicals or radiations).

39. In mildew diseases the pathogen is seen as white, grey, brownish or purplish growth on the host surface.

40. Herbicides and weedicide are the chemicals which are used to kill unwanted plants/weeds in the field.

41. Triazine, auxin derivatives e.g. 2,4-D, 2,4,5-T etc. are the example of herbicides/weedicides.

42. Violent reaction of the host tissues after attack of a pathogen in which there is prompt death of the invaded tissue preventing further spread of the pathogen is called hypersensitivity.

43. The initial invasion of the host by an organism is called penetration.

44. The establishment of the pathogen within the host following penetration is called infection.

45. In Downey mildew the superficial growth is a tangled cottony or downy growth while in powdery mildew enormous numbers of spores are formed on the superficial growth of the fungus giving the host surface a dusty or powdery appearance.

46. The interval between infection and appearance of disease is called incubation period.

47. *Orobanche cernua* or broomrape belongs to Angiospermic family Orobanchaceae, is a total root parasite and its root absorb nutrient from the roots of host plants. Generally it grows on the roots of *Nicotiana tabacum*, cabbage etc.

48. Plants that react to certain viruses or environmental conditions with production of specific symptoms and are used to detect and identify these factors called indicator plants.

49. Viruses that are acquired by their vectors through their mouth parts, accumulate internally, then are passed through their tissues and introduced into plants again through the mouth parts of the vectors are called circulative viruses.

50. The progression of external evidences of disturbance of the morphological and physiological processes of the host plant are called symptoms.

51. Receptive hypha is a specialized hypha protruding out of a pycnium and functioning as a female gamete; also called flexuous hypha.

52. The chain of events which occur between the time of infection and the complete expression of disease is called disease development.

53. Cross protection is a phenomenon in which plant tissues infected with one strain of a virus are protected from infection by other strains of the same virus.

54. The presence of pathogen in a population of plants or of pathogen in a position or material where it has the possibility of causing infection and producing disease is called infestations.

55. Insecticides are used to kill insects. D.D.T., Melathion, Aldrin etc. are the examples of insecticides.

56. Nematicides are used to kill nematodes; chloropicrin and methyl bromide are the examples of nematicides.

57. Diplanetism is a condition of having two periods of mobility in one life history as of zoospores in some oomycetes.

58. *Viscum* sp., *Dendrophthoe* and *Santalum* are the examples of partial stem parasites, the seeds of these plants germinate on the branches of a suitable host and finally the woody tissues of the two plants become continuous. E.g. Mango, Pear, Guava etc.

59. Latent infection is a state in which a host is infected but does not show symptoms and the period required by a virus to multiply and accumulate within the vector before the later becomes infective is called latent period.

60. The control of export and import of plant to prevent spread of disease and pest is called quarantine.

61. Resting spore is an inactive stage of a fungus, usually a thick walled spore that is resistant to unfavourable conditions, and often germinates after a rest period.

62. Warfarin, zinc sulphate and zinc phosphide are the most used rodenticides.

63. Organochlorines are the insecticides which are made up of hydrocarbon with chlorine. Some important examples of organochlorines are D.D.T., Aldrin, Dialdrin, Endosulphan, B.H.C etc.

64. Organochlorines are lipophilic and non biodegradable so these chemicals accumulate in the body and environment and cause harmful effects.

65. The removal and burning of infected plant parts or debris is called sanitation.

66. Dichlorodiphenyl tetrachloro-ethane (D.D.T.) is the world's most famous insecticide.

67. D.D.T. was discovered by Dr. Paul Muller. It is banned in India since 1985 and now B.H.C. is used as its substitute in India.

68. The softening, discolouration, and disintegration of a succulent plant tissue as a results of fungal or bacterial infection is called rotting.

69. In some viral diseases symptoms that are absent under certain environmental conditions but appear when the host is exposed to certain conditions of light and temperature are called masked symptoms.
70. Vein clearing is destruction of chlorophyll in the vein tissue, as a result of infection by a virus or other pathogen.
71. *Striga* is an example of partial root parasite; *S. lutea* is parasite on the roots of *Sorghum* and *Penisetum* sp.
72. Aldrin is used in the base of buildings to protect them from termites.
73. A parasite that in nature can grow and multiply only on living organisms are called obligate parasite. E.g. Rust fungi, powdery and downy mildew fungi etc.
74. Pustule is small blister like elevation of epidermis as spores emerge. E.g. White rust of cruscifers caused by *Albugo candida.*
75. The loss of rigidity and drooping of plant parts wholly or partially is called wilting.
76. Obligate saprophytes have no relation with living cells and obtained their food from dead organic materials or from inorganic materials.
77. The quantitative measurement of disease or the process of measuring diseases quantitatively is called disease assessment.
78. Organophosphate are the another group of insecticides, these are organic ester of phosphoric acid and triphosphoric acids.
79. The parasitic nature of nematodes in plants was first discovered by Berkeley in 1855.
80. Most of the nematodes known as plant pathogens fall under the order Tylenchida of class Nematoda.
81. Syndrome is a group of signs and symptoms that occur together and characterize the disease.
82. The most common used organophosphates are melathion, parathion, fenitrothion, trithion etc.
83. The diseases of animals and human caused by consumption of feed and foods invaded by toxin producing fungi are called mycotoxicosis.
84. The incidence of fields with diseased plants in a defined geographic area is called disease prevalence.
85. Sclerotium is resting, dormant, or winter stage of some fungi when they become a mass of hardened mycelium.
86. Virus indexing is the detection of the presence of viruses in plants using various techniques ranging from the development of symptoms in suitably inoculated susceptible genotypes of the host to serological techniques and DNA probes, it is essential while producing virus free plants through tissue culture or while introducing germplasm.

87. Melathion is the most commonly used organophosphates. It is an important component of Flit.

88. The coating of seed with a plastic substance in which a fungicide can be mixed is called pelleting.

89. Organophosphates act on central nervous system, it inhibits the action of cholinesterase enzyme, and it stops the breakdown of acetyl choline.

90. Vigorous biotype is biotype of an insect pest, which attacks both resistant and susceptible varieties of the host more susceptible than the other biotypes.

91. Vector is an organism (e.g. insect) that transmits a pathogen (e.g. virus).

92. Alternate host is one of the two kinds of plants on which a parasitic fungus (e.g. rusts) must develop to complete the life cycle or in other words it is a host of a pathogen growing during off-season for the main host crop.

93. Phytoalexins are the substances which inhibit the development of a fungus on hypersensitive tissue, formed when host plants cells come in contact with the fungus.

94. Carbamates are another group of insecticides; these are the salts of carbamic acid (NH_2COOH).

95. The effect of one or more meteorological factors which makes a plant vulnerable to attack by a pathogen is called predisposition.

96. Any infection caused by inoculums produced as a result of a primary infection or an infection caused by secondary inoculum is called secondary infection.

97. The structure of Carbamates is like acetyl choline so it binds with the enzyme acetylchlonesterase. This stops the normal functioning of nervous system.

98. Bacteriocin is bactericidal proteinaceous metabolites produced by certain strains of bacteria and active against one or more strains of the same or closely related species. The production of bacteriocin is controlled by extrachromosomal DNA (e.g. plasmid).

99. Carboryl, carbofuran or furadon, aldicarb or temic, propoxur or Baygon etc. are some the most commonly used Carbamates.

100. Dithiocarbamates (diethane M-22) is used as fungicide and Phenyl carbamates & thiocarbamates are used as herbicides.

101. Methyl isocinide is used as raw material in the preparation of carboryl.

102. Pyrethrin is an insecticide which is obtained from the inflorescence of *Chrysanthemum cinerariifolium* and *C. marshalii*.

103. The eruption or discharge of gum or other substances from diseased tissues is called exanthema.

104. Pesticides which are derived from urea are called triazines. E.g. atrazine, simazine etc.

105. The derivative of plant hormone auxins are also used as selective herbicides. E.g. 2,4-D, 2,4,5-T, delapal, trichloroacetic acid (TCA) etc.

106. Most of the herbicides inhibit the Photosystem-II of photosynthetic machinery of weeds. Because of this inhibition there is no photolysis of water and no oxygen release. Finally plants die due to lack of food.
107. Some herbicides block the translocation and movement of food by blocking sieve tubes.
108. Fasciation is a plant disease, due to cell injury in the bud, resulting in flattened and sometimes spirally curved shoots.
109. Most of the insecticides affect the nervous system of the insects but some block the respiratory system to kill them.
110. Increasing of concentration of pesticide in the body tissues of living organism is called biomagnifications or bioconcentration.
111. Due to long time of use of D.D.T. in India, the concentration of D.D.T. in the tissue of Indians is 13-30 ppm. It is highest in the world.
112. Fungistasis is a phenomenon in which fungus spores fail to germinate or hyphae fail to grow even if temperature and moisture conditions are favourable.
113. The first bioherbicide was Devine, it is actually mycoherbicide.
114. Devine was prepared from fungus *Phytophthora palmivora* in 1981 against a common weed *Morrenia odorata* (milk weed vine) which is present in citrus orchards.
115. Collego is another bioherbicide which is prepared from the spores of fungus *Colletotricum gloeosporiodes*.
116. *Cactoblastis cactorum* insect was used in India and Australia to fight against *Opuntia* or prickly pear cactus.
117. Fungus *Cercospora rodmani* was used in USA to control weed *Eichornia crassipes*.
118. The crop plants which inhibit the growth of weeds around them is called smoother crop. The examples of smoother crops are Sunflower, soybean, Barley, rye, sorghum etc.
119. Ladybug is used to control aphids, and Gambusia fish is used to control mosquito larva.
120. GABA (ã-aminobutyric acid), is a plant metabolite involved in induction of resistance to stress in plants.
121. Sporeine is the first bioinsecticide which was prepared in the Germany, it block the ion transport in the insect gut.
122. By sterilization of male partner with help of gamma rays method was used by Kipling in 1955 to eradicate Screw worm.
123. Gypsy moth was controlled by pheromone trap method.
124. Rotenone, nicotin, pyrethroides, azadirachtin, thurioside etc. are the examples of natural insecticides.

125. The fungi producing compatible male and female gametes on physiologically distinct mycelia are called heterothallic fungi and the phenomenon in which a fungus produces compatible male and female gametes on the same mycelium is called heterothallism.

126. Rotenone is obtained from the roots of *Derris elliptica.*

127. Nicotine is isolated from the leaves of *Nicotiana tobacum.*

128. Pyrethroide is obtained from the inflorescence of *Chrysanthemum cinerariifolium.*

129. Azadirachtin is extracted from the seeds of *Azadirachta indica.* The extract obtained from the leaves of neem tree used to control Japanese beetle and leaf eating pests.

130. The symptoms of certain viral diseases of plants characterized by intermingles patches of normal and light green or yellowish colour is called mosaic.

131. Thurioside is a bacterial originated insecticide which is obtained from *Bacillus thuringiensis* (microbial insecticide).

FIGURES

Figure 6.6 Smut disease symptoms on barley.

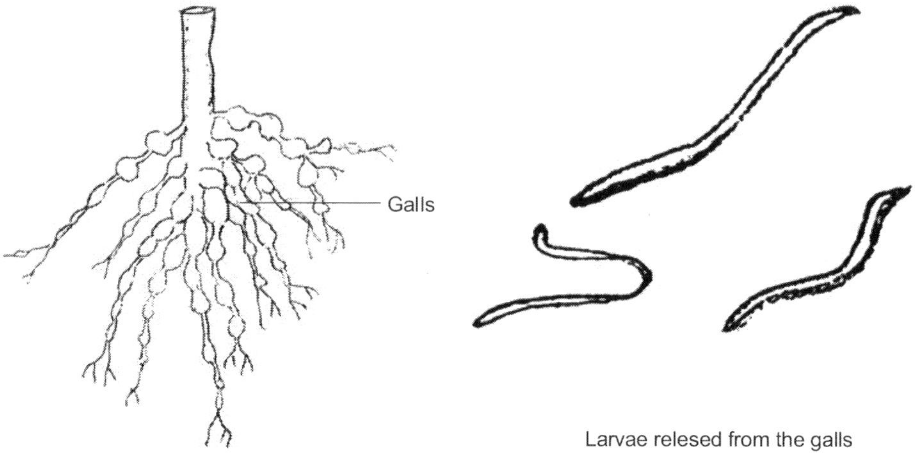

Figure 6.7 Symptoms of nematode infection on root and the causal organism.

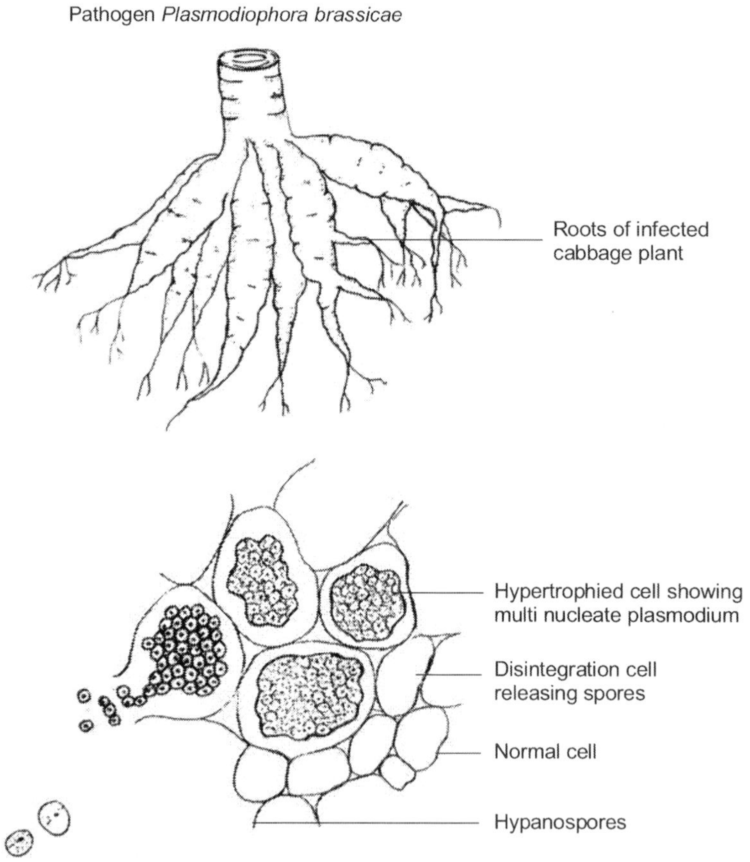

Figure 6.8 Symptoms of club root disease of crucifers and T.S of an infected root part.

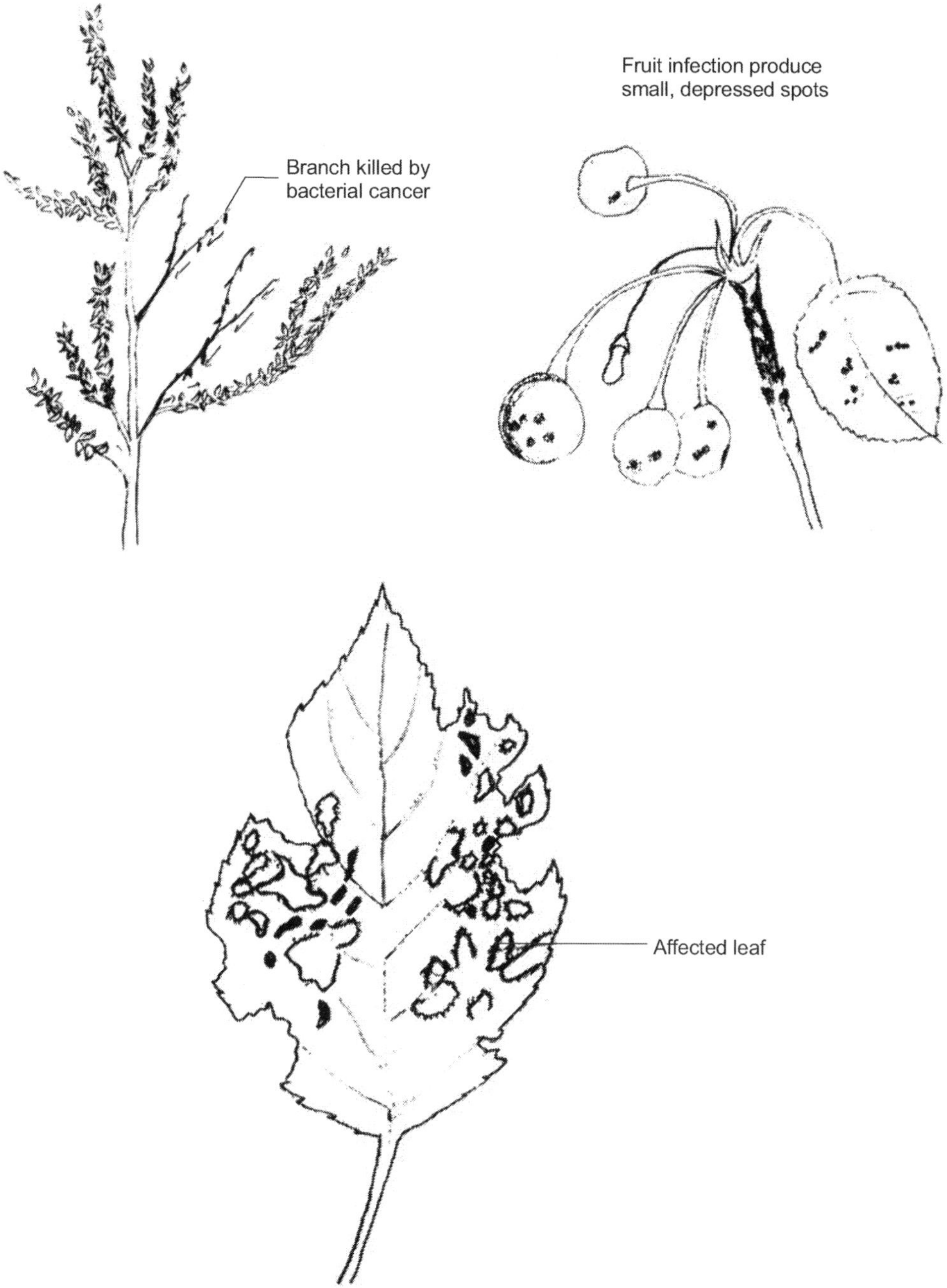

Figure 6.9 Affected plant and symptoms of citrus canker disease.

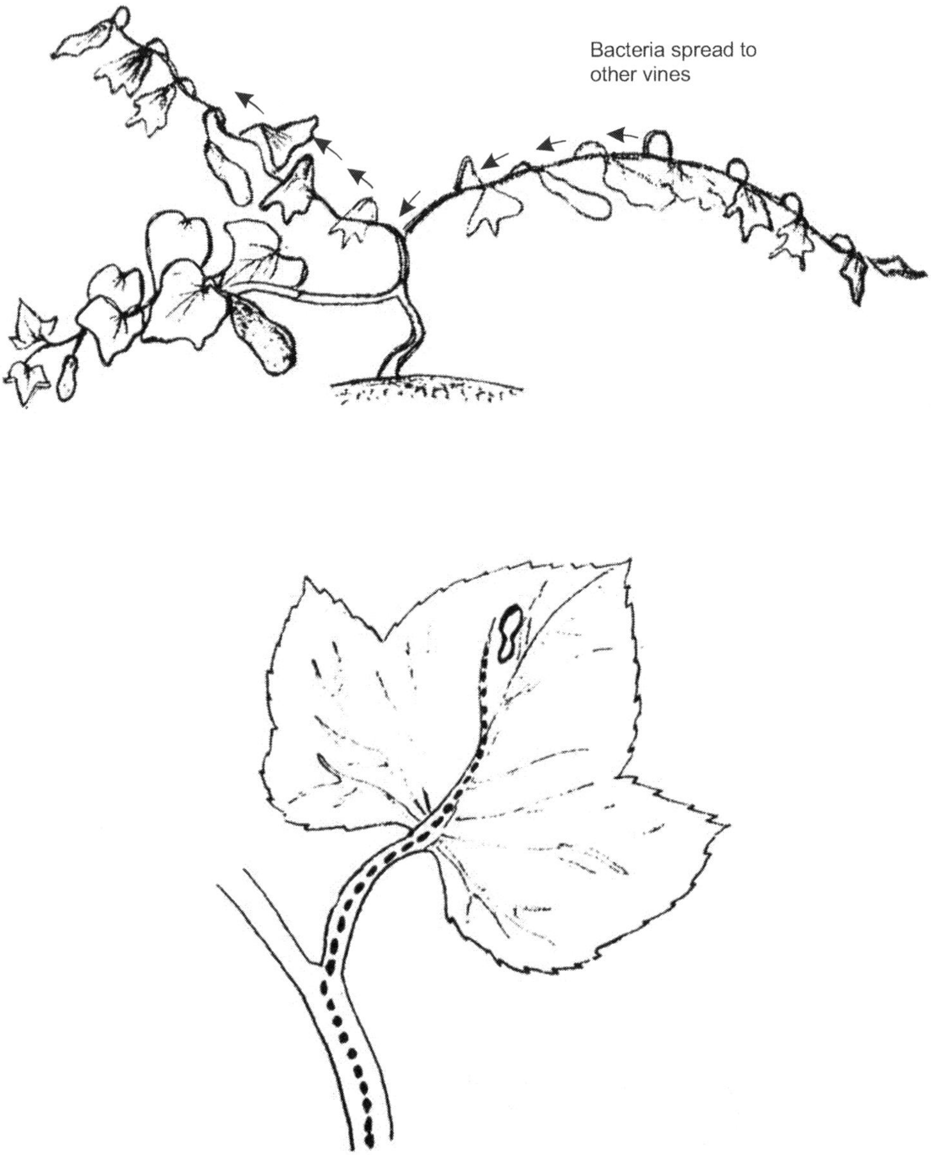

Figure 6.10 Wilt disease of cucurbits.

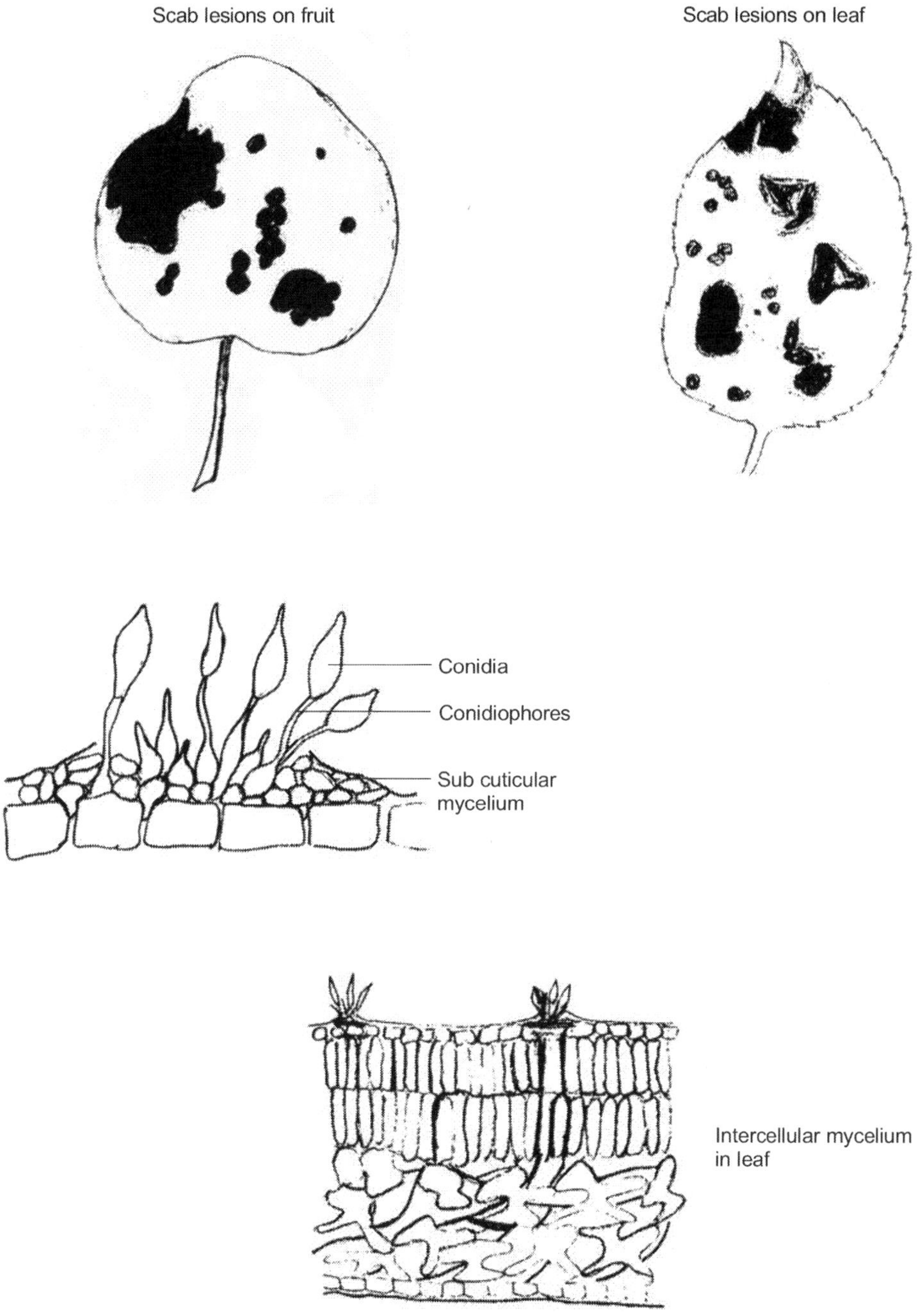

Figure 6.11 Symptoms of scab of apple on fruit and leaf, development of conidia on host surface.

ANSWERS

1. (b)	2. (d)	3. (c)	4. (c)	5. (d)	6. (a)
7. (b)	8. (b)	9. (a)	10. (a)	11. (d)	12. (d)
13. (c)	14. (d)	15. (b)	16. (b)	17. (d)	18. (c)
19. (b)	20. (b)	21. (b)	22. (d)	23. (c)	24. (b)
25. (d)	26. (d)	27. (d)	28. (c)	29. (c)	30. (c)
31. (d)	32. (a)	33. (c)	34. (a)	35. (b)	36. (a)
37. (a)	38. (b)	39. (a)	40. (b)	41. (d)	42. (d)
43. (d)	44. (c)	45. (d)	46. (b)	47. (d)	48. (b)
49. (b)	50. (c)	51. (d)	52. (a)	53. (b)	54. (b)
55. (c)	56. (d)	57. (b)	58. (b)	59. (d)	60. (a)
61. (b)	62. (b)	63. (a)	64. (c)	65. (c)	66. (d)
67. (c)	68. (b)	69. (a)	70. (c)	71. (c)	72. (b)
73. (d)	74. (a)	75. (a)	76. (b)	77. (c)	78. (b)

OLD QUESTION PAPER OF VARIOUS EXAMS

CSIR-NET Exam, June-2010 (Based on Memory)

Paper-I Part-B

1. The donor and acceptor of electron for cytb6/f are
   (a) PQ and UQ (b) UQ and Cyt (c) UQ and PC (d) PQ and PC
2. The chemical salicyl hydroxamic acid inhibits
   (a) Cytochrome oxidase (b) Alternate oxidase
   (c) ATP synthase (d) NAD-dehydrogenase
3. Which of the following DO NOT occur during seed development?
   (a) Accumulation of storage proteins (b) Synthesis of LEA proteins
   (c) Desiccation (d) Synthesis of gibberellic acid
4. In signal transduction pathway of which plant hormones genes 'rht', 'spy', 'gay' are associated
   (a) Auxin (b) Gibberellic acid
   (c) Abscisic acid (d) Ethylene
5. Which microbe has been used against insect plant pathogens?
   (a) *Agrobacterium tumefaciens* (b) *Bacillus thuringensis*
   (c) *Fusarium nudum* (d) *Agrobacterium rhizogenes*
6. In HIV diagnosis technique base on Western blotting, the patient sample is screened for
   (a) Antibody (b) Antigen
   (c) Virus (d) Virus outer coat protein
7. A person suffering from glucose-6-phosphate dehydrogenase deficiency are found to be resistant to
   (a) Fungus (b) Bacteria (c) Plasmodium (d) Leishmania
8. Dinosaurs become extinct during
   (a) 1.6 billion years ago (b) 6.5 million years ago

(c) 65 million years ago (d) 6.5 billion years ago

9. Which statement is correct for reselected species?
   (a) Large number of progeny with large size
   (b) Large number progeny with small size
   (c) Small number of progeny with large size
   (d) Small number of progeny with small size
10. During regulation of trp operon by attenuation there is
    (a) Immature termination of translocation (b) Immature termination of transcription
    (c) Termination of replication (d) Ribosome fails to read transcript
11. Polygalactourinidase (PG) antisense RNA is used in reference to which phenomenon
    (a) Viral resistance (b) Herbicide resistance
    (c) Seed setting (d) Fruit ripening
12. Bcl-2 and Bax proteins involved in apoptosis are
    (a) Proapoptosis and anti-apoptotic (b) Both proapoptic
    (c) Anti-apoptotic and pro-apoptotic (d) Both anti-apoptotic
13. Nick translation means
    (a) Translation by cytosolic ribosome
    (b) Translation of protein from stalled sites
    (c) Replication by DNA polymerase I after removing RNA primers
    (d) Replication of DNA by DNA polymerase I from Nicks produced by DNase treatment
14. For final processing of antigen macrophage after phagocytosis directs engulfed material to
    (a) Lysosome (b) Mitochondria (c) Proteosome (d) Peroxisome
15. Flow cytometer is used to measure the number of
    (a) Cells (b) DNA (c) RNA (d) Proteins
16. Which portion of antibodies is recognized by Fc receptor of macrophage?
    (a) λ-chain (b) κ-chain (c) Heavy chain (d) J-chain
17. Km for enzyme is equal to where
    (a) V_0 = Vmax (b) $2V_0$ = Vmax (c) $4V_0$ = Vmax (d) V_0 = 2Vmax
18. The precursor for most of phenolic secondary metabolites is
    (a) Phenyl alanine (b) Isopentyl pyrophosphate
    (c) Mevalonate (d) Geranyl phosphate
19. Which is NOT true in a process of succession from pioneering stage to climax stage?
    (a) Biomass increases (b) Food chain to food web
    (c) Wild niche (d) Dependency on detritus chain increases

20. Hydrostatic opening of stomata is due to
    (a) Movement of K ions into guard cells
    (b) Movement of water into guard cells
    (c) Low water potential in mesophyll cells
    (d) Decreased concentration of ABA
21. Which is correct explanation for sibling species?
    (a) Species which are morphologically different and reproductively isolated
    (b) Species which are morphologically alike and can interbreed
    (c) Species which look morphologically similar but reproductively isolated
    (d) Species which are morphologically different but can interbreed
22. Which statement is correct for archaebacteria?
    (a) Lacks histone protein
    (b) Lacks introns in genes
    (c) Have 9+2 arrangement in flagella
    (d) Cell wall is made up of peptidoglycan
23. Darwinian evolutionary fitness is measured in terms of
    (a) Good health
    (b) Ability to fight with others
    (c) Reproductive success
    (d) Lesser mutations
24. Incorrect blood transfer to person leads to death of a blood receptor. The major cause of death is
    (a) Precipitation of RBC in coronary arteries of heart
    (b) Failure of kidney due to excess breakdown product of RBC
    (c) Agglutination of RBC
    (d) Hyperthermia due to release of pyrogens
25. On the basis of statement given below the mode of inheritance is
    - Mostly males are sufferer of disease.
    - All male child developed from affected mother are diseased.
    - Female develop disease only when her father is diseased and mother is carrier.

    (a) X- linked recessive
    (b) Autosomal dominant
    (c) X-linked dominant
    (d) Autosomal recessive
26. In mRNA immature termination occurs from codon UAA. Which of the following transition mutation will lead to its reversal
    (a) Change at first A
    (b) Change at second A
    (c) Change at first U
    (d) All of these
27. The function of telomerase is
    (a) Synthesis of DNA at ends of chromosome
    (b) Synthesis at RNA primers
    (c) Reverse transcriptase of causing cancer
    (d) Replication of normal DNA

28. Which is NOT true in a process of succession from pioneering stage to climax stage?
    (a) Food chain to food web (b) Biomass increases
    (c) Wide niche (d) Dependency on detritus chain increases
29. Minimum resolution of electron microscope is
    (a) 0.1 mm (b) 0.2 nm (c) 20 nm (d) 100 nm
30. Vif gene is one of the important regulatory gene in lifecycle of HIV virus. It was observed that HIV virus with mutated Vif gene fails to culture of normal T-helper cells. the main reason is
    (a) Vif product targets the enzyme responsible for hypermutation to ubiquitination and cellular degradation
    (b) Vif gene help in insertion of viral genome into host genome
    (c) Vif gene help in reverse transcription process
    (d) Vif gene help in insertion of viral genome into host genome
31. A cross between hens with different comb shape was carried as shown in fig.

Rose × Pea

Walnut × Walnut

Walnut RosePea Single

The conclusion which can be drawn is
    (a) Single gene is involved for comb shape
    (b) Two independent segregating genes are involved
    (c) Four independent segregating alleles are involved
    (d) Two genes are showing epistasis
32. Which of the following is vascular plant pathogen?
    (a) *Fusarium* (b) *Claviceps* (c) *Botrychium* (d) *Verebarium*
33. Which of the following cell has been proved in elucidation of lipid bilayer of plasma membrane?
    (a) RBC (b) Virus (c) Bacterial cell (d) Kidney cell
34. Which antibiotic is responsible for premature termination of translation in bacteria?
    (a) Penicillin (b) Tetracycline (c) Chloramphenicol (d) Puromycin
35. Hardy-Weinberg law will NOT operate under conditions, when
    (a) Skewed sex ratio
    (b) Weak selection of one of the allele
    (c) 3 alleles are involved
    (d) Mutated allele is not involved in sexual selection
36. Motile zoospores occurs in
    (a) Ascomycetes (b) Basidiomycetes

(c) Zygomycetes (d) Chytridiomycetes

37. Mendel during his experiment on garden pea observed F_2 ratio of 3:1 on selfing round seed plants where 3 were round and 1 was wrinkled. If all round seed plants are selfed to raise the F3 generation the observed ratio will be
    (a) Of plants with round seed and with wrinkled seeds
    (b) All plants with round seeds
    (c) Half of plants with round seed and rest with wrinkled seeds
    (d) 1/3 of plants with round seeds and 2/3 of plants bearing round and wrinkled seeds

38. In an experiment bacterial cell was removed using enzymatic methods. It was observed that on culturing such bacteria, for many generations cell wall was not seen. It is proof to
    (a) Recessive mutation (b) Inheritance of acquired character
    (c) Reverse mutation (d) Adaptive mutation

39. Among the following which DNA polymerase lacks proof reading activity?
    (a) DNA polymerase ε (b) DNA polymerase δ
    (c) DNA polymerase α (d) DNA polymerase α and ε

40. Tigers do not occur in Srilanka while they are seen in India. While the leopards seen in both India and Srilanka. The main reason is
    (a) Tigers are not good swimmers
    (b) Srilankans have removed tiger due to excessive hunting
    (c) India care more for tigers
    (d) Leopard originated before separation of Indian and Srilanka due to plate shifting

41. Among the following maximum gross productivity is observed at
    (a) Cold deciduous forests (b) Boreal forests
    (c) Temperate conifers (d) Temperate deciduous

42. Which statement is correct for DG = 0
    (a) Reaction is in equilibrium (b) Operates only under constant P and V
    (c) Follows PdV equation (d) All of these

43. In prokaryotes which subunit of RNA polymerase is released after initiation process?
    (a) σ (b) β' (c) β (d) α

44. Which biogeographical region covers maximum part of Indian?
    (a) Semi arid (b) Himalaya
    (c) Deccan Peninsular (d) Gangatic plains

45. In prokaryotes IF2 binds to
    (a) Amino acyl tRNA and ATP (b) Initiator tRNA and GTP
    (c) Ribosome and mRNA (d) mRNA and 30S subunit of ribosome

46. ABC transporter are NOT involved in
    (a) Accumulation of pigment in plant vacuole
    (b) Secretion of mattering factor in yeast
    (c) Transport of chloroplast content for degradation in vacuole
    (d) Uptake of phosphate by bacteria
47. Major function of lymphatic system is
    (a) Phagocytosis (b) Innate immunity
    (c) Maintain fluid balance (d) Acquired immunity
48. Which of the following enzyme of nitrogen metabolism is located in plastid?
    (a) Aspartate synthetase (b) Nitrate synthetase
    (c) Nitrate reductase (d) Nitrogenase
49. Hypertrophy in anterior pituitary in adults leads to
    (a) Gigantism (b) Dwarfism (c) Acromegly (d) Cretinism
50. *E. coli* gal+ and bio+ genes are located very near to each other on genome. A bacterium with F' gal+ on conjugation with F will lead to diploid condition for
    (a) bio only (b) gal only
    (c) gal and bio both (d) *E. coli* will always remain haploid
51. What is basic assumption for using molecular clocks for estimation of molecular evolution?
    (a) Rate of evolution is constant over time
    (b) Rate of neutral mutation is constant over time
    (c) DNA do not show variations among organism
    (d) Proteins are better marker for phylogeny as compare to DNA
52. Uricotellism is an adaptation for
    (a) Marine habitats (b) Conserving salts (c) High altitudes (d) Conserving water
53. Which instrument is used to measure electrical activity of heart?
    (a) Sphygmomanometer (b) Electroencephalogram
    (c) Electrocardiograph (d) Electrocardiogram
54. In brain meninges are absent at
    (a) Dura mater (b) Pia mater (c) Grey mater (d) Arachnoid mater
55. Which molecule does not have SP3 hybridization?
    (a) CH_4 (b) NH_3 (c) CH_2O (d) H_2O
56. Phytochromes involved in red/far-red response are dimeric chromoproteins. The two sub-units of phytochromes are linked with
    (a) Kinase domain (b) PTB domain

(c) PAS domain (d) Hinge region

57. Which statement is NOT correct regarding brown fat?
    (a) It produces more heat
    (b) It is present in lesser amount in new born baby
    (c) It produces little ATP
    (d) In mitochondria electron transport chain and phosphorylation are uncoupled

58. Which of the following is a species protein in sea urchin which plays important role in reorganization during acrosomal reaction?
    (a) Bindin (b) Avidin (c) Fertilin (d) Cortical granule

59. Aggregation of myxamoeba into slime mould during formation of fruiting body is an example of
    (a) Multicellularity (b) Embryonic development
    (c) Patter formation (d) Social behavior

60. Secondary structure of RNA is stabilized by hydrogen boding between
    (a) GC and AU (b) GC and AT
    (c) GC, AU and GU (d) GC only

61. Most of DNA binding protein binds to DNA by particular motif to modulate gene expression. Genes which are under regulation of gibberellic acid have GRE where, GREB binds. The motif in GREB is
    (a) Leucine zipper (b) b ZIP (c) Zinc finger (d) Homeo domain

62. Explain origin of life the first biomolecule generated under lab condition were
    (a) Lipids (b) Amino acids (c) Carbohydrates (d) Nucleic acids

63. In an experiment approach trp operon and lac operon were fused. Under what condition there would be expression of β-galactosidase?
    (a) Low lactose and glucose (b) High lactose and glucose
    (c) High tryptophan (d) Low tryptophan

64. Origin of tropical weed *Eupatorium odoratum* is
    (a) Ethopia (b) Peru (c) Mexico (d) Brazil

65. When a new sapling is transferred to soil, roots have very loose contact with soil particles and it suffers from water stress because of
    (a) Loss of water from roots to soil (b) Cavitation
    (c) Positive hydraulic pressure (d) Capillary action

66. How many H atoms would be replaced CCl_4 in presence of ethanol during NMR spectrum
    (a) 1 (b) 2 (c) 3 (d) 4

67. In bioinformatics for structural proteomics, the structure of protein can be determined from
    (a) NIH (b) EMBL (c) PDB (d) Gene bank

68. Which selectable marker gene is routinely used for selection of transgenic plant?
   (a) Amphicillin (b) Tetracyclin (c) Hygromycin (d) Carbenicillin

69. Reversal of sexual dimorphism is observed under condition
   (a) Strong female choice
   (b) Skewed sex ratio
   (c) Male and female changes their roles
   (d) Number of males are very low

70. Which of the following diseases or pathogen have been completely eradicated from India?
   (a) Small pox, Polio
   (b) Small pox, Guinea worm
   (c) Yellow fever, Plaque
   (d) Sleeping sickness, Yellow fever

71. According to Holiday model if markers are present outside of crossover point, then recombinant molecules would be generated when
   (a) There is no resolution
   (b) Nicks on outer strand during resolution
   (c) Nick is on inner strand during resolution
   (d) Always recombinant would be produced

72. The lateral separation of amphibian embryo at two celled stage will result in
   (a) Single embryo
   (b) Two embryos joined at belly region
   (c) Identical twins
   (d) Two embryos missing various organs

73. In which cancer treatment erb-B antibodies are used
   (a) Prostate (b) Oral (c) Lung (d) Breast

74. In plants gene 'sepallata' is NOT involved in formation of organ
   (a) Sepals (b) Petals (c) Stamens (d) Carpels

75. If hydra is fragmented into various parts, separate group of cells repattern themselves into various small hydra. Such an mode of development is termed as
   (a) Regeneration
   (b) Morphylaxis
   (d) Epimorphogenesis
   (d) Morphylaxix and epimorphogenesis

76. *E. coli* cells were transformed with plasmid carrying tetracycline resistance gene and later on plated on LB medium and various colonies were observed but same number of colonies were observed for control non-transformed *E. coli* cells. the probable reason is
   (a) No transformation occurred
   (b) *E. coli* cells were already resistance to tetracycline
   (c) Bacteria are plated just after heat shock and incubation with plasmid
   (d) Wrong plasmid was used

77. An embryo lacking biocoid is injected with biocoid mRNA at middle portion. It will result into
   (a) Normal phenotype
   (b) Two heads and no tarsons
   (c) Head middle and tarsons at both end
   (d) No head and tarson at both ends

78. In an experimental population the birth rate is 18 per 1000 and death rate is 14 per 1000. Size of population is 10,000 at time, 't' then what will be size of population time 't + 1'

(a) 10,000 (b) 10,040 (c) 10,140 (d) 11,040

79. Which is the best method for screening of plants for bioprospecting?

(a) Ecological studies
(b) Ethno driven
(c) Comparison of DNA sequence
(d) Comparison of protein sequence

80. Which plant has been used in phytoremediation for uptake of cadmium from contaminated soil?

(a) *Brassica juncea*
(b) *Oscimum basalicum*
(c) *Silene vulgaris*
(d) None of these

81. The major function of surfactant protein secreted by lungs is

(a) Helps in absorption of oxygen
(b) Stops entry of dust particles into lungs
(c) Keeps alveoli inflated
(d) Do not allow growth of bacteria

82. Anterior-posterior limb axis and dorsal verntral nerve plate is determined by

(a) Bicoid
(b) Cactus
(c) Pax3
(d) Sonic hedgehog

83. Which is NOT a strategy for adaptation to dry environment?

(a) C_4/CAM cycle
(b) Low stomatal conductance
(c) High hydraulic conductance
(d) High root to shoot ratio

84. On a Ramchandran plot glycin, D-Alanine and D- Asp were plotted. Which statement is correct among the following

(a) D-alanine will occupy greatest area
(b) D- alanine and Asp will occupy similar area
(c) Smallest area would be occupied by D-Asp
(d) All amino acids will occupy greatest area

85. Which enzyme is used in estimation of glucose from blood sample using biosensors?

(a) Alkaline phosphatase
(b) Glucose oxidase
(c) Glycogen phosphatase
(d) Glucokinase

86. Most of nutrients are absorbed from intestine by co-transport. Which of the following mineral plays important role in transport of amino acids across intestine epithelium?

(a) K (b) Ca (c) Na (d) Na and Ca both

87. In which case statistical approach analysis of variance can be applied

(a) Body weight is dependent on person height
(b) Diabetes dependent on concentration of insulin
(c) Productivity of lake is not significantly different from other five lakes
(d) Seed weight is influenced by calcium in soil

88. Linking number of DNA is 9 with writhe -1, then number of twists is
    (a) 9 (b) 10 (c) 11 (d) 12
89. Which statement explains the phenomenon of character displacement?
    (a) Organism of different species look different where habitat does not overlap while looks entirely similar where their habitat overlaps
    (b) Organisms of different species are morphologically similar in different habitat
    (c) Organisms phenotypically similar are genetically different
    (d) Organism develops various new characters in different habitats
90. Root cap cells originated from
    (a) Root epidermis (b) Columella cells
    (c) Quiescent centre (d) Vascular cambium
91. Meiosis produce, n + 1 and n – 1 gametes. The probable reason is
    (a) Non-disjunction during metaphase I (d) Non-disjunction during metaphase II
    (c) Non-disjunction during metaphase I & II (d) Non-disjunction during anaphase I & II
92. Chromatids appear in form of dyads during
    (a) Metaphase of mitosis (b) Metaphase of meiosis
    (c) Anaphase of mitosis (d) Prophase of mitosis

CSIR-NET Exam. December – 2009 (based on memory)

1. Among the following which amino acid do not absorbs wavelength of 250-300 nm
   (a) Cystine (b) Histidine
   (c) Phenyl alanine (d) Tryptophan
2. Vinblastin has been extensively used for treating cancer. This is an example of
   (a) Heat therapy (b) Radiotherapy
   (c) Chemotherapy (d) Surgery
3. Transport of ions across membrane depends on
   (a) Concentration gradient
   (b) Membrane potential
   (c) Concentration of gradient and membrane potential
   (d) Independent of both
4. Among the following which is not a function of hydrogen peroxide release during plant stresses response?
   (a) Crosslinking glycans in cell wall (b) Production of jasmonic acid

(c) Lignin deposition (d) Production of ethylene and salicylic acid

5. Promoters for RNA polymerase III are located at

(a) +1 to +10 (b) -35 to -10
(c) Within transcribed sequence (d) Downstream after termination

6. When tryptophan in excess most of times RNA polymerase dismount after transcription of first 150 nucleotide in trp operon. This is termed as

(a) Feedback inhibition (b) Catabolite repression
(c) Antitermination (d) Attenuation

7. Which of the following is necessary for transport of mRNA from nucleus

(a) RNA editing (b) 5'-capping
(c) 3'-capping (d) Secondary structure

8. Which of the following can be regarded as programmed cell death?

(a) Death induced by toxin (b) Death by inflammation
(c) Death of cells during normal development (d) Death due to phagocytosis

9. Under which phase of bacterial growth bacteria increase is size but do not divide

(a) Log phase (b) Lag phase (c) Stationary phase (d) Death phase

10. 'Taq' enzyme utilized in PCR is a

(a) DNA polymerase (b) Reverse transcriptase
(c) RNA polymerase (d) Ligase

11. Among the following highest assimilation efficiency is observed in

(a) Carnivores (b) Omnivores (c) Herbivores (d) Microbivores

12. Which of the following is not an extracellular matrix protein?

(a) Lamin (b) Collagen (c) Fibronectin (d) Albumin

13. Among the following which radioisotope is not an emitter-

(a) H_3 (b) I_{125} (c) C_{14} (d) P_{32}

14. If organism is at very high risk of extinction according to IUCN, then it is kept in category of

(a) Endangered (b) Rare
(c) Vulnerable (d) Critically endangered

15. The best technique for analyzing total mRNA

(a) DNA hybridization (b) Northern analysis
(c) Southern analysis (d) RNA *in situ* hybridization

16. Which statement is correct regarding edge effect

(a) They are rich in diversity (b) They are poor in diversity
(c) Low competition (d) High predation pressure

17. Which statement is not true about *E. coli* DNA ligase?
    (a) NAD is source of AMP as cofactor (b) DO not link single stranded DNA
    (c) Links double stranded blunt ends (d) Requires ATP as energy source
18. Cell with rigid lignified cell wall and dead protoplasm is
    (a) Collenchyma (b) Sclerenchyma (c) Chlorenchyma (d) Companion cells
19. Which of the following is not a co-dominant marker
    (a) SNP (b) RFLP (c) RAPD (d) SMPLs
20. Among the following which is not an assumption of Hardy-Weinberg rule
    (a) Random mating (b) Small population size
    (c) No natural selection (d) No mutation
21. Which of the following are abiotic factors?
    (a) Temperature, rainfall, pH, parasites
    (b) Temperature, rainfall, pathogens
    (c) Temperature, rainfall, pH, viruses
    (d) Temperature, rainfall, ph, soil
22. During the process of succession arrival of late successional stage depends on environment modified by earlier successional stage. The process is referred as
    (a) Co-evolution (b) Inhibition (c) Facilitation (d) Tolerance
23. The ecosystem having longest energy transfer time is
    (a) Tropical rain forest (b) Desert
    (c) Temperate deciduous forest (d) Open ocean
24. Phonetic classification is based on
    (a) Phylogenetic relationship
    (b) Genetic relationship
    (c) Overall similarity of characters and gaps between variations
    (d) Anatomical and embryological characters
25. Which of the following molecule can be utilized for establishing early evolutionary process?
    (a) Mitochondrial DNA (b) Chloroplast DNA
    (c) Nuclear DNA (d) Ribosomal RNA
26. The term used for bubble like structure generated during early process of origin of life by Oparin is
    (a) Micelles (b) Probiont (c) Coacervates (d) Protobionts
27. Which gas was absent during pre-biotic environment?
    (a) CH_4 (b) SO_2 (c) CO_2 (d) O_2

28. Toll like receptors are a type of pattern recognition receptor (PRR) and recognize molecules that are broadly shared by pathogens but distinguishable from host molecules, collectively referred as pathogen-associated molecules patterns. They are
    (a) Present only in mouse
    (b) Present on membrane of ER
    (c) Present on cytosolic face of plasma membrane
    (d) Are transmembrane protein

29. According to the polymer trap hypothesis small sugars such as sucrose are converted to raffinose and other oligosaccharides is loaded in phloem. major site of synthesis of raffinose is
    (a) Transfer cells (b) Companion cells
    (c) Sieve tube (d) Intermediary cells

30. Uptake of mineral like zinc, magnesium and iron across the membrane in plants is by
    (a) ABC transporter (b) H^+ co-transporter
    (c) ATP dependent transporter (d) ZIP transporter

31. Major transport of nitrogen in xylem sap in form of
    (a) Glutamate (b) Allantoin (c) Glutamine (d) Ammonia

32. Plants dissipate excess excitation energy as heat so as to protect from photo-oxidative damage. The mechanism is known as
    (a) Photo chemical quenching (b) Non-photochemical quenching
    (c) Photoinhibition (d) Merven effect

33. Which lipid is found exclusively on one face of membrane?
    (a) Phosphotidylethanolamine (b) Phosphotidyl inositol
    (c) Cholesterol (d) Phosphotidyl choline

34. Chaperons (Hsp 70) are absent in
    (a) Mitochondria (b) Chloroplast
    (c) Golgi bodies (d) Endoplasmic reticulum

35. Prolamellar body are present in
    (a) Chloroplast (b) Leucoplast (c) Chromoplast (d) Etioplast

36. Three classes of genes A, B and C regulates the development of flower in *Arabidopsis*. If a loss-of-function mutation occurs in the B-type genes, what will be the composition of the flower whorls?
    (a) Sepals-petals-stamens-carpels (b) Sepals-sepals-stamens-carpels
    (c) Sepals-sepals-carpels-carpels (d) Petals-petals-stamens-stamens

37. In regulative development, the prospective potency of cells
    (a) Equal to prospective fate (b) Lesser than prospective fate
    (c) More than prospective fate (d) Not determined

38. T_4 bacteriophage after infecting *E. coli* generally hacks host machinery for transcription of its own genes. It is done by
    (a) Modifying host RNA polymerase (b) Degrading host RNA polymerase
    (c) Synthesis of own RNA polymerase (d) Degradation of host genome
39. Morphylaxis can be defined as
    (a) Reinitiation of cell division in existing cells, followed by repatterning of those cells
    (b) Production of lost organ by division in remaining cell
    (c) Production of complete organism by single cell
    (d) Movement of organism toward stimulus
40. Which thermodynamics property cannot be directly measured in cell?
    (a) Free energy (b) Enthalpy
    (c) Temperature (d) Entropy
41. Which of the following represents the gametophyte generation in plants?
    (a) Megaspore (b) Egg (c) Ovule (d) Embryo sac
42. Bacterial two component system includes
    (a) Sensory kinase and phosphotransferase (b) Signal and receptor
    (c) Sensory kinase and phosphotransferase (d) Stimulus and response
43. Different strains of viruses can be identified by
    (a) Observing symptoms of diseases in plants (b) Electron microscopy
    (c) Fluorescence microscopy (d) PCR
44. The virus inserted in genome can be recognized by
    (a) Microarray (b) FISH
    (c) Southern blot (d) Northern blot
45. The major function of type-III secretion by pathogenic bacteria is
    (a) Release signal for quorum sensing (b) Efflux of drugs
    (c) Release of virulence factors (d) Release of competence factors
46. If cell is not dividing (arrested in cell cycle) which repair mechanism will not occur
    (a) Excision repair mechanism (b) Recombination repair mechanism
    (c) Transcriptional coupled repair mechanism (d) DNA synthesis annealing repair
47. Which organelle has characteristic galactolipids in its membrane?
    (a) Golgi body (b) Endoplasmic reticulum
    (c) Mitochondria (d) Chloroplast
48. In formaldehyde the pure orbitals involved in bonding between C and O is
    (a) Only C (b) Only O
    (c) Both C and O (d) H, C and O

49. Which statement is correct for globular proteins?
    (a) Always contain α helix
    (b) Contain β sheets
    (c) Contain β pleated sheet
    (d) Turns

50. *Agrobacterium tumefaciens* causes crown gall disease in dicot plants. which phytohormone genes are present in T-DNA
    (a) Auxin only
    (b) Cytokinin only
    (c) Auxin and cytokinin
    (d) Cytokinin and brassicosteroids

51. *E. coli* based Humulin is a
    (a) Interferon
    (b) Insuline
    (c) Growth factor
    (d) Disacchride

52. An enzyme has Glu76 and Asp_{52} at active site. The pl for Glu is 5.6 and for Asp is 4.5. The enzyme function when Glu is in protonated form and Asp in deprotonated form. The pH where enzyme will show maximum activity will be
    (a) 4.5 (b) 5.2 (c) 5.05 (d) 10.1

53. If plant with genotype AaBb is self pollinated where the A and B are not linked, then the probability of getting AABB genotype will be
    (a) 1/2 (b) 1/4 (c) 1/8 (d) 1/16

54. During gamete formation alleles which do not undergo recombination segregate during
    (a) Mitosis (b) Meiosis-I (c) Meiosis-II (d) Cleavage

55. Two different mutant of Drosophila gives a black body colour. When these mutants are crossed all progeny have wild type colour. It means mutation are
    (a) Co-dominant (b) Allelic (c) Epistatic (d) Non allelic

56. Bending of coleoptiles tip of oat toward source of unilateral light wavelength 454 nm is due to
    (a) Polar transport of auxin
    (b) Degradation of auxin toward light
    (c) Lateral distribution of auxin toward shaded area
    (d) Synthesis of auxin in shaded area

57. Among the following which is terpene
    (a) Xanthophyll (b) Lycopene (c) Chlorophyll (d) Coumarin

58. Among the following which plant removes heavy metal from water
    (a) *Nymphia vishin*
    (b) *Pistia stratiotes*
    (c) *Eichornia crassipes*
    (d) *Salvia officinalis*

59. Polygenic traits in crops can be identified by
    (a) Cluster analysis
    (b) Gene mapping
    (c) Tandem array analysis
    (d) QTL mapping

60. A *Neurospora* strain has start and stop growth behavior. The mutated gene was found to be on mitochondria. If female *Neurospora* having stop mutation is crossed with wild type male *Neurospora*. Phenotype of progenies will be

    (a) All wild type (b) All start and stop mutant
    (c) Majority of wild type (d) Majority of start and stop mutant

61. Inversion is termed as crossover suppressor because

    (a) Crossing over within an inversion loop, when it does occur, leads to deleted and duplicated crossover chromosomes and mortality of zygotes carrying them
    (b) No crossing over in the inversion loop
    (c) Crossing over lead to formation of all acentric chromosomes
    (d) Segregation of chromosomes is not normal

62. Unique character of family Caryophylaceae is presence of

    (a) Terpenes (b) Glycosides (c) Saponins (d) Alkaloids

63. Which of the following is not a core angiosperm

    (a) Austobaileyales (b) Amborellales (c) Nymphaeales (d) Magniolales

64. Among the following which amino acid has two buffering zone?

    (a) Glycine (b) Alanine (c) Cysteine (d) Glutamine

65. Functional response of predators means

    (a) Killing of prey population
    (b) In response to prey, the killing strategy developed by predator
    (c) Regulation of predator population by availability of prey
    (d) Choosing prey depending on density of prey

66. Which of the following organism have been extensively used in generation of transgenic plants

    (a) *Bacillus thuringensis* (b) *Agrobacterium tumefaciens*
    (c) *E.coli* (d) *Baculo viruses*

67. For making transgenic animals in fertilized egg the best place to insert trans-gene is in

    (a) Female pronuclei (b) Male pronuclei
    (c) Cytoplasm (d) Cleavage cells

68. To determine variation in wing length of butterfly from five different places which would be best statistical test?

    (a) Chi-square (b) Student t-test
    (c) F-test (d) Regeneration analysis

CSIR-NET EXAM. JUNE - 2009 (BASED ON MEMORY)

1. Red wine and red grapes are important source of which anti-tumour agent
   (a) Bardysanin (b) Taxol (c) Cincristine (d) Resveratol
2. Molecular marker cannot be utilized for
   (a) Identifying the clones
   (b) Mapping of genes
   (c) Identifying the locus of gene on chromosome
   (d) Indentifying the expressed product
3. Sendai virus enters host cell by
   (a) Endocytosis (b) Phagocytosis
   (c) Cell fusion (d) Receptor mediated endocytosis
4. Which function is not related with Th1 cells?
   (a) Secretion of IL-2
   (b) Promoting antibody binding to soluble antigens
   (c) IFN-γ
   (d) Induce phagocytosis
5. Sxi genes of *Drosophila* regulate expression at
   (a) Transcriptional level (b) Translational level
   (c) Post translational level (d) Post transcriptional level
6. *Mycobacterium tuberculosis* is able to cause disease because it enters host cell it does not allow endosome to mature into
   (a) Lysosomes (b) Peroxisomes
   (c) Endoplasmic Reticulum (d) Golgi body
7. Cysteine Asparate protein kinases involved in process of apoptosis function as
   (a) Initiator and executioner (b) Initiator and inflammatory
   (c) Inflammatory and executioner (d) Initiator, inflammatory and executioner
8. Which of the following is a component of MAP kinase signal transduction pathway?
   (a) IP_3 (b) ERK (c) Protein kinase B (d) JAK kinase
9. If parents with genotype AABBccddeeFF and aabbCCDDEEff are crossed, the genotype of resulting progeny will be
   (a) AABBccDDeeFf (b) AaBbCcDdEeFf
   (c) aaBBccDDeeFF (d) AaBbCCddEeFf

10. Which kinase activity is associated with phytochrome photoreceptors responsible for Red/Far red response?

(a) Histidine (b) Ser/Thr kinase (c) Tyrosine (d) Aspartate

11. Fas protein involved in cell mediated immune response

(a) Have death domain (b) Act as inducer
(c) Generate G protein (d) Inhibit apoptosis

12. Bubonic plaque caused by *Yersinia pestis* cannot be eradicated completely because

(a) Causal organism does not express surface antigens
(b) Causal organisms cannot be cultured *in vitro*
(c) Antibodies are not generated by causal organism
(d) Y. pestis have broad host range

13. Small amount of lethal mutation always tend to remain in population is due to

(a) Positive selection (b) Negative selection
(c) Mutation selection balance (c) Frequency dependent selection

14. Distance between the two linked genes A and B is 20 cm on test cross of with recessive parent how many offspring will have genotype

(a) 10 (b) 20 (c) 40 (d) 80

15. Air inhaled during breathing contains principle gases in order N2>O2>CO2>H2. The gases in exhaled air would be in order

(a) $N_2>CO_2>O_2>H_2$ (b) $N_2>H_2>CO_2>O_2$
(c) $N_2>O_2>CO_2>H_2$ (d) $N_2>CO_2>H_2>O_2$

16. The glycocalyx around cell membrane can be determined by

(a) Iodine (b) Saffranin (c) Methylene blue (d) Lectins

17. In *Neurospora crossa* tetrad analysis showed following result + : m :: 6 : 2. The phenomenon involved for above result would be

(a) Branch migration (b) DNA replication
(c) Holiday junction (d) Strand exchange

18. During evolution increased ornamentation in male is a result of

(a) Directional selection (b) Co-evolution
(c) Sexual selection (d) Natural selection

19. Maximum possible isomers for glucose are

(a) 4 (b) 8 (c) 16 (d) 32

20. Which of the following is NOT a property of an enzyme?

(a) From complex with substrate (b) Increase rate of reaction
(c) Decrease Gibb's free energy (d) Decrease activation energy

21. Which era is characterized by dramatic diversification among eukaryotes?
    (a) Triassic (b) Devonian (c) Carboniferous (d) Cambrian
22. The Mendelian law of independent assortment is due to arrangement of chromosome during
    (a) Anaphase-I (b) Anaphase-II (c) S-phase (d) Cytokinesis
23. Among the following most variable stage of cell cycle is
    (a) G1 (b) S (c) M (d) G2
24. An organism influence the evolutionary pace of the other organism in
    (a) Co-evolution (b) Parallel evolution
    (c) Divergent evolution (d) Convergent evolution
25. It has been observed that during prolong animal cell culture and differentiation cell tends to stop dividing. They are said to be in
    (a) Senescence (b) Quiescent (d) Apoptosis (d) G1
26. Type of mutation which is most suitable for study of regulation of cell like DNA replication is
    (a) Gain of function (b) Loss of function
    (c) Suppressor mutation (d) Conditional mutation
27. A population of 200 is in Hardy- Weinberg equilibrium with allele frequency of 'A' = 0.7 and 'a' = 0.3. The number of carriers in population will be
    (a) 98 (b) 84 (c) 42 (d) 18
28. ABC transporter in plants which are responsible for detoxification of Xenobiotics and prevent oxidative damage are located at
    (a) Tonoplast (b) Peroxisome
    (c) Endoplasmic reticulum (d) Plasma membrane
29. Exponential growth in bacteria would be expected during
    (a) Log phase (b) Lag phase
    (c) Stationary phase (d) Deceleration phase
30. Which technique is most suitable to study transcription factor and its binding site?
    (a) Northern blotting (b) Western blotting
    (c) DNAse I foot printing (d) Microarray
31. Which of them is not utilized for comparison of operational taxonomic unit (OTU) in numerical taxonomy?
    (a) Unweighted pair group method (b) Percentage similarity
    (c) Jaccarrd coefficient (d) None of these
32. Thylakoid membrane has lateral asymmetrical positioning of photosystem in chloroplast. Which statement is correct?
    (a) PS-I in non appressed portion and PS-II in appressed portion
    (b) PS-II in non appressed portion and PS-I in appressed portion

(c) Both PS-I and PS-II in appressed portion
(d) Both PS-I and PS-II in non appressed portion on thylakoid

33. Which organelle requires intact membrane system for ATP synthesis?
(a) Mitochondria (b) Chloroplast
(c) Endoplasmic reticulum (d) Chloroplast and mitochondria

34. Which statement is correct regarding concogenes?
(a) They are viral genes
(b) They are mutated viral genes
(c) They are muted form of genes controlling cell division
(d) They suppresses tumors

35. The movement of chloroplast is mediated by
(a) Dynein (b) Kinesin (c) Actin (d) Myosin

36. A sample is no normal distribution ranging from (μ+1ó) to (μ+2ó). The data in range would be
(a) 17 (b) 67 (c) 50 (d) 98

37. The flagellin protein is associated with
(a) Bacteria (b) Protist (c) Virus (d) Eukaryotic cell

38. Starch filled plastids are responsible for geotropism in columella cell beneath the root cap. They are termed as
(a) Elioplast (b) Amyloplast (c) Chloroplast (d) Proplastid

39. Among the following which is not a result of acid rain
(a) Low amount of phosphate availability (b) Low amount of aluminium availability
(c) Low availability of nutrients to plants (d) Increased acidity of soil

40. Organ identity genes are responsible for correct positioning of floral organs on floral meristem. Mutation in them will lead to
(a) Loss of organs from certain whorls
(b) Appearance of organs at incorrect positions
(c) More number of organs in certain whorls
(d) No flower

41. Glycolysis of protein occurs in
(a) Endoplasmic reticulum (b) Golgi body
(c) Mitochondria (d) Nucleus

42. Generally organism tends to remain in realized niche. Under what condition realized niche can be greater than fundamental niche
(a) Abundance of resources
(b) One species helping other in utilization of resources

(c) Heterogeneity of resources
(d) Moving of organism from source to new sink area

43. Among the following which antibiotic will inhibit protein synthesis in chloroplast?
(a) Cyclohexamide (b) Chlorophenicol
(c) Rifampin (d) Ricin

44. Worldwide maximum cultivated transgenic crop is
(a) Insect resistance cotton (b) Herbicide resistance soybeans
(c) Growing plant for desired molecules (d) Edible vaccines

45. According to survival of fittest concept of natural selection one species out compete other species. Under such condition not two species can coexist in same niche but more than one species can live in same niche under condition
(a) High competition (b) Abundant resources
(c) Marginal overlapping (d) Utilization of different resources

46. The inbreeding coefficient of offspring on marriage between brother and sister sibling will be
(a) 0.5 (b) 0.05 (c) 0.25 (d) 0.75

47. Perennial habit among trees would be more preferred under conditions
(a) Low survival during sapling stage and high during adult
(b) Low survival during sapling stage and low during adult
(c) High survival during sapling stage and high during adult
(d) High survival during sapling stage and low during adult

48. Most of trees in Indian tropical forest belongs to family
(a) Arecaceae (b) Fabaceae
(c) Dipterocarpaceae (d) Bromeliaceae

49. Which could be suitable for constructing the genomic library 70 kb of DNA
(a) YAC (b) BAC (c) Cosmid (d) P1 based vector

50. Scientific names of bacteria, fungi, plants and animals are given by
(a) International Union of Biological Nomenclature
(b) There is different organization for naming plants and fungus
(c) Names of plants and animals are given by same organization
(d) There are three different organization for naming bacteria, plant and animals

51. If we want to obtain glycosylated protein from microbe. Suitable choice will be
(a) Bacteria (b) Yeast (c) Mycoplasma (d) Animal cell

52. Which is correct for termination of transcription in eukaryotes?
(a) Terminates prior to polyadenylation (b) Terminates during polyadenylation
(c) Terminates after polyadenylation (d) Forms hair pin loop

53. A gene consists of two introns and a 5' UTR region, the probable number of exon will be

(a) 2 (b) 3 (c) 5 (d) 4

54. During germination of seeds, after imbibitions of water first step would be

(a) Mobilization of reverse food (f) Cell division
(c) Transcription of specific genes (d) Embryo differentiation

55. Heritability due to genetic variance for a trait of importance is 0.2. Which would be most appropriate approach to select trait in next generation in a short time?

(a) Pedigree selection (b) Mass selection
(c) Family selection (d) Selection by progeny testing

56. Which statement is correct regarding functioning of topoisomerase

(a) Separate double stranded DNA (b) Act as primer
(c) Renaturate the single stranded DNA (d) Attach to super coiled DNA and relax it

57. A protein specially abundant in desiccated seeds and also help in osmotic adjustment

(a) LEA (b) Globin (c) Hsp (d) α amylase

58. What would be effect of photosynthesis in C_3 and C_4 plants on elevating the concentration of CO_2 under light saturated conditions?

(a) C_3 plants will saturate fast and C_4 plants remain unaffected
(b) C_4 plants saturate fast and C_3 plants remain unaffected
(c) Both types of plants will saturate fast
(d) No effect on both types of plants

59. The essential mineral required for cell adhesion protein cadherin is

(a) Iron (b) Sodium (c) Calcium (d) Magnesium

60. Common metabolites in nucleotide biosynthesis from glucose by pentose phosphate pathway is

(a) Glyceraldhyde -3-phasphate (b) Fructose-6-phosphate
(c) Phospho-ribosyl pyrophosphate (d) Di hydroxyl acetone phosphate

61. A gene consists of two introns and a 5' UTR region, the probable number of exon will be

(a) 5 (b) 4 (c) 3 (d) 2

62. Anticodon sequence lies in

(a) DNA (b) tRNA (c) mRNA (d) rRNA

63. Among the following which would be most suitable marker for selection for selection of animals with agronomic traits

(a) RFLP (b) RAPD (c) Mini-satellite (d) EST

64. In which technique O-Phenyl Diamine is used as chromogenic substrate

(a) ELISA (b) RIA (c) Southern blotting (d) Western blotting

65. Animal biologist generally uses Line-transect method for estimating density. It is based on assumption that
    (a) Organism will not move from marked transect
    (b) All organism are in straight line
    (c) Organism lack any competition
    (d) That animals on the line are seen

66. Which of the metabolite in nitrate assimilation is not located in the chloroplast?
    (a) Glutamine (b) Uric acid (c) Nitrite (d) Xanthin

67. In Northern hemisphere there is slow turnover of nutrient in terrestrial ecosystem as compare to southern hemisphere. The probable reason is
    (a) High rainfall in southern hemisphere
    (b) Soil is nutrient deficient in northern hemisphere
    (c) Plants are not good in uptake of nutrients in northern hemisphere
    (d) Temperature is low which is not suitable for nutrient recycling

68. In tissue culture experiment to initiated to shoots from undifferentiated mass of cell the medium must contain
    (a) Low auxin and high cytokinin (b) High auxin and high cytokinin
    (c) High auxin and low cytokinin (d) Low auxin and low cytokinin

69. In aroid plant the temperature of inflorescence rise around 8-10oC as compare to plant during maturation. It is due to activity of
    (a) Dehydrogenated (b) Cytochrome oxidase
    (c) Alternate oxidase (d) Peroxidase

70. Suppose a chromosomal aberration in a chromosome ABCDEF leads to ABCDEFGCDFEG. The probable reason is
    (a) Duplication and followed by EF inversion
    (b) Duplication followed by pericentric inversion
    (c) Only duplication
    (d) Only inversion

71. An aminopurine is attached to ribose sugar by N_9-C' glycosidic bond would be termed as
    (a) Nucleotide (b) Deoxyadenosine
    (c) Adenosine (d) Adenosine monophosphate

72. In knockout mice experiment germline transmission of gene A, null allele from a male chimera shows retarded growth of all mutant heterozygotes. On inbreeding animals produced the expected ratio of heterozygote pups but only 50% of heterozygote are with retarded growth of phenotype. These results are consistent with the following
    (a) Sex linked inheritance (b) Genomic imprinting
    (c) Cytoplasmic inheritance (d) Dominant effect

73. For constructing recombinant plasmid, plasmid and DNA to be inserted are digested with same restriction enzyme and kept in same reaction solution. To prevent self sealing of plasmids, which of the following enzyme is utilized?
    (a) Alkaline phosphate
    (b) Antibody for specific antigen tagged with toxin
    (c) Low immunogenic toxin
    (d) Anti-toxin
74. The model organism to study cell lineage is
    (a) *Xenopus* (b) Yeast
    (c) *Drosophila* (d) *Caenorhabdtis elegans*
75. What percentage of photo active radiation is actually utilized for photosynthesis by plants?
    (a) Lesser then 1% (b) 1-3% (c) 10-20% (d) >20%
76. Maximum absorption of UV light at wavelength 280 nm by a protein is due to
    (a) Aromatic amino acids
    (b) Aliphatic amino acids
    (c) Aromatic amino acids and peptide bond
    (d) Aromatic and aliphatic amino acid
77. Which molecule has property of self replication?
    (a) Protein (b) Carbohydrate (c) Lipids (d) Nucleic acid
78. If there are only 20 individuals in a population then as per IUCN would be kept under category
    (a) Extinct (b) Rare
    (c) Endangered (d) Critically endangered
79. Which of the following is negative regulator of trp operon
    (a) Lactose (b) C-AMP (c) Allolactose (d) Tryptophan
80. ? and ψ values for right handed á helix are expected to be
    (a) ? negative ψ negative (b) ? negative ψ positive
    (c) ? positive ψ negative (d) ? positive ψ positive

DBT – JRF – 2010 Exam (Based on Memory)

1. What happens first in case of cerebral damage?
   (a) Loss of proper posture (b) loss of smell
   (c) Loss of taste (d) None of these
2. Which of the following does not favour DNA denaturation?
   (a) Adding salt (b) Decreasing ionic strength
   (c) Heating (d) Treatment with alkali
3. A good inhibitor should have
   (a) Low enzyme-inhibitor dissociation constant
   (b) High enzyme- inhibitor dissociation constant
   (c) Both (a) and (b)
   (d) None of these
4. Which vitamin deficiency causes the symptoms of hemolytic anemia?
   (a) Vit. E (b) Vit. D (c) Vit. A (d) Vit.K
5. A discrete portion of a protein assumed to fold independently of the rest of the protein and possessing its own function-
   (a) Motif (b) Fold (c) Helix (d) Domain
6. If plant is used as a source of vitamin production, production would be maximum in
   (a) Organ (b) Callus
   (c) Suspension (d) Anther culture
7. What is the promoter used in pET vectors?
   (a) pET 32 is used (b) M13 helper phage
   (c) T7 promoter is used (d) None of these
8. Restriction enzymes in E. coli does not cut self DNA because they are
   (a) REC- (b) REC+ (c) DAM- (d) DAM+
9. Among the following seed pairing occurs in
   (a) tRNA (b) mRNA (c) rRNA (d) Micro RNA
10. Zooxanthallae is?
    (a) Blue Green Algae (b) Algae of corals
    (c) Both (a) and (b) (d) None of these
11. Which of the following protozoan parasite replicates inside lysosome?
    (a) Plasmodium (b) Taxoplasma
    (c) Trypanosoma (d) Leishmania

12. Phenyl acetic acid in penicillin production added as
    (a) Inducer (b) Inhibitor (c) Substrate (d) Precursor
13. In which of the following fermentation processes an inhibitor is used
    (a) Penicillin (b) Riboflavin (c) Tetracycline (d) None of these
14. Which of the enzyme is used in solid sequencing?
    (a) Ligases (b) Sequanases (c) Both (a) and (b) (d) None of these
15. Huntington's disease is caused by an expansion of the
    (a) Triplet sequence of amino acids GAG (b) Triplet sequence of amino acids CAG
    (c) Triplet sequence of nucleotides CAG (d) Triplet sequence of nucleotides GAC
16. In CATH 'C' stands for
    (a) Configuration (b) Conformation (c) Classification (d) Class
17. Which type of fishes reproduces in sea water and live rest of their life in fresh water?
    (a) Catadromous (b) Anadromous
    (c) Fresh water fish (d) Sea water fish
18. Which programme is used for phylogenetic analysis?
    (a) Bootstrapping (b) Probability
    (c) Ch square test (d) Null hypothesis
19. Proportion of A:T:C:G is 0.4:0.85:1.56:1 therefore it is
    (a) Purine rich DNA (b) Cruciform DNA
    (c) Double stranded DNA (d) Single stranded
20. Real time studies in plants is done with help of
    (a) YFP (b) GFP (c) GUP (d) Luciferase
21. Which of the following organism is used in Ames test?
    (a) *E. coli* (b) *Staphylococcus aureus*
    (c) *Pseudomonas aerugenosa* (d) *Salmonella typhimurium*
22. Pyrosequencing is so named because
    (a) It releases pyrophosphate while sequencing
    (b) It generated programs
    (c) Both (a) and (b)
    (d) None of these
23. In resting membrane the potential according to Nerst equation is closer to ions that are
    (a) Impermeable (b) Permeable
    (c) Least abundant (d) Most abundant

24. What factor does not affect oxygen transfer rate?
    (a) Ph (b) Agitation (c) Temperature (d) Antifoams
25. Which of the following process takes place during the formation of Disulphide Bridge?
    (a) Oxidation of sulfahydroxyl group (b) Reduction of sufahydroxyl group
    (c) Both reduction and oxidation (d) None of these
26. What is Marine snow?
    (a) Continuous detritus fall from upper layers
    (b) Ice crystals in upper layers
    (c) Due to marine algae
    (d) None of these
27. Which of the following should be used when the partition coefficient is very low
    (a) Counter current (b) Co-current
    (c) Multi batch extraction (d) Single batch extraction
28. Which of the following is not found in hnRNA?
    (a) Introns (b) Exons
    (c) Polycistronic coding (d) None of these
29. Which antibody can pass through placenta?
    (a) IgG (b) IgA (c) IgE (d) IgM
30. Which of the following is involved in RNA editing?
    (a) si RNA (b) hn RNA (c) sn RNA (d) mn RNA
31. Marker free plants can be generated by
    (a) Inversion (b) Deletion
    (c) Insertion (d) Co-transformation
32. What is serine protease activity site?
    (a) Glu-Asp-Met (b) His-Asp-Ser (c) Ser-Ser-His (d) His- Asp- Glu
33. Which algorithm uses once a gap is always a gap?
    (a) Smith Waterman (b) Needle Wnusch
    (c) FASTA (d) Clustlaw
34. During the treatment of Diabetes milletus why they are not providing glucose-6-phosphate?
    (a) It is unable to cross the cell membrane
    (b) Exogenous glucose-6-phosphate is toxic to cells
    (c) It is able to cross the cell membrane
    (d) None of these

35. What is the disadvantage involved in use of agitator in a bioreactor?
    (a) Increases bubble escape from the reactor
    (b) Increase residence time
    (c) Both (a) and (b)
    (d) None of these
36. What is the relationship between actual growth mass and apparent growth mass?
    (a) Actual growth mass is equal than apparent growth mass
    (b) Actual growth mass is less than apparent growth mass
    (c) Actual growth mass is greater than apparent growth mass
    (d) Actual growth mass is no way related to apparent growth mass
37. Which of the form irreversible enzyme substrate complex?
    (a) Uncompetitive and competitive inhibitor (b) Competitive inhibition
    (c) Non competitive inhibitor (d) Uncompetitive inhibition
38. If we use DNA Pol. instead of Taq Pol. in PCR then what can be the situation?
    (a) Add fresh enzyme after every denaturation step
    (b) Add DNA Pol. after every denaturation step
    (c) Both (a) and (b)
    (d) None of these
39. Which of the following RNA work in RNA editing?
    (a) rRNA (b) mRNA (c) tRNA (d) gRNA
40. Mice attain immunological maturity in
    (a) 2 weeks (b) 4 weeks (c) 6 weeks (d) 12 weeks
41. Which of the following lipids will work as signaling molecule?
    (a) Phospho 4-5 bis inositol (b) Phospholipase a2
    (c) Phospholipid sphingomylein (d) Phospholipid ethanolamine
42. Polyploidy is induced by
    (a) Auxin (b) Ornithine
    (c) Colchicine (d) None of these
43. Among the following which will not favour denaturation of DNA?
    (a) High Ca salts (b) High pH
    (c) Formide (d) Heat
44. Philadelphia chromosome is because of
    (a) Translocation (b) Deletion (c) Duplication (d) Inversion

45. Which of the following interaction is not found in Ab/Ag receptor/ligand interaction?
    (a) Hydrogen bond (b) Ionic bond
    (c) Covalent bond (d) Vander Waals interaction
46. What is the percentage of Head Space volume found in the reactor?
    (a) 50% (b) 20-25% (c) 10-15% (d) None of these
47. Dye injection in the plant cell will go in to the adjacent cells through
    (a) Tight junctions (b) Plasmodesmata
    (c) Desmosomes (d) Mesosomes
48. Which carbon in DNA is deoxygenated?
    (a) 1' position (b) 2' position (c) 3' position (d) 5' position
49. What is the sequence of enzyme used in RT PCR?
    (a) RNA dependent DNA Pol followed by RNA dependent DNA Pol
    (b) RNA dependent DNA Pol followed by DNA dependent DNA Pol
    (c) DNA dependent RNA Pol followed by DNA dependent DNA Pol
    (d) None of these
50. What is used for stabilization of the protoplast during protoplast fusion?
    (a) Mannose (b) Mannitol (c) Sucrose (d) Glucose
51. Which of the following cannot get digested by the gut flora?
    (a) Lignin (b) Pectin (c) Starch (d) Cellulose
52. Somatic mutation results in
    (a) Affinity maturity (b) Allelic exclusion
    (c) Class switching (d) Antibody diversity
53. Homology modeling is used for
    (a) Paralogs (b) Analogs (c) Homologs (d) Orthologs
54. Which pathway helps to protect cell from O_2 redicles?
    (a) TCA (b) C_4
    (c) Calvin (d) Pentose phosphate
55. The heat required for evolution of 1 kg saturated liquid is
    (a) Latent heat (b) Sensible heat (c) Specific heat (d) None of these
56. Which virus is well known for its antigenecity?
    (a) HIV (b) Influenza virus
    (c) Hepatitis C virus (d) None of these
57. Which antibiotics bind to 50S subunit of ribosome?
    (a) Penicillin (b) Puromycin

(c) Erythromycin (d) Chloromphenicol

58. Which of the following does not determine protein-protein interactions?
   (a) Far Western blotting
   (b) Chromatin immunoprecipitation
   (c) Bimolecular fluorescence complementation
   (d) Co-immunoprecipitation
59. If the operator region and the repressor regions are mutated in lac operon
   (a) b- galactosidase is produced even in the presence of lactose
   (b) b- galactosidase is not produced in presence of lactose
   (c) b- galactosidase is not produced in presence of glucose
   (d) b- galactosidase is produced even in the absence of lactose
60. Gene polymorphism detected by
   (a) Northern hybridization (b) Slot
   (c) Dot blot (d) Southern hybridization
61. How can you distinguish between linear and circular form of DNA?
   (a) Nuclease digestion (b) Exonuclease digestion
   (c) Endonuclease digestion (d) None of these
62. How many ATP involved in Na+ K+ pump and enzyme is
   (a) 2 ATPs and enzyme is tetramer (b) 2 ATP and enzyme is monomer
   (c) 1 ATP and enzyme is tetramer (d) 1 ATP and enzyme is monomer
63. What plant tissue culture technique is used to produce hybrid plants from male sterile plant species?
   (a) Pollen culture (b) Anther culture (c) Meristem culture (d) Organ culture
64. SDS binds to which domain of protein?
   (a) Negative charged amino acid (b) Positive charged amino acid
   (c) Hydrophobic side chain (d) No binding
65. What is clean gene technology?
   (a) Removing markers after use (b) Marker free transgenic plants
   (c) Conventional breeding without markers (d) None of these
66. Which is not a naturally competent bacterium?
   (a) *Haemophilus* (b) *Streptococus*
   (c) *Bacillus subtilis* (d) *E. coli*
67. When HFr strain is conjugated with F-, the result is
   (a) Always F^+ (b) Always HFr^+
   (c) Rarely F^+ (d) None of these

68. A type of structural chromosomal aberration is
    (a) Cri du chat syndrome (b) Edwards syndrome
    (c) Down's syndrome (d) Patau syndrome
69. In penicillin recovery pH is lower
    (a) To avoid contamination (b) For precipitation
    (c) For induction (d) None of these
70. What is genotype of Klinerfelter's syndrome?
    (a) XXX (b) XXY (c) XYY (d) XXO
71. Marker free plants can be generated by
    (a) Deletion (b) Inversion
    (c) Insertion (d) Co-transformation
72. The giant microorganism commonly called as 'sulfur pearl' is
    (a) *Dialister* (b) *Epulopiscium fischelsoni*
    (c) *Thiomargarita namibiensis* (d) None of these
73. What is the cause of aroma in rice plants?
    (a) 2-venyl 1-pyrroline (b) 2-acetyl 1-pyrimidine
    (c) 2-acetyl 1-pyrroline (d) 3-acetyl 1-pyrroline
74. Rifampicin interferes with
    (a) RNA polymerase I (b) RNA polymerase II
    (c) Bacterial RNA polymerase (d) RNA polymerase III

JAWAHARLAL NEHRU UNIVERSITY COMBINED BIOTECNOLOGY ENTRANCE EXAMINATION PAPER, 1995

1. Which one of the following transports water in higher plants?
   (a) Xylem (b) Phloem (c) Hypha (d) proton pump
2. Colchicine can block one of the following stages
   (a) Cell division (b) Spindle formation (c) Anaphase (d) Interphase
3. The female reindeer uses its horns for
   (a) Digging out food underneath the ice
   (b) Fight with other animals
   (c) Attract the male
   (d) No specific purpose

4. By differentially labeling the coat protein and the DNA of phage T2 Hershey and Chase demonstrated that
   (a) Only the protein enters the infected cell
   (b) The entire virus enters the infected cell
   (c) The phage DNA enters the infected cell
   (d) The phage coat protein directs synthesis of new progeny phage
5. Which of the following is not associated with protein synthesis?
   (a) Ribosomes (b) Proteosomes (c) mRNA (d) tRNA
6. Ribozyme is a
   (a) Protein (b) Carbohydrate (c) Lipid (d) RNA
7. One of the following is an amino acid
   (a) Proline (b) Histidine (c) Tryptophan (d) Maleic acid
8. A heteropolysaccharide is
   (a) Starch (b) Cellulose (c) Glycogen (d) Chitin
9. The fruit of mango is
   (a) Achene (b) Drup (c) Nut (d) Caryopsis
10. A natural plant hormone controlling fruit ripening is
    (a) CO_2 (b) Ethylene (c) Acetylene (d) CO
11. Plants perceive duration of day length by photoconversion of the following photoreceptor
    (a) Cytochrome (b) Chlorophyll (c) Bacteria (d) Blue green algae
12. One of the following is not an autotroph
    (a) Animals (b) Plants (c) Bacteria (d) Blue green algae
13. Bacteria causing crown gall formation in plants are
    (a) *Rhizobium* (b) *Pseudomonas*
    (c) *Mycobacterium* (d) *Agrobacterium*
14. Enterotoxin is produced by a species
    (a) *Streptococcus* (b) *Corynibacterium*
    (c) *Pseudomonas* (d) *Lactobacillus*
15. In biogas production the type of gases evolved are
    (a) Methane and nitrogen (b) Nitrogen and CO_2
    (c) Methane and CO_2 (d) H_2 and CO_2
16. Plant cells lack one of the following-
    (a) Centriole (b) Chloroplasts (c) Peroxisomes (d) Vacuoles

JAWAHARLAL NEHRU UNIVERSITY COMBINED BIOTECH-NOLOGY

ENTRANCE EXAMINATION PAPER, 1996

1. The high frequency of recombination in bacterial conjugation is associated with
   (a) The large F factor (b) A multicopy F factor
   (c) F factor integrated in the chromosome (d) F factor bound to the membrane
2. The proof of the double helical structure of DNA was obtained initially through
   (a) Electron microscope (b) X-ray diffraction studies
   (c) Absorption photometry (d) Fluorescence studies
3. Assuming that half-life 32P, the radioactive isotope of phosphorous is 4 days, how many atoms of ^{32}P would be still there after 42 days, if at day zero, there are 2×106 atoms?
   (a) 3×0^6 (b) 4×0^6 (c) 15×0^6 (d) None
4. Respiration and photosynthesis both require one of the following
   (a) Chloroplasts (b) Sunlight (c) Cytochromes (d) Mitochondria
5. The greatest degree of genetic variability would be expected among the offspring of a plant that reproduces
   (a) Asexually (b) Sexually with obligate self-pollination
   (c) Sexually with obligate cross pollination (d) Vegetative propagation
6. Barbara Mc Clintock won Nobel Prize for her work on
   (a) Split genes (b) Jumping gene
   (c) Translocation in Maize (d) Cell division in maize
7. Which one of the following d-amino acid is a diamino monocarboxylic acid?
   (a) Leucine (d) Lysine
   (c) r- carboxy glutamic acid (d) Glycine
8. An ecosystem consists of
   (a) Producers, consumers and decomposers in particular locality
   (b) Carnivores and herbivores in an area
   (c) A living community and its environment
   (d) All the plants and animals in an area
9. If a man of blood group 'AB' marries a woman of blood group 'A', whose father was of blood group 'O', their children will be
   (a) AB, A, B, O (b) AB, A, B only (c) A, B only (d) A, B, O only

10. One of the following in autotrophic prokaryote

    (a) *Chlamydomonas* (b) *Aspergillus* (c) *Oscillatoria* (d) *Clostridium*

11. Polyploidy in plants can be induced by

    (a) X-rays (b) Colchicine (c) Nitrous acid (d) Mustard gas

12. Biologically apple is a

    (a) True fruit (b) False fruit
    (c) Modified stem (d) Modified leaf

13. Citric acid is industrially produced by

    (a) *Penicilium notatum* (b) *Saccharomyces cerevisiae*
    (c) *Aspergillus niger* (d) *Pseudomonas aeruginosa*

14. The most abundant species of RNA in an eukaryotic cell is

    (a) mRNA (b) rRNA (c) tRNA (d) hn RNA

15. What is the common component among ATP, FAD and NAD?

    (a) ATP unit (b) ADP unit (c) A haeme group (d) Flavin unit

16. In a culture of *E. coli,* the cell population increased from 1.5×10^6 cells/ml to 12×10^6 cells/ml in 36 minutes the generation times is

    (a) 6 min (b) 9 min (c) 12 min (d) 18 min

17. Which of the following(s) has/have impaired activity in case of vitamin B1 deficiency?

    (a) Succinate dehydrogenase (b) Pyruvate dehydrogenase complex
    (c) Isocitrate dehydrogenase (d) Glucose-6- phosphate dehydrogenase

18. Which of the following proteins does not bind to DNA?

    (a) DNA polymerase (b) RNA polymerase
    (c) Lac repressor (d) Permease

UNSOLVED MODEL QUESTION PAPERS - I

1. When did life originated on earth (approximately)?
   (a) 4.5 billion years ago
   (b) 3.2 billion years ago
   (c) 2 billion years ago
   (d) 500 million years ago
2. Which kind of plants life appeared first on earth?
   (a) Filamentous green algae
   (b) Coccoid blue green algae
   (c) Lichens
   (d) Red algae
3. When did Pteridophytes originate on earth?
   (a) Cambrian
   (b) Carboniferous
   (c) Upper Silurian
   (d) Devonian
4. The theory of Natural Selection was proposed by
   (a) Lamarck (b) Darwin (c) De Vries (d) Mendel
5. In plants evolution, *Archaeopteris* links
   (a) Bryophyta with Pteridophyta
   (b) Pteridophyta with Angiosperms
   (c) Pteridophyta with Gymnosperms
   (d) Gymnosperms with Angiosperms
6. The Theory of Mutation was proposed by
   (a) De Vries (b) Wallace (c) Tschermak (d) Correns
7. In lysosome, the enzyme which helps in digestion is
   (a) Peroxidase (b) Carboxylase (c) Oxidase (d) Hydrolytic
8. Peroxisomes are
   (a) Spherical and rich in oxidative
   (b) Spherical and rich in hydrolytic
   (c) Rod-shaped and rich in fats
   (d) Rod-shaped and rich in amino acids
9. Which structure disappears at the time of cell division?
   (a) Plasma membrane
   (b) Centriole
   (c) Nuclear membrane
   (d) Plastids
10. In which organisms the cell wall is made up of only proteins?
    (a) Algae (b) Mycoplasma (c) Bacteria (d) Lichens
11. Single membrane encloses
    (a) Plastid
    (b) Nuleolus

(c) Lysosome (d) Endoplasmic reticulum

12. One similarity between prokaryotic and eukaryotic cells is
    (a) Cell division (b) Cell wall
    (c) rRNA and RNA synthetases present (d) Protein synthesis
13. The cell wall of Gram-negative bacteria has
    (a) 30-40 peptidoglycan layers
    (b) 10-15 peptidoglycan layers
    (c) 1-3 peptidoglycan layers
    (d) No peptidoglycan layers
14. In which stage of cell division the DNA becomes double?
    (a) Interphase (b) Anaphase (c) Metaphase (d) Prophase
15. In meiosis crossing over takes place at what stage
    (a) Leptotene (b) Zygotene (c) Pachytene (d) Diplotene
16. How many meiosis will produce 128 pollen grains?
    (a) 16 (b) 56 (c) 128 (d) 32
17. Phloem of gymnosperms does not have
    (a) Companion cells (b) Phloem parenchyma
    (c) Phloem fibres (d) Sieve cells
18. Vessel-less angiosperm is
    (a) *Magnolia* (b) *Trochodendron*
    (c) *Nymphea* (d) *Nilumbium*
19. Bicollateral vascular bundle is found in
    (a) *Dracaena* stem (b) *Cucurbita* stem
    (c) *Ficus* root (d) *Tinospora* stem
20. An amphivassal vascular bundle has
    (a) Phloem surrounded by xylem
    (b) Xylem surrounded by phloem
    (c) Xylem and phloem in a line or radius
    (d) Xylem and phloem in separate radii
21. Polyarch xylem without secondary thickening is found in
    (a) Monocot stem (b) Monocot root
    (c) Dicot stem (d) Dicot root
22. Monocot stem bearing vascular bundles in a ring
    (a) *Ruscus* (b) *Tricticum* (c) *Tamus* (d) *Dracaena*

23. In a dihybrid cross phenotype ratio of progeny will be
    (a) 3 : 1 (b) 1 : 1 (c) 9 : 3 : 3 : 1 (d) 1 : 1 : 1 : 1
24. If a homozygous dwarf plant is crossed with a heterozygous tall plant, the percentage proportion of dwarf progeny will be
    (a) 50% (b) 75% (c) 25% (d) 100%
25. 'One gene one enzyme' was suggested by
    (a) Khurana and Holley (b) Lynen and Bloch
    (c) Watson and Crick (d) Beadle and Totum
26. Cell division is enhanced by
    (a) Colchicine (b) Gibberellic acid
    (c) Kinetin (d) Morphactin
27. Breeding of F1 with the recessive parent is a
    (a) Back cross (b) Purity of gametes
    (c) Test cross (d) Reciprocal cross
28. A test cross is useful for study of
    (a) Phenotype (b) Interbreeding capacity
    (c) Genotype (d) Fertility ratio
29. Which one is known as 'pond silk'?
    (a) *Ulothrix* (b) *Spirogyra* (c) *Cladophora* (d) *Volvox*
30. Agar agar is obtained from
    (a) *Gracilaria* (b) *Batrachospermum*
    (c) *Chlorella* (d) *Fucus*
31. Cystocarp in red algae is
    (a) Haploid (b) Diploid (c) Triploid (d) Polyploid
32. Indian Botanist who suggested the utility of blue green algae as biofertilizer
    (a) H.D. Kumar (b) M.O.P. Iyangar
    (c) R.N. Singh (d) Y. Bhardwaja
33. Cap cells are present in
    (a) *Vaucheria* (b) *Oedogonium* (c) *Bulbochaete* (d) *Cladophora*
34. After fertilization is produced in *Coleochaete* a structure called
    (a) Cystocarp (b) Spermocarp (c) Mericarp (d) Pericarp
35. Litmus is a natural dye obtained from
    (a) Algae (b) Fungi (c) Lichen (d) Mycoplasma

36. Who discovered plasmids?
    (a) Lederberg (b) Fleming
    (c) Alexopoulos (d) Rober Hooke
37. Nitrogen fixing bacteria in the roots of Legumes are
    (a) *Azotobacter* (b) *Rhizobium* (c) *Pseudomonas* (d) *E. coli*
38. In Ascomycotina karyogamy occurs in
    (a) Ascogonium (b) Ascus
    (c) Antheridium (d) Ascogynous hyphae
39. Powdery Mildew disease is caused by
    (a) *Phytophthora* (b) *Rhizopus* (c) *Erysiphe* (d) *Aspergillus*
40. Downy Mildew disease is caused by
    (a) *Erysiphe* (b) *Peronospora* (c) *Phytophthora* (d) *Sclerospora*
41. Early blight of Potato is caused by
    (a) *Alternaria* (b) *Helminthostachys*
    (c) *Cercospora* (d) *Plasmopara*
42. Bordeaux mixture contains
    (a) Lime and calcium sulphate (b) Sulphur and lime
    (c) Copper sulphate and lime (d) Copper sulphate and sulphur
43. Yellow vein mosaic of Bhindi is a
    (a) Bacterial disease (b) Fungal disease
    (c) Viral disease (d) Physiological deficiency disease
44. Tobacco mosaic virus was crystallized first by
    (a) Ivanowsky (b) Stanley (c) Beiternick (d) Stabin
45. There is a single chloroplast in every cell of the thallus
    (a) *Riccia* (b) *Marchantia*
    (c) *Sphagnum* (d) *Anthoceros*
46. Thickness of the neck of archegonium in *Polytrichum* is
    (a) 5 cells (b) 3 cells (c) 6 cells (d) 4 cells
47. In mosses elaters are
    (a) Single celled with spiral thickening (b) 4-celled with spiral thickening
    (c) 4-celled without thickening (d) Not present
48. An aquatic Bryophyte is
    (a) *Sphaerocarpus* (b) *Riella* (c) *Pellia* (d) *Porella*

49. The book 'Liverworts of Western Himalaya and Punjab Plains' is written by
    (a) P.N. Mehra (b) S.K. Pande (c) R.S. Chopra (d) S.R. Kashyap
50. In the evolution of thallus of Marchantiales the retrogressive theory was suggested first by
    (a) K. Goebel (b) S.R. Kashyap (c) D.H. Campbell (d) G.M. Smith
51. Heterospory originated during the period
    (a) Upper Silurian (b) Middle Devonian
    (c) Lower Carboniferous (d) Upper Devonian
52. Female gametophyte of *Marsilea* has archegonia
    (a) 2 (b) 4 (c) 1 (d) 8
53. A fern which grows along with mangroves
    (a) *Marsilea* (b) *Certopteris* (c) *Acrostichum* (d) *Salvinia*
54. A pteridophyte which has the maximum number of chromosomes in a cell
    (a) *Lycopodium clavatum* (b) *Equisetum arvense*
    (c) *Ophioglossum reticulatum* (d) *Marsilea minuta*
55. Rhizome of fern with secondary growth is
    (a) *Cyathea* (b) *Osmunda* (c) *Adiantum* (d) *Botrychium*
56. Gametophytes of *Equisetum* may be
    (a) Male gametophyte (b) Female gametophyte
    (c) Bisexual gametophyte (d) All
57. The tallest living cycad is
    (a) *Stangeria* (b) *Dioon* (c) *Microcycas* (d) *Cycas*
58. In the life cycle of Pinus how much time is involved between origin of ovule and coming out of a mature seed form the cone?
    (a) 6 months (b) 9 months (c) 2 years (d) 3 years
59. 'Taxol' is extracted from
    (a) *Cephalotaxus* (b) *Pinus* (c) *Taxus* (d) *Thuja*
60. A C.S.I.R. monograph on *Pinus* has been written by
    (a) P. Maheshwari and C. Biswas (b) P. Maheshwari and R.N. Konar
    (c) P. Maheshwari and V. Vasil (d) P. Maheshwari and H.P. Singh
61. Seed is multicotyledonous in
    (a) *Cycas* (b) *Ephedra* (c) *Pinus* (d) *Gnetum*
62. *Williamsonia* plant survived during the Geographical period
    (a) Permo-carboniferous (b) Jurassic-Lower Cretaceous

(c) Upper Crretaceous (d) Eocene

63. Caruncle (aril) is present in the seed of

(a) Pea (b) Castor (c) Maize (d) Cotton

64. Hypanthodium inflorescence is present in

(a) *Riccinus* (b) *Euphorbia* (c) *Acacia* (d) *Ficus*

65. Anatropous ovule is found in

(a) *Polygonum* (b) *Lathyrus*
(c) *Chenopodium* (d) *Brassica*

66. Respiratory roots are present in

(a) Orchids (b) *Cuscuta*
(c) *Ficus bengalensis* (d) *Avicennia*

67. Cystolith is made of

(a) Calcium oxalate (b) Calcium oxide
(c) Calcium carbonate (d) Silica

68. Ascending imbricate estivation is found in

(a) *Cassia* (b) *Prosopis* (c) *Arachis* (d) *Lathyrus*

69. *Usnea* is an example of

(a) Crustose lichen (b) Foliage lichen
(c) Fruticose lichen (d) An alga

70. Mycoplasma has

(a) Cell wall of cellulose (b) Ribosomes
(c) Nucleus (d) Plastids

71. Generally a plant virus has

(a) DNA (b) RNA
(c) Ribosomes (d) DNA + RNA both

72. Bentham and Hooker kept gymnosperms in their classification

(a) After Dicotyledonae
(b) In between Monocotyledonae and Dicotyledonae
(c) After Monocotyledonae
(d) Separate from Monocots and Dicots

73. In which plant of Momosaceae family the inflorescence has flowers of two colours?

(a) *Acacia* (b) *Albizia*
(c) *Dicrostachys* (d) *Mimosa*

74. KEW herbarium is in
(a) U.S.A. (b) Scotland (c) England (d) Russia

75. South zone herbarium of Botanical Survey of India is at
(a) Chennai (b) Mysore (c) Coimbatore (d) Calicut

76. In which family has the ovary a swollen and oblique placentaiton
(a) Malvaceae (b) Cruciferae
(c) Solanaceae (d) Rutaceae

77. All plants which perform photosynthesis have
(a) Chlorophyll a (b) Chlorophyll b (c) Chlorophyll c (d) Chlorophyll d

78. C_4 plant is
(a) Papaya (b) Wheat (c) Maize (d) Pea

79. During guttation water drops come out of the plant through
(a) Stomata (b) Hydathode (c) Lenticel (d) Wound

80. Both photosynthesis and respiration need
(a) Chlorophyll (b) CO_2 (c) Water (d) Cytochrome

81. Link between glycolysis and Krebs cycle is
(a) Citric acid (b) Acetyl Coenzyme
(c) Malic acid (d) Fumaric acid

82. Number of ATP evolved during glycolysis is
(a) 2 (b) 6 (c) 4 (d) 8

83. Number of oxidations during Krebs cycle is
(a) 2 (b) 3 (c) 5 (d) 6

84. Micronutrient is
(a) Iron (b) Zinc (c) Sodium (d) Potassium

85. Epigeal seed germination occurs in
(a) Gram (b) Maize (c) Castor (d) Pea

86. An example of self-pollination is
(a) Maize (b) *Vallisnaria* (c) *Slavia* (d) Sunflower

87. In an ecosystem the source of energy is
(a) Decomposition of plants and animals (b) Photosynthesis by plants
(c) Fermentation (d) Sunlight

88. Which one is not a halophyte?
(a) *Trapa* (b) *Atriplex* (c) *Haloxylon* (d) *Salsola*

89. Food chains are not met with in
    (a) Sea (b) City (c) Forest (d) Everywhere
90. The term ecosystem was coined by
    (a) Huxley (b) Tensley
    (c) Weaver and Clements (d) Odum
91. Red rot of sugarcane is caused by
    (a) *Cercospora* (b) *Helminthosporium*
    (c) *Fusarium* (d) *Colletotrichum*
92. Tikka disease of groundnut is caused by
    (a) *Alternari solanii* (b) *Cercospora personata*
    (c) *Colletotrichum falcatum* (d) *Sclerospora oryzae*
93. Protoplast culture requires
    (a) A tissue (b) Cell with cell wall
    (c) Cell with plasma membrane (d) Cell with a large vacuole
94. Tissue culture technique is not very successful in
    (a) Medicinal herbs (b) Crop plants
    (c) Fruit plants (Papaya, Banana) (d) Large woody trees
95. For tissue culture technique utensils required are made up of
    (a) Sand (b) Metal
    (c) Ordinary glass (d) Borosil glass
96. Utensils used in tissue culture technique are sterilized by
    (a) Sand (b) Ash
    (c) Soap (d) Ovan or autoclave
97. Endangered plants produce
    (a) Abundant seeds (b) Small seeds
    (c) Multiply vegetatively (d) Seeds mostly non-viable
98. Quinine is extracted from
    (a) Leaves (b) Fruit (c) Root (d) Stem bark
99. ‘Hing’ is a latex of
    (a) *Papaver* (b) *Digitalis* (c) *Artemisia* (d) *Ferula*
100. Sweet potato is
    (a) *Solanum* (b) *Ipomoea* (c) *Raphanus* (d) *Daucus*

UNSOLVED QUESTIONS PAPER - II

1. Who gave the name 'protoplasm'?
   (a) Brown (1831)
   (b) Mirbel (1802)
   (c) Schwann (1838)
   (d) Purkinje and Von Mohl (1839)
2. Which is the fat metablic and storage organelle?
   (a) Lysosome (b) Dictyosome (c) Sphaerosome (d) Glyoxysome
3. Extranuclear mitotic spindle is formed in
   (a) Plantae (b) Prokaryotes (c) Viruses (d) Cyanobacteria
4. Plasma membrane is
   (a) Non-permeable
   (b) Semi-permeable
   (c) Permeable
   (d) Neither permeable nor impermeable
5. Lysosome is
   (a) Single membraned
   (b) Double membrane
   (c) Triple membrane
   (d) Naked
6. Glycosidation occurs in
   (a) Lysosomes (b) Dictyosomes (c) Sphaerosomes (d) Peroxisomes
7. $F_0 - F_1$ are particles of
   (a) Mitochondria (b) Chloroplasts (c) Dictyosomes (d) Mesosomes
8. Cellular endoskeleton is formed in
   (a) Endoplasmic reticulum
   (b) Dictyosomes
   (c) Ribosomes
   (d) Chloroplasts
9. Cilia are formed from
   (a) Lysosomes (b) Ribosomes (c) Peroxisomes (d) Centrioles
10. Energy is stored in
    (a) Chloroplasts
    (b) Mitochondria
    (c) Ribosomes
    (d) Golgi bodies
11. The cell organelle essential for photoreception is
    (a) Glyoxysome (b) Peroxisome (c) Dictyosome (d) Sphaerosome
12. Which is a naked cell organelle
    (a) Nucleus (b) Nucleolus (c) Vacuole (d) Chloroplast
13. Cross over between homologous chromosome parts occurs during
    (a) Leptotene (b) Zygotene (c) Pachytene (d) Diplotene

14. Which is a secondary meristem in dicot stem
    (a) Fascicular cambium (b) Inter fascicular cambium
    (c) Phelloderm (d) Corck
15. The loose aerating pores in the bark are
    (a) Stomata (b) Lenticels (c) Hydathodes (d) Holes
16. Phellogen forms on the outside of
    (a) Secondary cortex (b) Phellem
    (c) Phelloderm (d) Wood
17. The origin of root laterals is
    (a) Endogenous (b) Exogenous (c) Cortical (d) Epidermal
18. Mutations were demonstrated by De Vries in
    (a) Garden pea (b) *Mirabilis jalapa*
    (c) *Oenothera lamarkiana* (d) *Drosophila*
19. The cross between pure dominant and pure recessive parents with two allelic characters is called
    (a) Monohybrid (b) Dihybrid (c) Trihybrid (d) Tetrahybrid
20. The purity of genotype is checked by
    (a) Self crossing (b) Back crossing
    (c) Hybrid crossing (d) Not crossing
21. The inheritance of linked characters was reported by
    (a) Bateson and Punnet (1906) (b) Morgan (1910)
    (c) Bridges (1916) (d) Beadle and Tatum (1948)
22. Experimental proof of cross over during synapsis at four strand stages came through studies on
    (a) *Penicillium* (b) *Mucor* (c) *Neurospora* (d) *Aspergillus*
23. Who was the propounder of one gene-one enzyme hypothesis?
    (a) Zinder (1950) (b) Zinder and Lederberg (1952)
    (c) Beadle and Tatum (1958) (d) Tatum (1940)
24. Gene change in base sequence of one triplet is called
    (a) Continuous mutation (b) Point mutation
    (c) Whole mutation (d) Mass mutation
25. Cistron is
    (a) One gene and one peptide chains (b) Two gene and two peptide chains
    (c) One gene and two peptide chains (d) One gene and three peptide chains

26. Artificial mutations through X-rays and UV rays were induced by
    (a) Hugo De Vries (1903) (b) Muller (1909)
    (c) Darwin (1859) (d) Morgan (1910)
27. The incomplete multiplication of genome set during polyploidy is called
    (a) Euploidy (b) Heteroploidy (c) Homoploidy (d) Hypoploidy
28. Sex linked gene in humans are related with
    (a) Tallness (b) Dwarfness
    (c) Skin colour (d) Colour blindness
29. Which is a parasitic algae?
    (a) *Trentepohlia* (b) *Fritschiella* (c) *Cephaleuros* (d) *Chondrus*
30. Algae having golden brown pigment, leucosin stored food, naked pyrenoide, silicious bivalve cell wall are classified as
    (a) Phaeophyceae (b) Rhodophyceae
    (c) Cryptophyceae (d) Bacillariophyceae
31. Fungi having no sexual reproduction are classified as
    (a) Basidiomycetes (b) Ascomycetes
    (c) Zygomycetes (d) Deuteromycetes
32. Which fungus has the sexuality between two compatible nuclei of vegetative cells
    (a) *Rhizopus* (b) Yeast (c) *Agaricus* (d) *Mucor*
33. First antirabies vaccine was prepared by
    (a) Lister (b) Pasteur (c) Koch (d) Waksman
34. Which is a photo-autotrophic bacterium?
    (a) *Chromatium* (b) *Rhizobium*
    (c) *Clostridium* (d) *Lacto bacillus*
35. Which is a symbiotic, aerobic, nitrogen fixing bacterium
    (a) *Rhizobium* (b) *Azotobacter* (c) *Streptomyces* (d) *Salmonella*
36. Which bacterium converts ammonia to nitrite?
    (a) *Rhizobium* (b) *Azotobacter* (c) *Nitrosomonas* (d) *Escherichia*
37. The phenomenon of transduction in bacteria was discovered by
    (a) Zinder and Lederberg (1952) (b) Griffith (1928)
    (c) Tatum (1944) (d) Avery and Coworkers (1941)
38. The bacterial viral shape is
    (a) Tadpole like (b) Rod shaped (c) Filamentous (d) Spherical

39. The length and width of T_2 phage is
    (a) 1000 × 500 Å (b) 1200 × 850 Å
    (c) 750 × 750 Å (d) 750 × 850 Å
40. Which is a DNA containing virus
    (a) TMV (b) Tobacco satellite virus
    (c) Turnip yellow virus (d) T_2 phage
41. The vegetative reproduction organs of Lichens are
    (a) Pycnidia (b) Isidia (c) Conidia (d) Sporidia
42. Litmus dye is obtained from
    (a) *Rocella* (b) *Usnea* (c) *Cladonia* (d) *Peltigera*
43. N-acetyl Muramic acid and N-acetyl Glucosamine are present in the cell wall of
    (a) *E. coli* (b) *Mucor* (c) *Parmelia* (d) TMV
44. The rhizoids are obliquely septate in
    (a) *Ferns* (b) *Marchantia* (c) *Funaria* (d) *Riccia*
45. Heterospory is present in
    (a) *Selaginella* (b) *Dryopteris* (c) *Lycopodium* (d) *Equisetum*
46. The largest, motile, multiflagellate sperm is present in
    (a) *Pinus* (b) *Cycas* (c) *Ginkgo* (d) *Cedrus*
47. The archegonia are absent in
    (a) Ferns (b) *Gnetum* (c) Buttercup (d) *Cycas*
48. Smallest organism, growing and reproducing autonomously is
    (a) Virus (b) T_2 phage (c) Mycoplasma (d) Pneumococcus
49. Early blight of potatoes is controlled by spraying
    (a) Lime sulphur (b) Sulphur dust (c) Bordeaux mixture(d) DDT
50. Red streaks with ashen-gray centre are produced on the midrib of sugarcane leaves by the infection of
    (a) *Cystopus candidus* (b) *Alternaria solani*
    (c) *Puccinia gramins* (d) *Colletotrichum falcatum*
51. Which disease can be effectively controlled through temperature and its physical manipulations?
    (a) Loose smut of wheat (b) Wheat ruts
    (c) Tikka disease of groundnut (d) White rust of crucifers
52. Which of the following herbaria has least number of plant sheets
    (a) Royal Botanical Gardens and Herbarium, London
    (b) British Museum, London

(c) Jorden des Plantes, Paris
(d) Harvard University Cambridge, Massachusets

53. A specimen exactly like holotype among original collection is called
(a) Lectotype (b) Neotype (c) Isotype (d) Paratype

54. According to International Code of Botanical Nomenclature a homonym is considered as
(a) Tautonym (b) Synonym (c) Basonym (d) Heteronym

55. The barrier between anther and stigma is called
(a) Dichogamy (b) Herkogamy (c) Homogamy (d) Dicliny

56. Plant pollination by birds is called
(a) Entomophily (b) Anemophily (c) Chiropterophily (d) Ornithophily

57. Production of seedless fruits in plants is induced by
(a) Lycomarasmine (b) Gibberellins (c) Tabtoxinin (d) Victoxinin

58. The concept of cellular totipotency was proposed by
(a) Vasil and Hildebrandt (1965) (b) Steward and coworkers (1964)
(c) Koner and Natraja (1965) (d) Guha and Maheshwari (1965)

59. Easily available water to plants from soil is
(a) Hygroscopic (b) Capillary
(c) Gravitational (d) Hygroscopic and capillary

60. Plants absorb solutes from soil water through
(a) Diffusion against gradient pressure (b) Diffusion not against gradient pressure
(c) Imbibition (d) Osmosis

61. Water moves upto top of trees through xylem by
(a) Osmosis (b) Diffusion (c) Imbibition (d) Capillary action

62. Ascent of sap in plants best explained by the theory of
(a) Cohesion (b) Bolhm
(c) Root pressure (d) Physical forces

63. National Botanical Research Institute is situated at
(a) Calcutta (b) Delhi (c) Lucknow (d) Coimbatore

64. The Royal Botanical Gardens Kew, one of the most well-known International Herbarium and Botanical Garden located in
(a) U.S.A (b) Germany (c) England (d) Australia

65. Dwarfism, curling of leaves and death of apical meristem are caused by the deficiency of
(a) Iron (b) Potassium (c) Copper (d) Boron

66. In C_3 plants CO_2 acceptor is
    (a) Phospho-enolpyruvic acid (b) Ribulose diphosphate
    (c) Pyruvic acid (d) Maleic acid
67. PS II system of light reaction is absent in
    (a) C_3 plants (b) C_4 plants
    (c) CAM plants (d) CAM and C_3 plants
68. Cyclic electron transfer occur only in
    (a) PS I system (b) PS II system
    (c) PS I and PS II systems (d) None of the PS systems
69. Light reaction occurs in
    (a) Cytoplasm (b) Mitochondria (c) Quantosomes (d) Ribosomes
70. Glycolysis occurs in
    (a) Cytoplasm (b) Mitochondria (c) Chloroplast (d) Vacuole
71. Total ATP produced during Krebs cycle is
    (a) 28 (b) 30 (c) 32 (d) 38
72. Which organism fixes atmospheric nitrogen to nitrates
    (a) *Nostoc* (b) *Chlorella* (c) *Microcytis* (d) *Ulothrix*
73. Which is a growth retardant in plants
    (a) Auxins (b) Gibberellins (c) Cytokinins (d) Abscisic acid
74. Florigen is made up of
    (a) Anthesis (b) Abscisic acid
    (c) IAA (d) Anthesis and Gibberellins
75. The influence of cold on flowering is called
    (a) Photoperiodism (b) Vernalization
    (c) Photoreception (d) Chilling
76. Antagonism between two organisms through gases or allelochemicals is called
    (a) Commensalism (b) Amensalism
    (c) Mutualism (d) Parasitism
77. A level not eaten by any in a food chain is called
    (a) Primary consumers (b) Secondary consumers
    (c) Tertiary consumers (d) Top consumers
78. Which plant is used in paper industry?
    (a) Teak (b) Bamboo (c) Sal (d) Mango

79. Which is a drug yielding plant?
    (a) *Lichi sinensis* (b) *Cinchona officinalis*
    (c) *Mangifera indica* (d) *Dalbergia sissoo*
80. Penicillium is obtained from
    (a) *Penicillium vermiculatum* (b) *Penicillium roqueforti*
    (c) *Penicillium notatum* (d) *Penicillium fumigatus*
81. Vinegar is synthesized by
    (a) *Acetobacter aceti* (b) *Rhizopus oryzae*
    (c) *Aspergillus niger* (d) *Penicillium notatum*
82. Riboflavin is synthesized by
    (a) *Ascoidea* (b) *Aspergillus niger*
    (c) *Penicillium fumigatus* (d) *Eremothecium ashbyii*
83. Calder 1937 divided British-India into how many regions on the basis of physical conditions and floral features
    (a) 5 (b) 6 (c) 9 (d) 11
84. A plant culture from roots in the nutrient medium is called
    (a) Callus culture (b) Suspension culture
    (c) Organ culture (d) Embryo culture
85. Which is the sanctuary in Rajasthan state?
    (a) Chilka lake (b) Madumalai wild life
    (c) Dachigam (d) Ranthambor tiger project
86. Which is a method of conservation of Genetic Resources?
    (a) Offsite collection (b) Reducing crop variety
    (c) Deforestation (d) Reducing crop number
87. Which crop in India gave better yields by crossing with wild germplasm?
    (a) Sorghum (b) Sugarcane (c) Maize (d) Potatoes
88. Which is a leading crop of green revolution in India?
    (a) Maize (b) Wheat (c) Milletes- Bajra (d) Sorghum
89. Ginkgo is now almost out of endangered species because of
    (a) Offsite collection (b) Gene bank
    (c) *Ex situ* conservation (d) *In situ* conservation
90. Banana germplasm has been stored through
    (a) Seed bank (b) Tissue culture
    (c) Special orchards (d) Cryopreservation

91. Potent plant for harvesting germplasm is
    - (a) *Grevia asiatica*
    - (b) *Cola nitida*
    - (c) *Pyrus sylvestris*
    - (d) *Achrus sapota*
92. Which is the most beneficial product from industrial biotechnology
    - (a) Wine and brandy
    - (b) Soap
    - (c) Detergents
    - (d) Talc powder
93. Biologically active and highly useful compound produced through applied biotechnology in medicines is
    - (a) Proteases
    - (b) Sterptokinase
    - (c) Amylases
    - (d) Cellulases
94. Which of the following is a reserve biosphere in India?
    - (a) Naina devi
    - (b) Valley of flowers
    - (c) Mansarovar
    - (d) Vrindavan
95. Van Mahotsava week is celebrated in India during
    - (a) First week of June
    - (b) Second week of June
    - (c) First week of May
    - (d) Second week of May
96. Which is a family having superior oblique ovary and swollen axile placentation?
    - (a) Malvaceae
    - (b) Solanaceae
    - (c) Brassicaceae
    - (d) Rutaceae
97. In *Euphorbia* inflorescence is
    - (a) Spadix
    - (b) Hypanthodium
    - (c) Capitulum
    - (d) Cyathium
98. Life supporting atmospheric zone is called
    - (a) Troposphere
    - (b) Thermosphere
    - (c) Ionosphere
    - (d) Mesosphere
99. The theory of inheritance of acquired characters was proposed by
    - (a) Darwin
    - (b) Lamarck
    - (c) De Vries
    - (d) Mendel
100. The binomial nomenclature was given by
    - (a) Hooker
    - (b) Linnaeus
    - (c) Strasburger
    - (d) Hutchinson

SOLVED MODEL QUESTION PAPER -I

1. Pyrenoids are
   (a) Starch surrounded by protein grains (b) Starch surrounded by oil droplets
   (c) Protein surrounded by oil droplets (d) Protein surrounded by starch grains
2. Storage (reserve food) product of most of the algae is
   (a) Glycogen (b) Fat (c) Starch and oil (d) Cellulose
3. Sexual reproduction in *Chlamydomonas* is
   (a) Isogamous (b) Anisogamous (c) Oogamous (d) All of these
4. From which of the following alga Iodine is commercially obtained
   (a) *Ectocarpus* (b) *Laminaria* (c) *Oedogonium* (d) *Gelidium*
5. Red rust of tea is caused by
   (a) Red alga (b) Green alga (c) Fungus (d) All the above
6. All fungi are always
   (a) Parasites (b) Autotrophs (c) Heterotrophs (d) Saprophytes
7. Heterothallism in fungi was discovered by
   (a) Bessey (b) Robert Hooke (c) Saccardo (d) Blakeslee
8. Yeast is abundantly found in
   (a) Organic substances rich in fats (b) Organic substances rich in sugar
   (c) Moist bread (d) Horse dung
9. Fungi imperfecti (Deuteromycete) lack
   (a) Hyphae (b) Spores
   (c) Asexual reproduction (d) Sexual reproduction
10. Ergot is obtained from
   (a) *Rhizobium* (b) *Claviceps* (c) *Phytomonas* (d) *Albugo*

11. Which of the following is litmus yielding lichen
    (a) *Cladonia rangifera* (b) *Rocella montaignei*
    (c) *Lecanora esculenta* (d) *Cetaria islandica*
12. A common phycobiont in lichen is
    (a) *Microcystis* (b) *Trebouxia* (c) *Euglena* (d) *Citraria*
13. Lichens are good indicators of
    (a) Air pollution (b) Water pollution
    (c) Soil pollution (d) All the above
14. Lichens participate in a
    (a) Xerosere (b) Halosere
    (c) Hydrosere (d) Psammosere
15. The fungal part of lichens generally belongs to the class
    (a) Phycomycetes (b) Basidiomycetes
    (c) Ascomycetes (d) Deuteromycetes
16. Gametophytic generation is dominant in
    (a) Angiosperms (b) Pteridophytes (c) Bryophytes (d) Gymnosperms
17. Simplest sporophyte among Bryophytes is of
    (a) *Marchantia* (b) *Riccia* (c) *Sphagnum* (d) *Funaria*
18. Protonema is a stage in the life cycle of
    (a) All Bryophytes (b) *Riccia* (c) *Marchantia* (d) *Funaria*
19. Peristome of *Funaria* has
    (a) 8 teeth in one ring (b) 32 teeth in two rings
    (c) 16 teeth in one ring (d) 16 teeth in two rings
20. In *Marchantia* female sex organs are formed on stalked structures termed as
    (a) Carpel (b) Archegoniophore
    (c) Gynophore (d) Megasporophyll
21. Heterospory is production of
    (a) Haploid and tetraploid spores (b) sexual and asexual spores
    (c) Large and small spores (d) Haploid and diploid spores
22. Actinostele is related to
    (a) Dictyostele (b) Siphonostele (c) Solenostele (d) Protostele
23. Trabeculae in *Selaginella* are modified
    (a) Endodermal cells (b) Pericycle cells
    (c) Cortical cells (d) Pericycle and endodermal cells

24. Gametophytic generation in fern plant is represented by
    (a) Antheridium (b) Archegonium (c) Zygote (d) Prothallus
25. Sporangia of ferns are situated on
    (a) The base of sporophyll (b) Adaxial side of sporophyll
    (c) Abaxial side of sporophyll (d) Apophysis
26. Xylem in Gymnosperms lacks
    (a) Xylem fibers (b) Tracheids
    (c) Vessels (d) Xylem parenchyma
27. Wings of Pinus seeds develop from
    (a) Ovuliferous scale (b) Bract scale (c) Cone axis (d) Seed coat
28. Compared to other seed plants, *Cycas* is marked by largest
    (a) Egg (b) Sperm (c) Ovule (d) All the above
29. The girdle leaf trace in *Cycas* goes to leaves
    (a) After making a semicircle (b) After making a complete circle
    (c) After making two circles (d) Non of these
30. In Gymnosperms the endosperm is
    (a) Haploid (b) Diploid
    (c) Triploid (d) None of the above
31. The branch of botany which deals with extinct plants is
    (a) Palaeobotany (b) Extant Botany
    (c) Extra-terrestrial Botany (d) Economic botany
32. Coal is a type of
    (a) Compressed fossil (b) Petrified fossil
    (c) Impression (d) None of these
33. Which is said to be age of ferns
    (a) Carboniferous (b) Jurassic (c) Triassic (d) All of these
34. Which type of fossil allows to study the internal structure of a fossil plant
    (a) Compression (b) Impression (c) Pertified (d) Psedofossil
35. Continuity of cytoplasm from cell to cell is maintained through
    (a) Middle lamella (b) Plasmodesmata
    (c) Endoplasmic reticulum (d) Cell membrane system
36. Which of the following cell organelle is considered to be rich in oxidative enzymes
    (a) Lysosome (b) Endoplasmic reticulum
    (c) Golgi Complex (d) Mitochondria

37. Rough endoplasmic reticulum is especially well developed in cells actively engaged in
    (a) Secretory function (b) Lipid synthesis
    (c) Protein synthesis (d) Nucleotide synthesis
38. Ribosomes are made up of
    (a) RNA and DNA (b) DNA and Proteins
    (c) RNA and proteins (d) DNA and carbohydrates
39. Human eyes resolving power is
    (a) 50μ (b) 150μ (c) 100μ (d) 200μ
40. Puffs in the salivary gland chromosomes represent the site of
    (a) Protein synthesis (b) RNA synthesis
    (c) DNA synthesis (d) None of these
41. During cell division chiasmata is seen in
    (a) Interphase (b) Zygotene (c) Pachytene (d) Diplotene
42. An exception to Mendel's Law is
    (a) Independent assortment (b) Dominance
    (c) Purity of gametes (d) Linkage
43. Test cross is defined as
    (a) Cross between parents to find out the genotype of the parents
    (b) Cross between the hybrid and the dominant parent
    (c) Cross between the hybrid and the recessive parent
    (d) Cross between the hybrid and any one of its parents
44. Down's syndrome is an example of
    (a) Monosomy (b) Trisomy (c) Triploidy (d) Polyploidy
45. Germinating barley seeds are employed in the preparation of
    (a) Cheese (b) Wine (c) Beer (d) Lactic acid
46. Hybridomas are result of fusion of
    (a) Male reproductive cells
    (b) Female reproductive cells
    (c) Normal antibody producing cells with myeloma
    (d) Abnormal antibody producing cells with myeloma
47. The nuclease enzyme which begins its attack from a free end of a polynucleotide chain is
    (a) Polymerase (b) Exonuclease (c) Endonuclease (d) Ligase
48. The first transgenic crop was
    (a) Pea (b) Flax (c) Tobacco (d) Cotton

49. Flavr Savr is a transgenic
    (a) Tomato (b) Potato (c) Brinjal (d) Grape
50. Interferon is a glycoprotein which is
    (a) Antiviral (b) Antibacterial (c) Antipathogenic (d) Antioxidant
51. BLAST is
    (a) Basic Local Alignment Sequence Tool
    (b) Basic Local Arrangement Sequence Tool
    (c) Basic Local Alignment Search Tool
    (d) Basic Local Arrangement Search Tool
52. Which among the following give information about gene sequence database
    (a) NCBI (b) ENVIS (c) MALDI (d) None of these
53. MIAME is concerned with
    (a) DNA Microarray (b) Protein Microarray
    (c) Expression analysis (d) All the above
54. What do present day computers use for processing information
    (a) Vacuum tube (b) Transistors
    (c) Microprocessors (d) All of these
55. Which of the following is system software
    (a) DOS (b) Word (c) Power Point (d) Fox pro
56. Victorin is
    (a) Phytotoxin (b) Vivotoxin (c) Pathotoxin (d) None of these
57. Downy mildew refers to a group of fungi and the diseases they cause and
    (a) They are obligate (b) They belong to Oomycetes
    (c) Characterized by 'downy growth' (d) All of these
58. Smut diseases are caused by the members of
    (a) Basidiomycetes (b) Oomycetes
    (c) Myxomyetes (d) Ascomycetes
59. Which Indian Plant pathologist contributed extensively to elucidate the role of enzymes and toxins in plant disease?
    (a) T.S. Sadasivam (b) H.C. Dube
    (c) C.V. Subramanian (d) K.C. Malhotra
60. Brown leaf spot of rice is caused by
    (a) *Puccinia graminis* (b) *Pyricularia oryzae*
    (c) *Xanthomonas oryzae* (d) *Helminthosporium oryzae*

61. Tunica corpus theory was proposed by
    (a) Hofmeister (1857) (b) Nageli (1878)
    (c) Schmidt (1924) (d) Hanstein (1868)
62. Which of the following is not a feature of sclerenchyma cells?
    (a) Secondary cell wall is made of lignin
    (b) Pits are absent from the wall
    (c) Intercellular spaces are absent
    (d) Major function is providing mechanical support
63. In which type of stele, vascular bundles are irregularly dispersed in ground tissue?
    (a) Protostele (b) Eustele (c) Atactostele (d) Solenostele
64. Among the plant families mentioned below the one that does not have 'Kranz anatomy'
    (a) Gramineae (b) Amaranthaceae (c) Compositae (d) Pinaceae
65. The type of stomata that has guard cells surrounded by three unequal subsidiary cells
    (a) Anisocyte type (b) Anomocyte type
    (c) Paracyte type (d) Diacyte type
66. Which of the following organisms can serve as biofertilizer for rice crop
    (a) Blue-green algae (b) *Frankia sp.*
    (c) Mycorrhizal fungi (d) *Azotobacter*
67. Organisms that thrive in cold are known as
    (a) Mesophile (b) Basophile (c) Psychrophile (d) Thermophile
68. The terms amoeboid movement, plasmodium and capillitium are concerned with
    (a) Zygomycetes (b) Acomyceytes
    (c) Myxomycetes (d) Actinomycetes
69. The bacteria not involved in the process of nitrification
    (a) *Nitrosomonas* (b) *Clostridium*
    (c) *Nitrosococcus* (d) *Nitrobacter*
70. Appropriate size of bacteria
    (a) 0.01-0.25μ (b) 0.25 - 0.5μ (c) 0.1 - 10μ (d) 2 -1mm
71. Who revealed the actual role of pollen in fertilization
    (a) Johri (1992) (b) Maheshwari (1950)
    (c) Amici (1824) (d) Hofmeister (1849)
72. The mode of embryo sac development in majority of flowering plant is
    (a) Polygonum type (b) Allium type
    (c) Endymion type (d) Peperomia type

73. A mature embryo of *Triticum* has a single cotyledon called
   (a) Coleorhiza (b) Epiblast (c) Scutellum (d) Coleoptile
74. Filiform apparatus is formed by
   (a) Synergid (b) Antipodal (c) Egg cell (d) Tapetum
75. The type of ovule where its curvature affects the nucellus so that the ovule becomes horse-shoe shaped is
   (a) Orthotropous ovule (b) Amphilotropous ovule
   (c) Anatropous ovule (d) Hemianatropous ovule
76. Chipko movement is associated with
   (a) Protection of wastelands (b) Protection of wetlands
   (c) Protection of forests (d) Protection of dams and rivers
77. Which source of energy is the least source in India?
   (a) Atomic energy (b) Biomass energy
   (c) Hydro energy (d) Petroleum energy
78. Who proposed the concept of ecosystem
   (a) Weaves and Clements (b) Tansley
   (c) Kormandy (d) Cunningham
79. Choose the unit of Ozone levels
   (a) dB (b) Ergs (c) Watt (d) Dobson
80. CITES is concerned with the protection of endangered
   (a) Fauna only (b) Flora only
   (c) Whole ecosystem (d) Flora and Fauna
81. Release of domestic sewage into a water body leads to
   (a) Bioaccumulation (b) Bioremediation
   (c) Biomagnification (d) Eutrophication
82. Pick out the hot - spot region of India
   (a) Western Ghats (b) Kashmir Valley
   (c) Eastern Ghats (d) Vindhya - Satupura range
83. Who proposed the concept of Sea - floor spreading?
   (a) Alfired Wegener (b) Hermen Hess
   (c) N.I. Vavilov (d) Robert Myers
84. Where are wild relatives of a species concentrated?
   (a) Centre of origin (b) Centre of diversity
   (c) Mega diverse nation (d) Hot spots

85. Endemism is richest in
    (a) Desert ecosystem (b) Mountain ranges
    (c) Agricultural plains (d) Islands
86. Which is the primary acceptor of light energized electrons in chloroplasts?
    (a) Plastoquinones (b) Pheophytins
    (c) Benzoquinones (d) Cytochromes
87. Which enzyme shows affinity to O_2 and CO_2
    (a) PEP carboxylase (b) Glycolic Acid Oxidase
    (c) Rubisco (d) Citrate synthetase
88. Tall trees receive water at the canopy. It is best explained by
    (a) Cohesion - tension theory (b) Mass - flow hypothesis
    (c) Capillary theory (d) Root - pressure theory
89. Which is active in dual forms?
    (a) Chlorophyll (b) Carotene (c) Xanthophyll d) Phytochrome
90. Ethylene stimulates
    (a) Rooting and ripening (b) Senescence and cell division
    (c) Epinasty and ripening (d) Bud dormancy and cell elongation
91. Binomial system of nomenclature was conceived by
    (a) Andreo Caesalpino (b) Caspard Bauhin
    (c) Carl Linnaeus (d) Joseph Otto
92. Tree habit was considered primitive by
    (a) Takhtajan (b) Hutchinson
    (c) Engler and Prantl (d) Bantham and Hooker
93. Which is the most advanced family?
    (a) Ranunculaceae (b) Poaceae
    (c) Cucurbitaceae (d) Magnoliaceae
94. Choose the Angiosperm family with vessel less taxa
    (a) Chloranthaceae (b) Pinaceae
    (c) Cycadaceae (d) Hydrocharitaceae
95. Gynostegium is present in flowers of
    (a) *Vanda* (b) *Gynandropsis* (c) *Nerium* (d) *Calotropis*
96. A. Cybrid will have
    (a) The complete genome of two taxa
    (b) The cytoplasmic genes of two taxa

(c) Only the nuclear genes of two taxa
(d) Cytoplasmic genes of two and nuclear gene of one

97. Who is considered the "Father of Tissue culture"?
(a) Murashighe (b) Steward (c) White (d) Haberblandt

98. Professor Maheshwari's Delhi School of Plant Morphology is credited with the discovery of
(a) Haploids (b) Somatic embryo
(c) Cybrid production (d) Anther culture

99. Which method is not micropropagation?
(a) Axillary branch induction (b) Mericloning
(c) Synthetic seed production (d) Cybrid Production

100 Plastochorn index is calculated for
(a) Shoot meristem (b) Root meristem
(c) Callus proliferation (d) Axillary bud induction

MODEL QUESTION PAPER-II

1. Bacteria that prefer cold, thriving at temperatures between 0°C to 25°C
(a) Mesophile (b) Thermophile
(c) Psychrophile (d) Facultative bacteria

2. A disease which occurs occasionally or in scattered instances
(a) Endemic (b) Sporadic (c) Epidemic (d) Pandemic

3. A microbe that can live only in the presence of oxygen
(a) Strict (obligate) anaerobe (b) Strict (obligate) aerobe
(c) Strict (obligate) parasite (d) Strict (obligate) saprophyte

4. The prevention of sepsis by preventing or inhibiting the growth of causative microorganism
(a) Immunoglobulin (b) Disinfection (c) Antisepsis (d) Sterilization

5. A prokaryotic unicellular microorganism of the Kingdom Monera, existing as free living organisms or as parasites, multiplying by binary fission and having a large range of biochemical properties
(a) Virus (b) Fungi (c) Bacteria (d) Protozoa

6. A large group of non-motile, gram negative intracellular parasites are
(a) Fungi (b) Protozoa (c) *Chlamydia* (d) Virus

7. Genera of bacteria that are gram positive organisms occurring in pairs; also called streptococcus
(a) Diplobacilli (b) Streptobacilli (c) Diplococcus (d) Coccus

8. Appearance of an infectious disease or condition that attacks many people at a time in the same geographical area-
   (a) Pandemic (b) Endemic
   (c) Epidemic (d) General infection
9. One in which the organisms are originally confined to one area but enter the blood or lymph vessels and spread to other parts of the body
   (a) Local infection (b) General infection
   (c) Mechanical vector (d) Focal infection
10. A substance that destroys microorganisms
    (a) Viricide (b) Fungicide (c) Bactericide (d) Germicide
11. The temperature above which bacterial growth cannot take place
    (a) Maximum temperature (b) Minimum temperature
    (c) Optimum temperature (d) Obligate temperature
12. A genus found in human beings has no cell wall. They are the smallest free living organisms, known as
    (a) Mycoplasmas (b) Fungi (c) Protozoa (d) Virus
13. The state of producing or being able to produce pathological changes and disease is called
    (a) Infectious (b) Virulent
    (c) Pathogenicity (d) Strict obligate
14. A genus of gram negative, pathogenic, intracellular parasitic bacteria
    (a) Mycoplasmas (b) *Rickettsia*
    (c) *Chlamydia* (d) *Microaerophillic*
15. A genus of gram positive, non motile, opportunistic bacteria which tend to aggregate in irregular grape like clusters
    (a) Streptobacilli (b) Diplobacilli
    (c) Coccus (d) *Staphylococcus*
16. Pyruvic acid is produced at the end of
    (a) Calvin cycle (b) Krebs cycle
    (c) EMP pathway (d) PP pathway
17. The structure that regulates the entry and exit of a material or molecule in cell is
    (a) Plasma membrane (b) Cell wall
    (c) Nuclear membrane (d) Tonoplast
18. Heterothallism is common in
    (a) *Rhizopus* (b) *Cycas* (c) *Mangifera* (d) *Pisum*

19. Oxidative phosphorylation means formation of
    (a) ATP in photosynthesis (b) ATP in respiration
    (c) Use of O_2 in photosynthesis (d) Excretion of O_2 in respiration
20. Cystolith is made up of
    (a) Calcium oxalate (b) Calcium carbonate
    (c) Calcium phosphate (d) Calcium silicate
21. In a lake ecosystem the dominant second trophic level is of
    (a) Phytoplanktons (b) Zooplanktons (c) Snakes (d) Birds
22. Exchange of chromosome segment between maternal and paternal chromatids during meiosis is called
    (a) Linkage (b) Crossing over (c) Dominance (d) Meiosis
23. Two similar holotypes are called
    (a) Mesotype (b) Syntype (c) Isotype (d) Neotype
24. Girdling experiment cannot be performed in sugarcane because
    (a) Vascular bundles are scattered (b) Vascular bundles are in ring
    (c) It is unbranched (d) It is delicate cannot withstand injury
25. The histogen layer present at the apex of root tip is
    (a) Dermatogen (c) Periblem (b) Calyptrogen (d) Plerome
26. Vivipary is found in
    (a) *Cicer* (b) *Ricinus* (c) *Rhizophora* (d) Maize
27. A Polygonum type of embryo sac is
    (a) 7 celled, 8 nucleate (b) 8 celled, 7 nucleate
    (c) 7 celled, 7 nucleate (d) 8 celled, 8 nucleate
28. Cytokinin stimulate cell for
    (a) Elongation (b) Division (c) Wall thickening (d) Increase
29. The rate at which light energy is converted into chemical energy of organic molecules is known as ecosystems
    (a) Net primary productivity (b) Gross primary productivity
    (c) Net secondary productivity (d) Gross secondary productivity
30. Damage of O_3 layer is due to the harmful effect of
    (a) CO_2 (b) SO_2
    (c) O_2 (d) Chlorofluorocarbons
31. Oxysomes are present on
    (a) Cell wall (b) Outer membrane of mitochondria

(c) Inner membrane of mitochondrial cristae (d) Cytoplasmic membrane

32. The chief function of parenchymatous tissue is to
    (a) Synthesize food (b) Store food (c) Absorb moisture (d) Conduct food
33. Low temperature treatment is called
    (a) Pasteurization (b) Vernalization (c) Photomorphism (d) Fermentation
34. Which of the following type of RNA molecule is essential for protein synthesis?
    (a) nRNA (b) tRNA (c) rRNA (d) sRNA
35. Salt-loving plants are known as
    (a) Glycophytes (b) Halophytes (c) Xerophytes (d) Mesophytes
36. The stage of meiosis in which pairing of homologous chromosomes starts is called
    (a) Leptotene (b) Zygotene (c) Diplotene (d) Pachytene
37. In a dihybrid cross, the ratio of F2 generation is
    (a) 9:3:3:1 (b) 3:1 (c) 1:2:1 (d) 1:1
38. The book "Origin of Species" was written by
    (a) Linnaeus (b) Lamarck (c) Mendel (d) Darwin
39. The interaction between algae and fungi to form lichen is called
    (a) Parasitism (b) Mutalism (c) Commensalism (d) Amensalism
40. The actual location or place where an organism lives is called
    (a) Habitat (b) Ecosystem (c) Niche (d) Biome
41. A sequence of changes in the community structure of an ecosystem over a period of time is called
    (a) Natural selection (b) Succession (c) Neo-Darwinis (d) Lamarckism
42. Which of the given plant grows from stem cutting?
    (a) Sweet potato (b) *Bryophyllum* (c) *Hibiscus* (d) Carrots
43. What is the kind of soil in which cotton is grown?
    (a) Clayey soil (b) Black soil (c) Sandy soil (d) Moist soil
44. Which of the following is the seat of the origin of the lateral roots is
    (a) Epidermis (b) Cortex (c) Endodermis (d) Pericycle
45. Resin duct of a Gymnosperm us stem is an example of
    (a) Schizogenous cavity (b) Schizolysigenous cavity
    (c) Lysogenous cavity (d) Intercellular space
46. *Cycas* stem contains
    (a) Pycnoxylic wood (b) Manoxylic wood
    (c) Sap wood (d) Heart wood

47. Membrane, surrounding vacuole is called
    (a) Tonoplast (b) Nucleoplast (c) Protoplast (d) Vacuoplast
48. Clinging roots are found in
    (a) Orchid (b) Trapa (c) *Podostemon* (d) *Acacia*
49. *Datura* is a classical example of
    (a) Triploidy (b) Trisomy (c) Monoploidy (d) Monosomy
50. Ethyl alcohol is commercially manufactured from
    (a) Bajra (b) Grapes (c) Maize (d) Sugarcane
51. Which of the following species are restricted to an area?
    (a) Sibling species (b) Endemic species
    (c) Allopathic species (d) Sympatric species
52. Crop rotation is used by farmers to increase
    (a) Soil fertility (b) Community area
    (c) Organic content of soil (d) Nitrogenous content in the soil
53. Which of the following organism form the decomposers?
    (a) *Pteris* (b) Bacteria
    (c) Saprophytic fungi (d) Both (b) and (c)
54. Which of the following is an edible 'Fungi'?
    (a) *Mucor* (b) *Penicillium* (c) *Agaricus* (d) *Rhizopus*
55. Coir which is obtained from fruit of coconut is
    (a) Endocarp (b) Periderm (c) Mesocarp (d) Epicarp
56. 'Pneumatophores' are present /common in
    (a) Xerophytes (b) Hygrophytes (c) Mesophytes (d) Halophytes
57. The inhibition of one bacterial organism by another in a relationship is called as
    (a) Synergism (b) Antagonism (c) Symbiosis (d) Parasitism
58. Glycoprotein substance developed in response to, and interacting specifically with an antigen is known as
    (a) Germicide (b) Immunity (c) Virus (d) Antibody
59. Any rod shaped microbial organism refers to oen of the followings
    (a) Bacteria (b) Coccus (c) *Spirillum* (d) *Bacillus*
60. A method of asexual reproduction in bacteria in which the cell splits into two parts, each of which develops into a complete individual is called
    (a) Myosis (b) Binary fission
    (c) Vectored splitation (d) Inverse kinematics

61. The act of introducing disease germs or infectious material into an area or substance is
    (a) Attenuation (b) Infection (c) Contamination (d) Virulence
62. A disease that occurs continuously in a particular region but has low mortality
    (a) Endemic (b) Epidemic (c) Pandemic (d) Exodemic
63. A toxin produced by a microorganism and excreted into its surrounding medium, generally protein in nature is
    (a) Endotoxin (b) Exotoxin (c) Lipotoxin (d) Prototoxin
64. An agent that kills fungi and their spores is called
    (a) Fungicide (b) Sporicide (c) Fungo-sporicide (d) Bactericide
65. The state or condition in which the body or a part of it is invaded by pathogenic agents, that under favourable conditions multiplies and produces injurious effects
    (a) Contamination (b) Virulence (c) Acute (d) Infection
66. A microorganism that requires very little free oxygen for its growth
    (a) Mesophile (b) Macroaerophillic (c) Microaerophillic (d) Heteroaerophillic
67. A relationship in which organisms of two different species live in close association with each other for their benefit
    (a) Antagonism (b) Parasitism (c) Synergism (d) Mutualism
68. A disease affecting the majority of the population of a large region or one that is epidemic at the same time in many different parts of the world is known as
    (a) Epidemic (b) Endemic (c) Sporadic (d) Pandemic
69. A condition characterized by the multiplication of bacteria in blood
    (a) Septicemia (b) Bacteremia (c) Bulimia (d) Anaemia
70. A gram positive cocci that occurs in chains
    (a) Streptococci (b) Staphylococcus (c) *Rickettsia* (d) Streptobacilli
71. Infection caused by germs lodging and multiplying at one point in a tissue and remaining there is called
    (a) Local infection
    (b) General infection
    (c) Mixed infection
    (d) Endemic infection
72. Study of relationship between communities and environment is called
    (a) Ethology (b) Autecology (c) Synecology (d) None of these
73. Which one of the following is actually not a flower?
    (a) Shoe flower (b) Rose (c) Pea (d) Sunflower

74. Which of the following is the axis of the floral shoot and is the direct prolongation of the pedicel?
    (a) Corolla (b) Calyx (c) Thalamus (d) Root
75. The essential parts of a flower are
    (a) The sepal and petals (b) The sepals and stamens
    (c) The stamens and pistils (d) The petals and pistils
76. The primary function of stem is
    (a) To anchor plant in the soil
    (b) To help in storage of food
    (c) To give support to the leaves and flowers and spread them out on all sides
    (d) To help in vegetative reproduction
77. Match column 'A' (types of sub-aerial modifications for vegetative propagation) with Column 'B' (examples) then select the correct answer from the options given below

(Column A) (Types of sub-aerial modifications for vegetative propagation)	(Column B) (Examples)
A. Runner	1. *Chrysanthemum*
B. Stolon	2. *Mentha*
C. Sucker	3. *Pistia*
D. Offset	4. *Oxalis*

	A.	B.	C.	D.
(a)	4	2	3	1
(b)	4	1	2	3
(c)	4	3	2	1
(d)	4	2	1	3

78. Bladders are found in
    (a) *Utricularia* (b) Nepenthes (c) Mango (d) Neem
79. Silviculture refers to the
    (a) Cultivation of forest trees (b) Cultivation of ornamentals
    (c) Cultivation of rare plants (d) Cultivation of flower plants
80. Mint multiplies by
    (a) Bulbs (b) Suckers (c) Root cuttings (d) Rhizome
81. Branches of stem are
    (a) Endogenous in origin
    (b) Exogenous in origin

(c) Superficial
(d) Partly exogenous and partly endogenous

82. Match column 'A' (name of the fruit) with Column 'B' (edible part) then select the correct answer from the options given below

(Column A)	(Column B)
(Name of the fruit)	(Edible part)
A. Rice	1. Seed
B. Pea	2. Epicarp and mesocarp
C. Peach	3. Endosperm and embryo
D. Lady's finger	4. Entire fruit

	A.	B.	C.	D.
(a)	4	2	3	1
(b)	3	1	2	4
(c)	4	3	2	1
(d)	4	2	1	3

83. Select the edible stem from the following
(a) Potato (b) Sweet potato (c) Egg plant (d) None of these

84. Which is the recent branch of taxonomy among the following
(a) Chemotaxonomy (b) Karyotaxonomy
(c) Morphotaxonomy (d) None of these

85. Phylloclade is a modification of
(a) Stem (b) Root (c) Leaf (d) Petiole

86. Unbranched, erect, cylindrical stout axis with distinct nodes and internodes with jointed appearance is called
(a) Runner (b) Sucker (c) Stolon (d) Clum

87. Which of the following is known as 'Pond Scum'
(a) *Spirogyra* (b) *Ulothrix* (c) *Nostoc* (d) *Anabaena*

88. Both Chlorophyta and higher plants share the following characteristics
(a) Grana in chloroplasts (b) Starch as a storage fuel
(c) Cellulose in cell walls (d) All the above

89. Pyrenoid is found in one of the following organelles of *Spirogyra*
(a) Nucleolus (b) Vacuole (c) Mitochondria (d) Chloroplast

90. *Spirogyra* resembles Rhizopus in having
(a) Stored starch (b) Single celled gametangia
(c) Cellulose cell wall (d) Single nucleus in each cell

91. Which one of the following statement about *Spirogyra* is true
    (a) Asexual reproduction occurs by zoospore
    (b) Filaments showing scalariform conjugation are homothallic
    (c) Filaments showing lateral conjugation are homothallic
    (d) Filaments showing lateral conjugation are heterothallic

92. Heterothallism in fungi was discovered by
    (a) De Bary (b) Alexander Fleming
    (c) A.F. Blakeslee (d) Felix Dujardin

93. Penicillin was discovered by
    (a) Elie Metchnikoff (b) Alexander Fleming
    (c) A.F. Blakeslee (d) Felix Dujardin

94. All fungi are
    (a) Symbionts (b) Heterotrophs (c) Saprophytes (d) Parasites

95. Why are zygospores not generally formed in cultures of *Rhizopus* developed from a single spore?
    (a) Due to presence of + and - strains of hyphae
    (b) Due to absence of + and - strains of hyphae
    (c) Due to absence of light
    (d) Due to shortage of oxygen

96. The coprophilic fungi inhabit
    (a) Dead wood (b) Decaying leaves
    (c) Food articles (d) Dung substratum

97. First land inhabiting plants are
    (a) Pteridophytes (b) Bryophytes (c) Gymnosperms (d) Angiosperms

98. Gametophyte is predominant in
    (a) *Cycas* (b) Pea (c) Mosses (d) Ferns

99. The centre of the moss axis is occupied by
    (a) The xylem (b) A large pith
    (c) Red coloured cells (d) The so called conducting tissue

100. In moss, the middle sterile part of capsule is called
    (a) Spore case (b) Protonema (c) Columella (c) Foot

Answers

Model Question Paper-I

1. (a)	2. (c)	3. (d)	4. (b)	5. (b)	6. (c)
7. (d)	8. (b)	9. (d)	10. (b)	11. (b)	12. (b)
13. (a)	14. (a)	15. (c)	16. (c)	17. (b)	18. (d)
19. (b)	20. (b)	21. (c)	22. (d)	23. (a)	24. (d)
25. (c)	26. (c)	27. (a)	28. (d)	29. (a)	30. (a)
31. (a)	32. (a)	33. (a)	34. (c)	35. (b)	36. (d)
37. (c)	38. (c)	39. (c)	40. (b)	41. (d)	42. (d)
43. (c)	44. (b)	45. (c)	46. (c)	47. (b)	48. (c)
49. (a)	50. (a)	51. (c)	52. (a)	53. (d)	54. (c)
55. (a)	56. (c)	57. (d)	58. (a)	59. (a)	60. (d)
61. (c)	62. (b)	63. (c)	64. (d)	65. (a)	66. (a)
67. (c)	68. (b)	69. (c)	70. (b)	71. (c)	72. (a)
73. (c)	74. (a)	75. (b)	76. (c)	77. (a)	78. (b)
79. (d)	80. (d)	81. (d)	82. (a)	83. (b)	84. (b)
85. (d)	86. (b)	87. (c)	88. (d)	89. (d)	90. (c)
91. (b)	92. (c)	93. (b)	94. (d)	95. (d)	96. (d)
97. (d)	98. (a)	99. (d)	100. (a)		

Model Question Paper-II

1. (c)	2. (b)	3. (b)	4. (c)	5. (c)	6. (c)
7. (c)	8. (c)	9. (d)	10. (d)	11. (a)	12. (a)
13. (c)	14. (b)	15. (d)	16. (c)	17. (a)	18. (a)
19. (b)	20. (b)	21. (b)	22. (b)	23. (c)	24. (a)
25. (b)	26. (c)	27. (c)	28. (b)	29. (b)	30. (d)
31. (c)	32. (b)	33. (b)	34. (a)	35. (b)	36. (b)
37. (a)	38. (d)	39. (b)	40. (a)	41. (b)	42. (c)

43. (b)	44. (d)	45. (a)	46. (b)	47. (a)	48. (a)
49. (b)	50. (d)	51. (b)	52. (a)	53. (d)	54. (c)
55. (c)	56. (d)	57. (b)	58. (d)	59. (d)	60. (b)
61. (c)	62. (b)	63. (b)	64. (a)	65. (d)	66. (c)
67. (d)	68. (d)	69. (a)	70. (a)	71. (a)	72. (c)
73. (d)	74. (c)	75. (c)	76. (c)	77. (d)	78. (a)
79. (a)	80. (b)	81. (b)	82. (b)	83. (a)	84. (b)
85. (a)	86. (d)	87. (a)	88. (d)	89. (d)	90. (b)
91. (c)	92. (c)	93. (b)	94. (b)	95. (b)	96. (d)
97. (b)	98. (c)	99. (d)	100. (c)		

GATE-2009 PAPER (BOTANY)

1. Kyoto protocol is related to

(a) Acid rain
(b) Photochemical smog
(c) Ozone hole
(d) Global warming

2. Phagotrophs are

(a) Organisms that feed on dead organic matter
(b) Organisms that absorb dissolved organic matter
(c) Organisms that ingest other organisms or particulate organic matter
(d) Organisms that manufacture food from simple inorganic substances

3. Identify the INCORRECT statement

(a) 2,4-Dichlorophenoxyacetic acid (2,4-D) is the most common used chemical analogue of indole-3-acetic acid
(b) In somatic embryogenesis, embryo initiation needs a high concentration of 2,4-D
(c) Crown-gall disease depends on the presence of Ti-plasmid in Agrobacterium tumefaciens
(d) Agrobacterium tumefaciens responsible for hairy root formation in plants

4. Choose the correct relation between Angstrom (Å) and nanometer (nm)

(a) 1 Å = 101 nm (b) 1 Å = 10-1 nm (c) 1 Å = 10-2 nm (d) 1 Å = 102 nm

5. In hypgynous flower

(a) Ovary occupies the highest position on the thalamus
(b) Ovary may be partially sunken in the thallamus
(c) Ovary is completely sunken in the thalamus
(d) Ovary is naked

6. Fill up the blanks with appropriate maches.

The main axis of the inflorescence is known as----------and the stalk of the individual flower is called ----------.

(a) Pedicil and panicle
(b) Panicle and pedicel
(c) Pedicel and peduncle
(d) Peduncle and pedicel

7. Microorganisms responsible for nitrification
(a) Nitrosomonas and Nitrobacter
(b) Nostoc and Anabaena
(c) Rhizobium and Azotobacter
(d) Clostridium and Pseudomonas

8. Identify the CORRECT statements
(a) Vernalization is the process where flowering is promoted by heat shock
(b) The four different types of floral organs are initiated as separate whorls
(c) The flowering stimulus is transported to the meristem via the xylem
(d) Abscisic acid synthesis occurs via the carotenoid biosynthetic pathway

9. Which of the following statements are INCORRECT?
1. The frequency of recombination is a measure of linkage between genes on the same chromosome.
2. DNA polymerase I is the true DNA replicas in E. coli.
3. The conserved element closest to the transcription initiation site is called the CAAT box.
4. The introns in the nuclear pre-m-RNA are excised by spliceosomes.

(a) 1 and 2
(b) 2 and 3
(c) 1 and 3
(d) 1 and 4

10. Which of the following statements are TRUE for transposable genetic elements?
1. IS elements can be inserted at many different sites in bacterial chromosomes and plasmids.
2. In Tn5 the flaking IS elements are in the same orientation.
3. The Ac and Ds elements in maize were discovered by Barbara McClontock.
4. Tn10 consists of two IS elements flaking a gene for streptomycin resistance.

(a) 1 and 2
(b) 1 and 3
(c) 1 and 4
(d) 3 and 4

11. Floral formula is a method of describing a plant specimen in which the morphological characters are represented by symbols. Indentify the correct matches

(Group-I)	(Group-II)
1. B	A. Zygomorphic
2. O+	B. Male flower
3. C A	C. Epipetalous
4. (_)	D. Actinomorphic
	E. Superior ovary
	F. Inferior ovary

(a) 1-B, 2-D, 3-C, 4-E
(b) 1-A, 2-F, 3-E, 4-B
(c) 1-B, 2-D, 3-C, 4-F
(d) 1-F, 2-D, 3-A, 4-C

12. Identify the pathways where the following reactions occur

(Group-I) (Group-II)

P. Fatty acid + GTP + Co A = Acyl-CoA + GDP + Pi
Q. NH_3 + Glutamate + ATP = Glutamine + ADP + Pi
R. Succinate + E-FDA = Fumarate + E-$FADH_2$
S. Malonyl-S-CoA + ACP-SH = Malonyl-S-ACP + CoA-SH

1. Fatty acid synthesis
2. Fatty acid oxidation
3. Oxidative phosophorylation
4. Citric acid cycle
5. Gluconeogenesis
6. Amino acid biosynthesis

(a) P-5, Q-3, R-4, S-6 (b) P-1, Q-2, R-5, S-6
(c) P-3, Q-2, R-6, S-1 (d) P-2, Q-6, R-4, S-1

13. (Group-I)

P. *Ricinus cummunis*
Q. *Jatropha curcas*
R. *Pongamia pinnata*
S. *Madhuca indica*

(Group-II)

1. Rice bran oil
2. Mahua oil
3. Sun-flower oil
4. Castor oil
5. Karanja oil
6. Jatropha oil

(a) P-4, Q-6, R-5, S-2 (b) P-1, Q-6, R-4, S-5
(c) P-3, Q-6, R-5, S-2 (c) P-1, Q-6, R-5, S-2

14. (Group-I)

P. Forster mechanism
Q. Ping-Pong reaction
R. Feed-back inhibition
S. DNA recombination

(Group-II)

1. The initial product inhibit the initial reaction
2. The accumulation of end product inhibit the initial reaction
3. Process of exciton transfer in chromophore molecule
4. Double displacement reaction
5. Michealis-Menten enzyme kinetics
6. Holliday model

(a) P-5, Q-4, R-2, S-6 (b) P-2, Q-3, R-4, S-6
(c) P-3, Q-4, R-2, S-6 (d) P-6, Q-5, R-4, S-1

15. (Group-I)

P. AG ↓ CT
TC ↑ GA
Q. GTPy ↓ PuAC
CAPu ↑ PyTG
R. C ↓ CGG
GGC ↑ C
S. A ↓ AGCTT
TTCGA ↑ A

(Group-II)

1. *E. coRI*
2. *AluI*
3. *HpaII*
4. *HindII*

5. *PstI*
6. *HincII*

(a) P-1, Q-5, R-6, S-3 (b) P-2, Q-6, R-3, S-4
(c) P-1, Q-6, R-4, S-2 (d) P-1, Q-2, R-6, S-4

16. (Group-I) (Group-II)

(Group-I)		(Group-II)	
P.	Linnaeus	1.	Flora Indica
Q.	William Roxburgh	2.	Genera Plantarum
R.	Bentham-Hooker	3.	Artificial Sexual System
S.	Engler	4.	Sero Diagnostic System
		5.	Phylogenetic System
		6.	The Families of Flowering Plants

(a) P-2, Q-4, R-3, S-5 (b) P-1, Q-6, R-5, S-2
(c) P-3, Q-1, R-2, S-5 (d) P-3, Q-1, R-2, S-5

17. (Group-I) (Group-II)

(Group-I)		(Group-II)	
P.	Monarch butterfly	1.	Gene-gene matching
Q.	Horizontal resistance	2.	Bt. crops
R.	Supper weed	3.	Polygenic trait
S.	Biolistics	4.	Tungsten/gold
		5.	Herbicide resistance
		6.	Pusztai affair

(a) P-2, Q-3, R-5, S-4 (b) P-2, Q-6, R-5, S-4
(c) P-2, Q-6, R-5, S-1 (d) P-6, Q-4, R-1, S-3

18. (Group-I) (Group-II)

(Group-I)		(Group-II)	
P.	Water transport	1.	Hydathodes
Q.	Guttation	2.	Xylem
R.	Harting net	3.	Arbuscules
S.	Secondary active transport	4.	Ectotrophic micorrhizal fungi
		5.	Proton motive force
		6.	Lenticels

(a) P-2, Q-1, R-4, S-5 (b) P-1, Q-6, R-5, S-2
(c) P-2, Q-6, R-4, S- 5 (d) P-6, Q-4, R-1, S-3

19. (Group-I) (Group-II)

(Group-I)		(Group-II)	
P.	Tobacco leaf curl *bigeminivirus*	1.	*Culex tarsalis*
Q.	Cauliflower mosaic *caulimovirus*	2.	*Rhopalosiphum maidis*
R.	Rice tungro baciliform *badnavirus*	3.	*Bemisia tabaci*

S. Wheat yellow leaf *closterovirus*

4. *Periplaneta americana*
5. *Nephotettis sp.*
6. *Brevicoryne brassicae*

(a) P-2, Q-3, R-6, S-4 (b) P-3, Q-6, R-5, S-2
(c) P-3, Q-4, R-6, S-5 (d) P-2, Q-4, R-5, S-6

20. Identity which of the following statements are INCORRECT?
    1. *Rhizoctonia solani* causes seedling blight of wheat.
    2. *Gibberella zeae* causes root rot of wheat.
    3. *Phytopthora infestans* causes late blight of potato.
    4. *Tilletia caries* causes blunt of wheat.

    (a) 1 and 2 (b) 1 and 3 (c) 1 and 4 (d) 2 and 4

GATE-2009- (BIOTECHNOLOGY)

1. The method used for prediction of three dimensional structure of a protein for known structure(s) of one or more related proteins is

   (a) Multiple sequence alignment (b) Homology modeling
   (c) Phylogeny (d) Docking

2. To produce plants that are homozygous for all traits, the best choice is

   (a) Protoplast culture (b) Cell suspension culture
   (c) Anther and pollen culture (d) Apical meristem culture

3. Restriction endonucleases from two different organisms that recognize the same DNA sequence for cleavage are called

   (a) Isoschizomers (b) Idozymes
   (c) Concatamers (d) Palindromes

4. Caspases are involved in the process of

   (a) DNA replication (b) Mutation and recombination
   (c) Antibody synthesis (d) Apoptosis

5. Baculovirus expression system is used to express heterologous genes in

   (a) Mammals (b) Plants (c) Insects (d) Yeasts

6. Culture vessel in which physical, physicochemical and physiological conditions, as well as cell concentration, are kept constant is known as

   (a) Cell concentrator (b) Biostat
   (c) Batch bioreactor (d) Incubator

7. Virus resistant transgenic plants can be developed by the expression of
   (a) Cowpea trypsin inhibitor (b) Crystalline toxin protein
   (c) Snowdrop lectin (d) Defective movement protein
8. Which of the following are commonly used as reporter genes?
   1. NPT gene
   2. Luciferase gene
   3. CFTR gene
   4. GFP gene

   (a) 2 and 4 (b) 3 and 4 (c) 1 and 3 (d) 1 and 4
9. Which of the following statements are true about glyphosate tolenrant transgenic plants?
   1. Transgenic plants detoxify glyphosate.
   2. Transgenic plants produce an altered enzyme that is not affected by glyphosate.
   3. Transgenic plants sequester glyphosate in vacuoles.
   4. Transgenic plants overcome the inhibition of aromatic amino acid biosynthesis.
10. Match the items in Group I with an appropriate description in Group II

(Group-I)	(Group-II)
P. UPGMA	1. Protein sequence database
Q. CLUSTAL	2. Phylogenetic analysis
R. SWISS-PORT	3. 3-D structure visualization
S. RasMol	4. Multiple sequence alignment

   (a) P-4, Q-1, R-2, S-3 (b) P-2, Q-4, R-1, S-3
   (c) P-2, Q-3, R-1, S-4 (d) P-2, Q-1, R-4, S-3
11. Match the properties in Group I with the downstream operation in Group II

(Group-I)	(Group-II)
P. Size	1. Extraction
Q. Density	2. Distillation
R. Volatility	3. Filtration
S. Solubility	4. Sedimentation

   (a) P-3, Q-4, R-2, S-1 (b) P-4, Q-1, R-2, S-3
   (c) P-4, Q-3, R-1, S-2 (d) P-3, Q-2, R-4, S-1
12. Match the items in Group I with their functions in Group II

(Group-I)	(Group-II)
P. *rol* genes	1. Food and energy source
Q. Opines	2. Tumor formation
R. Virulence gene	3. Hairy root induction

S. *Aux* and *cyt* gene — 4. T-DNA transfer and integration

(a) P-4, Q-3, R-2, S-1 (b) P-3, Q-2, R-4, S-1
(c) P-1, Q-3, R-4, S-2 (d) P-3, Q-1, R-4, S-2

13. Which of the following statements hold true for pluripotent stem cells (PSCs) under *in vitro* conditions?
P. PSCs can be maintained in an undifferentiated state.
Q. PSCs exhibit abnormal and unstable karyotypes.
R. PSCs can differentiate into a wide variety of cell types.
S. PSCs cannot be passaged continuously.
(a) P, Q (b) P,R (c) Q, R (d) Q, S

14. Determine the correctness or otherwise of the following Assertion (a) and Reason (r)-
Assertion (a): IPTG (Isopropylthiogalactoside) is a gratuitous inducer of *lac* operon.
Reason (r): IPTG is an efficient inducer, but not a substrate of *lac* operon.
(a) Both are true and (r) is the correct reason for (a).
(b) Both are true but (r) is not the correct reason for (a).
(c) (a) is true but (r) is false.
(d) (a) is false but (r) is true.

15. Which one of the following statements are true about bioreactors?
1. Continuous bioreactors provide less degree of control and uniform product quality than batch bioreactors.
2. Batch bioreactors are ideally suited for reaction with substrate inhibition.
3. Choice of a bioreactor is dictated by kinetic considerations.
4. Fed batch bioreactors are also called semibatch bioreactors.
(a) 1 and 2 (b) 2 and 4 (c) 3 and 4 (d) 1 and 3

16. Match the items in Group I with correct options in Group II

(Group-I)	(Group-II)
P. DNA foot printing	1. Protein-protein interaction
Q. Yeast two-hybrid system	2. VNTR
R. DNA finger printing	3. DNA binding protein
S. SAGE	4. Transcriptome analysis

(a) P-1, Q-2, R-4, S-3 (b) P-3, Q-1 R-2, S-4
(b) P-3, Q-4, R-1, S-2 (d) P-4, Q-2, R-1, S-3

17. A RNA polymerization assay was performed using ^3H UTP as the labeled nucleotide with a specific activity of 500 $\mu Ci = 2.2 \times 10^6$ counts per min. after 10 min incubation, the trichloroacetic acid- insoluble radioactivity was found to be 692521 counts per min as determined in a liquid scintillation counter working at 60% efficiency for 3H. The amount of UTP incorporated into the RNA will be

(a) 15 nmol (b) 105 nmol (c) 150 nmol (d) 50 nmol

18. One unit of glucoamylase enzyme activity is defined as the amount of enzyme required to produce 1 μmol of glucose per min in a 4% solution of Lintner starch at pH 4.5 and 60o C. If in a reaction mixture with 1 ml of the crude enzyme preparation containing 8 mg protein and 9 ml of 4.44% starch, 0.6 μmol of glucose/ml-min is produced, what will be the specific activity of the crude enzyme preparation?

(a) 1 unit/mg protein (b) 1.5 units/mg protein
(c) 0.25 units/mg protein (d) 0.75 units/mg protein

19. Match the products in Group I with their possible applications in Group II

(Group-I)	(Group-II)
P. Erythropoietin	1. Blood clot
Q. Anti-fibrin 99	2. Binding and transport of iron
R. Collagenase	3. Anaemia
S. Transferrin	4. Animal cell separation

(a) P-3, Q-1, R-4, S-2 (b) P-3, Q-4, R-1, S-2
(b) P-2, Q-3, R-1, S-4 (d) P-2, Q-1, R-4, S-3

20. Match the products in Group I with their producer organisms in Group II

(Group-I)	(Group-II)
P. Ethanol from glucose	1. *Aspergillus niger*
Q. Probiotics	2. *Leuconostoc mesenteroides*
R. Citric acid	3. *Saccharomyces cerevisiae*
S. Sauerkraut	4. *Bifidobacterium*

(a) P-2, Q-3, R-2, S-4 (b) P-3, Q-4, R-1, S-2
(b) P-3, Q-4, R-2, S-1 (d) P-1, Q-4, R-3, S-2

GATE-2009 (MICROBIOLOGY)

1. Which of the following scientists are associated with the discoveries listed below?

(Group-I)	(Group-II)
P. Identification of DNA as the genetic material	1. Fleming
Q. Discoveries of penicillin	2. Avery
R. Demonstration of bacterial transformation	3. Beijerink
S. Demonstration of filterable infectious agent	4. Griffith

(a) P-2, Q-1, R-4, S-3 (b) P-2, Q-3, R-1, S-4
(b) P-2, Q-4, R-1, S-3 (d) P-2, Q-1, R-3, S-4

2. Density of cells of a bacterial culture is routinely measured using Spectrophotometer. This is based on the principle of

   (a) Light absorption (b) Light diffraction
   (c) Light scattering (d) Light reflection

3. Which one of the following life styles does *Rhizobium* species adopt to fix molecular nitrogen to ammonia?

   (a) Only symbiotic (b) Only non-symbiotic
   (c) Both symbiotic as well as non-symbiotic (d) Only commensalism

4. A tryptophan auxotrophic Hfr strain of *E. coli* is mixed with F- histidine auxotroph and plated on to minimal medium to select for prototrophs. The prototrophs are formed due to

   (a) Conjugation (b) Transduction (c) Transformation (d) Transfection

5. Identify the correct pair of amino acids which differ only by an atom in the side chain (R) group

   (a) Gly-Ala (b) Ser-Cys (c) Met-Leu (d) Ser-Asp

6. Which one of the following antibiotics inhibits protein biosynthesis precisely by blocking the peptidyl transfer stage?

   (a) Bleomycin (b) Rifampicin (c) Chloramphenicol (d) Tetracycline

7. Pseudopeptidoglycan is present in the cell wall of

   (a) *E. coli*
   (b) *Bacillus subtilis*
   (c) *Saccharomyces cerevisiae*
   (d) *Methanococcus jannaschii*

8. The high frequency of recombinants obtained from a cross between Hfr and F- strains is because

   (a) The F factor integrated into the chromosome mediates transfer of large regions of chromosome into the recipient
   (b) The strain bears high number of F pilli on its surface.
   (c) There are high numbers of copies of the F plasmid per cell
   (d) It induces recombination at high rate

9. Which one of the following bacterial toxins does NOT have ADP ribosylation activity?

   (a) Diphtheria toxin (b) Cholera toxin
   (c) S. aureus α-toxin (d) Pseudomonas exotoxin

10. Which one of the following approaches would be appropriate if one wants to cultivate microorganisms under constant physiological conditions?

    (a) Fed batch culture (b) Batch culture
    (c) Continuous culture (d) Discontinuous culture

11. A wild type of strain of E.coli growing on a mixture of glucose and acetate shows a typical biphasic diauxic growth. A mutant strain was isolated and was found not to exhibit the second phase of diauxic growth. This mutant is likely to be defective in
    (a) Pyruvate decarboxylase
    (b) Isocitrate lyase
    (c) Glucose-6-phosphate dehydrogenated
    (d) Phosophoglucose isomerase
12. A silent mutation is one that
    (a) Result in a truncated polypeptide
    (b) Replaces an amino acid with an equivalent amino acid in a polypeptide
    (c) Does not change the amino acid sequence of the polypeptide
    (d) Changes the reading frame of the m-RNA leading to an altered polypeptide
13. A rotator motion of bacterial flagellum is driven by
    (a) Energy obtained through hydrolysis of ATP
    (b) Trans-membrane electrocheimical potential lined to proton pumping
    (c) Direct uptake of extra-cellular nutrients
    (d) Concentration gradient of nutrients in the environment
14. Which one of the following metabolic intermediates is involved in the biosynthesis of Phenylalanine and Tyrosine in bacteria?
    (a) Chorismate (b) Pantothenate
    (c) α-ketobutyrate (d) Indole-3-phosphate
15. The DNA mixture containing unlabelled and N^{15} labelled DNA from phage T4 was denatured and allowed to reanneal. How many bands one would observed upon $CsCl_2$ density gradient centrifugation of the above mixture?
    (a) 0 (b) 2 (c) 2 (d) 3
16. A temperate phage differs from a virulent phage in that, temperate phage
    (a) Can exhibit only lytic cycle
    (b) Can choose between lytic and lysogenic cycle
    (c) Can exhibit only lysogenic cycle
    (d) Engages in lysogenic cycle only at high temperature
17. RecA is a protein involved in
    (a) Recombinational repair (b) Mismatch repair
    (c) Nucleotide excision repair (d) Base excision repair
18. In which of the following bacterium, chemical energy is converted to both mechanical as well as light energy?
    (a) *Vibrio fisheri* (b) *Bacillus subtilis*

(c) *E. coli* (d) *Pseudomonas fluorescence*

19. Initial density of a culture of bacteria with a generation time of 30 minutes was 1×10^5 cells/ml. After 5 hours of incubation, what serial dilution will you have to plate out to get 100 colonies per ml.?

(a) 10^3 (b) 10^4 (c) 10^5 (d) 10^6

20. Effective chemotherapeutic agents are difficult to develop for the treatment of fungal infections because

(a) Fungi have cell wall
(b) Fungi have better mechanisms to inactivate drugs
(c) Fungi are eukaryotic cells and their cellular machinery is similar to that of the host
(d) Fungi pathogens typically infect organs inaccessible for antibiotic treatment

21. Which one of the following is NOT the criterion for using ribosomal RNAs as evolutionary chronometers?

(a) The large size of ribosomal RNAs, gives large number of possible sequence combinations
(b) Ribosomal RNAs are functionally invariant
(c) Ribosomal RNAs are universally distributed
(d) Ribosomal RNAs have enzyme activity

IISC-2009- BIOLOGICAL SCIENCE

1. Which process does NOT occur in mitochondria?

(a) Fatty acid biosynthesis (b) Protein synthesis
(c) β-oxidation (d) DNA synthesis

2. Which test is commonly used to identify a reducing sugar?

(a) Iodine test (b) Benedict's test (c) Biuret test (d) Emulsion test

3. Using site-directed mutagenesis, four mutants were generated. Which mutant has the largest difference in the number of atoms compared to the wild type?

(a) Ser → Cys (b) Tyr → Phe (c) Lys → Ala (d) Arg → Lys

4. From the following sets, pick the one that contains exlusively hydrophobic amino acids.

(a) Asp, Glu, Lys, Arg, Ser (b) Arg, Phe, Tyr, Trp, Asn
(c) Ala, Ile, Leu, Phe, Val (d) Ala, Arg, Phe, Leu, Tyr

5. An enzyme (Molecular weight = 5,500 Da) has a concentration of 93. 5 g/litre and maximum velocity of 0.875 M^{-1} Sec^{-1}. What is the specificity constant for the enzyme and substrate if K_m is 0.438 M?

(a) 22.54 M^{-1} Sec^{-1} (b) 117.5 M^{-1} Sec^{-1}

(c) 6.54 M^{-1} Sec^{-1} (d) 225.4 M^{-1} Sec^{-1}

6. The major storage polysaccharide of mammals is found in highest amounts in which organ?
   (a) Kidney (b) Liver (c) Pancreas (d) Spleen
7. Which one of the following statements about a circular double stranded DNA genome having 21% adenosine is TRUE?
   (a) It has 10.5% guanosine (b) It has 29% guanosine
   (c) It has 21% guanosine (d) It has 58% guanosine
8. The peptide, Ala-Arg-Gln-Met-Thr-Trp-Lys-Val, was digested with cyanogens bromide to produce
   (a) Ala-Arg-Gln-Met + Thr-Trp-Lys-Val (b) Ala-Arg-Gln-Met-Thr-Trp + Lys-Val
   (c) Ala-Arg + Gln-Met-Thr-Trp-Lys-Val (d) Ala-Arg-Gln + Met-Thr-Trp-Lys-Val
9. Deficiency of which vitamin causes inefficient collagen hydroxylation?
   (a) Vitamin C (b) Vitamin A (c) Vitamin B9 (d) Vitamin B1
10. Statins reduce cholesterol levels by inhibiting which enzyme?
    (a) NADPH oxidase (b) HMG-CoA reductase
    (c) HMG-CoA synthetase (d) Mevalonate kinase
11. Which enzyme is involved in non-template dependent nucleic acid synthesis?
    (a) DNA polymerase 3 (b) DNA polymerase 1
    (c) RNA polymerase (d) Poly (a) polymerase
12. Telomerase adds 6 nucleotide repeats to the ends of chromosomes called telomers. The telomeric repeat sequence of Tetrahymena is TTGGGG. The telomeric repeat sequence of yeast is GTGTGT. Propagation of Tetrahymena chromosomes in yeast will result in extension as follows
    (a) TTGGGG TTGGGG TTGGGG (b) GTGTGT GTGTGT GTGTGT
    (c) GTGTGT TTGGGG GTGTGT (d) TTGGGG GTGTGT TTGGGG
13. If one λ phage particle infects a cell lysogenic for λ, which of the following will probably happen?
    (a) l DNA will circularize but will not replicate (b) Cell will die
    (c) l prophage will be excised (d) l DNA will not be injected
14. The stringency of hybridization of two strands of DNA can be increase by
    (a) Increasing the salt concentration (b) Decreasing the DNA concentration
    (c) Increasing the temperature (d) Addition of inert components such as polyethylene glycol
15. Three polypeptides (A,B and C) whose masses are 55 kDA, 50 kDa and 75 kDA with pI of 6.5, 7.0 and 8.0, respectively, were subjected to standard reducing SDS-PAGE. The order of their separation from top to bottom would be
    (a) A, B and C (b) B, A and C (c) A, C and B (d) C, A and B

16. In a sequencing reaction, the dATP was left out of the tube to which ddATP was added. What would be the consequence?
    (a) No DNA synthesis would occur
    (b) Synthesis would always stop at the position at which the first A was incorporated
    (c) Synthesis would terminate randomly regardless of the nucleotide incorporated
    (d) Normal DNA synthesis would occur
17. Which part of chromatin contains genes that are being expressed?
    (a) Euchromatin (b) Centromere (c) Heterochromatin (d) Telomere
18. The time required for two replication forks travelling in opposite directions to traverse the entire E. coli chromosome at 37ºC is 40 min, regardless of the culture conditions. However, in a rich medium, the cells divide every 20 min. which of the following statements is true for cells growing in rich medium?
    (a) Half the daughter cells are nonviable
    (b) There is an average of four replication forks per chromosome
    (c) There is an average of six replication forks per chromosome
    (d) One fourth of the daughter cells are non-viable
19. Bacterial genomes undergo compaction due to the action of which of the following?
    (a) Helicase (b) Ligase
    (c) Condensins (d) Histone like proteins
20. Rho-dependent transcription termination involves
    (a) Rho binding to stem-loop structure in RNA
    (b) Rho binding to nascent RNA and then contacting paused RNA polymerase
    (c) Rho binding termination sites in DNA and creating a block to RNA polymerase
    (d) Rho binding tightly to unstructured RNA and inhibiting translation
21. Which one of the following enzymes is responsible for class switch recombination during immunoglobulin generation?
    (a) Activation-induced deaminase (b) RAG
    (c) Telomerase (d) MRN complex
22. Which one of the events occur every early during T cell activation?
    (a) IL-2 production
    (b) High affinity IL-2 receptor expression
    (c) Rise in cytosolic calcium
    (d) Production of granzymes
23. Which cell surface receptor is directly required for anti-viral immunity?
    (a) IL-2 receptor (b) IL-4 receptor
    (c) TNF-α receptor (d) Interferon-α/β receptor

24. Hormone treatment often results in a decreased response to subsequent treatment with the same hormone. This is due to
    (a) Decrease in the synthesis of receptors (b) Increase in degradation of the hormone
    (c) Decrease in the surface receptors accompanied by decreases in signal transduction
    (d) Decrease in the affinity between hormone and receptor
25. Which hormone binds to intracellular receptors?
    (a) Insulin (b) Growth hormone
    (c) Tri-iodothyronin (d) Thyroid stimulating hormone
26. Which receptor is an ion channel?
    (a) Acetylcholine receptor (b) EGF receptor
    (c) Corticosteroid receptor (d) Growth hormone receptor
27. Which one of the following is NOT implicit in the primary structure of a protein?
    (a) Covalent bonds in the side chains (b) Covalent bonds involving hydrogen atoms
    (c) Covalent bonds between sulphur atoms
    (d) Covalent bonds at the amino acid carboxyl ends of the polypeptide
28. The rate constant of a reaction increase by a factor of 1.23 when the temperature is increased from 20°C to 27°C. what is the activation energy of the reaction?
    (a) 216 kj (b) 21.6 kj (c) 2.16 kj (d) 0.216 kj
29. Let us consider a sphere of certain radius and a cone of same radius. The height of the cone is twice the diameter of its base. What is the relationship between volume of sphere (S) and cone (c)?
    (a) S>C
    (b) C>S
    (c) S=C
    (d) No definite relationship exists between S and C
30. An individual performs a certain task correctly 90% of the time. What is the probability that if there are two such individuals who function independently, at least one of them will perform the task correctly?
    (a) 99.99% (b) 99.9% (c) 99% (d) 90%
31. There are two single stranded DNA of 10 nucleotides each. What is the probability that they form a double stranded DNA with all the 10 base pairs in Watson-Crick pairing?
    (a) $(1/4)^{10}$ (b) $(1/10)^4$ (c) $(1/4)^5$ (d) $(1/5)^4$
32. A Michaelis-Menten enzyme with sugar isomerase activity was purified from a species of Streptomyces. The steady state parameters were measured for several different sugars.

Sugar	K_M (in M)	K_{cat} (in Sec.-1)
Arabinose	5×10^{-3}	20

Xylose	1×10^{-3}	10
Ribose	1×10^{-1}	100
Glucose	5×10^{-4}	50

Which sugar is used up the tastes when it is present at low concentrations?

(a) Glucose (b) Arabinose (c) Ribose (d) Xylose

33. When a DNA molecule is heated, the strands come apart. Which one of the following is the possible explanation for strand separation?
    (a) The solubility of adenine and guanine is greatly decreased at high temperature and solubility of the deoxyribose is unchanged
    (b) The energy of thermal vibrations is greater than the energy of the weak bonds stabilizing the double stranded structure
    (c) The solubility of the bases is unchanged, but the solubility of the deoxyribose is increased
    (d) The solubility of the bases is greatly increased at high temperature

34. When Glutamine at the sixth position in the β-chain of the human hemoglobin is replaced by Valine
    (a) Secondary structures are significantly altered
    (b) Tertiary structure is significantly altered
    (c) Quaternary structure is significantly altered
    (d) No significant structural change takes place

35. Which factor secreted by the hypothalamus serves as the neurosecretory link for the control of reproduction in mammals?
    (a) GHRH (b) GnRH (c) CRH (d) TRH

36. The presence of metal ions in a protein solution can be found by
    (a) Atomic absorption spectroscopy (b) Infrared spectroscopy
    (c) Thin layer chromatography (d) Fluorescence spectroscopy

37. How many chlorine atoms surround a sodium atom in sodium chloride?
    (a) 4 (b) 6 (c) 8 (d) 12

38. Vibrational spectra of molecules correspond to
    (a) Ultraviolet radiation (b) Visible radiation
    (c) Infrared radiation (d) Microwaves

39. Optical density of 1 means
    (a) 1% of the incident light is absorbed
    (b) 10% of the incident light is transmitted
    (c) 10% of the incident light is absorbed
    (c) 90% of the incident light is transmitted

40. Which parameter is NOT necessary when finding the equilibrium potential of an ion using the Nernst equation?
    (a) The temperature
    (b) The ratio of external and internal ion concentrations
    (c) The charge of the ion
    (d) The permeability of the ion channel
41. N-linked glycosylation can occur in a protein containing
    (a) Asn-Thr-Thr-Gly
    (b) Asn-Asn-Asn-Gly
    (c) Glu-Asn-Thr-Gly
    (d) Ser-Ser-Ser-Gly
42. In which cellular compartment does membrane protein biogenesis initiate?
    (a) Rough endoplasmic reticulum
    (b) Golgi bodies
    (c) Smooth endoplasmic reticulum
    (d) Endosomes
43. The inverted repeat sequence in a single stranded DNA forms
    (a) G-quatrets (b) Cruciform DNA (c) Hairpin DNA (d) Triplex DNA
44. Triskelions are components of
    (a) Ribosomes (b) Proteasomes (c) Clathrins (d) Integrins
45. In order for H-Ras to be functional, it needs to associate with the cell membrane. The process involves which post-translational modification?
    (a) Glycosylation
    (b) Hydroxylation
    (c) Farneslylation
    (d) Dephosphorylation
46. Molecules that pass through a cell membrane most easily are
    (a) Small and hydrophobic
    (b) Large and polar
    (c) Ions
    (d) Monosaccharides
47. The resting membrane potential has a positive charge on the outside of the membrane. The most important contributor to this positive charge is the
    (a) Outward movement of K^+
    (b) Inward movement of $Cl/^-$
    (c) Action of the Na^+/K^+ ATPase pump
    (d) Hihg permeability of Na^+ across the membrane
48. All secretory proteins in mammalian cells contains a
    (a) Single sequence for translation into the lumen of the endoplasmic reticulum
    (b) Cleavage site for a specific caspase
    (c) Membrane anchor signal
    (d) Motif for tyrosine phosphorylation
49. A mammalian cell typically has 1.2 meters of double stranded DNA when completely outstretched. The total time to duplicate the DNA is 5 hours. If the rate of duplication is 16 μm/min. how many origins of replication will be formed?

(a) 250 (b) 1500 (c) 1 (d) 500

50. A mutant of E. coli with a heat-sensitive DNA ligase (25°C permissive, 37°C non-permissive) has been used to show DNA synthesis is discontinuous. Examination of DNA replication in the presence of [^{3}H]-Thymidine in the mutant would demonstrate the accumulation of short segments of

(a) Unlabeled DNA at 25°C but not at 37°C
(b) Unlabeled DNA at 25°C and at 37°C
(c) Radioactive DNA at 25°C but not at 37°C
(d) Radioactive DNA at 37°C but not at 25°C

51. Upon activation of the translational machinery in a cell, 2000 amino acids are joined together to form a protein molecule. What changes will occur with the synthesis of this protein molecule?

(a) The entropy of 2000 amino acid molecules will increase
(b) The entropy of 2000 amino acid molecules will decrease
(c) The entropy of the universe will decrease
(d) The entropy of 2000 amino acid molecules will decrease and the entropy of the universe will increase

52. The diffusion coefficient for a molecule at very low concentrations in a non-interacting solvent depends upon

(a) The shape of the molecule
(b) The viscosity and shape of the molecule
(c) Viscosity, shape and temperature
(d) Independent of these parameters at infinite dilution

53. Membranes with unsaturated fatty acids are more flexible and fluid because

(a) Unsaturated fatty acids pack closely together to form ordered arrays
(b) Unsaturated fatty acids bend at the double bond (*cis*) preventing close packing
(c) Saturated fatty acids have a 'kink' that produces more fluid aggregates
(d) Unsaturated fatty acids have *cis* double bonds that prevent formation of stable bilayers

54. A pulse chase experiment is used to measure the half life of a protein inside a bacterial cell. Which of the following isotopically labeled compounds is best suited for the experiment?

(a) Water with O^{18} (b) Methionine with S^{35}
(c) Phosphotyrosine with P^{32} (d) Ammonium acetate with N^{15}

55. The Ramchandaran map was calculated assuming which of the following?

(a) Non bonded atoms in a polypeptide cannot come closer than the sum of their Van der waals radii
(b) Non bonded atoms in a polypeptide cannot come closer than the sum of their covalent radii
(c) Non bonded atoms in a polypeptide cannot come closer than 5 Å
(d) Non bonded atoms in a polypeptide can come closer than a distance greater than in (b) but less than in (A)

56. The ratio of the average mass of an amino acid to the average mass of a nucleotide is approximately

(a) 1 (b) 2 (c) 3 (d) 4

57. The net charge of the dipeptide Glu-Lys at physiological PH is

(a) 0 (b) 1 (c) 2 (d) -1

58. Absorbance of 0.22 mM ATP solution at 260 nm with a cuvette having a path length of 0.5 cm is 0.154. the molar absorption coefficient of ATP is

(a) 15400 (b) 7700 (c) 770 (d) 1540

59. Considering the thickness of cellular membrane, roughly how many residues would you expect in a transmembrane α-helix?

(a) 5 (b) 10 (c) 20 (d) 40

60. Xeroderma pigmentosum is caused due to defect in which of the following DNA repair pathways?

(a) Homologous recombination (b) Mismatch repair
(c) Nucleotide excision repair (d) Nonhomologous end joining

61. Mutations in Glyoxylate amino-transferase enzyme leads to primary hyperoxaluria Type 1, a rare genetic disease in which calcium oxalate 'stones' accumulate in the kidney. The mutation mapped at N-terminus of the enzyme altered 'Gly' to 'Arg'. The biochemical analysis indicates that the mutant enzyme is expressed and folded properly. The phenotype associated with such patients is due to

(a) Mutation leading to dimerization of the enzyme
(b) Turnover rate of mutant enzyme is increased
(c) Mutation phenotype not directly linked to the enzyme
(d) Mutation leading to mistargeting of the protein

62. In an operon, positively controlled gene expression occurs because

(a) Represor is inactivated by activator binding to initiate transcription
(b) RNA polymerase is able to bind to the promoter and transcribe the gene on its own
(c) Activator is needed for high level transcription initiation
(d) Represor removal facilitates RNAP binding to the promoter to facilitate transcription

63. Location of a gene in a genome to a narrow region can be found out by-

(a) Generalized transduction (b) Conjugation mapping
(c) Specialized transduction (d) Transpositional insertion

64. A mother and a father, both Cystic Fibrosis (CF) carriers, have two children who do not suffer from CF. The chance of a third pregnancy producing a child with the disease is

(a) 25% (b) 33.33% (c) 50% (d) 100%

65. Evidence indicating that chloroplasts were originally free-living prokaryotes that subsequently evolved a symbiotic relationship with a eukaryotic host includes all of the following EXCEPT

(a) Similarities in structure between chloroplasts and prokaryotes
(b) Presence of circular DNA, in chloroplasts and prokaryotes
(c) Susceptibility of chloroplasts to prokaryotic protein synthesis inhibitors
(d) Ability of chloroplast to synthesize all their own proteins

66. Which technique is commonly used to study epigenetic alterations?
(a) Electrophoretic mobility shift assay
(b) Chromatin immunoprecipitation
(c) Isoelectric focusing
(d) Western blotting

67. A point mutation in newly identified receptor tyrosine kinase renders it constitutively active. Which of the following is NOT true?
(a) It is gain-of-function mutation
(b) Mutant kinase can function in a ligand-independent fashion
(c) It can contribute to carcinogenesis
(d) The mutation is recessive

68. The cross between a mutant mother and a normal father gives rise to offspring that are all mutant; the reciprocal cross gives rise to offspring that are all normal. This means that the mutation is likely to be
(a) In the mitochondrial genome (b) In the nuclear genome
(c) Dominant (d) Recessive

69. Fruit flies have a diploid number of 8 and honey bees have a diploid number of 32 chromosomes. Assuming no crossing over takes place in these two species, the genetic variation among offsprings from the same two parents is likely to be
(a) Same in both the species (b) More in fruit flies than in honey bees
(c) More in honey bees than in fruit flies
(d) Cannot be determined from the information given

70. Which cellular processes produce the carbon dioxide that is exhaled?
(a) Electron transport system and glycosis
(b) Electron transport system and citric acid cycle
(c) Citric acid cycle and glycolysis (d) Glycolysis and gluconeogenesis

71. Which enzyme is involved in epigenetic inheritance?
(a) Mitogen activated protein kinase (b) Acetyl coenzyme A carboxylase
(c) Telomerase (d) Histone methyl transferase

72. Which of the following is an enterovirus?
(a) Foot and mouth disease virus (b) Hepatitis C virus
(c) Rotavirus (d) Coxsackie virus

73. What is the broad mechanism of action of the anti-viral drug Ribavarin?
    (a) It interferes with viral protein synthesis
    (b) It interferes with processing of viral poly-proteins
    (c) It interferes with duplication of viral genetic material
    (d) It blocks the binding of virus to receptors on host cells

74. Mating type switching process that occurs in yeast is an example of which of the following?
    (a) Alternative splicing (b) Change in feedback loop
    (c) Change in DNA methylation pattern
    (d) Change in physical rearrangement of the genome

75. Which one of the following is correct?

1.	*E. coli*	A.	Spore formation
2.	*Stahylococcus aureus*	B.	Obligate intracellular parasite
3.	*Chlamydia*	C.	Lipopolysaccharide
4.	*Clostridium*	D.	Lipoteichoic acid

    (a) 1-C, 2-D, 3-B, D-A (b) 1-C, 2-B, 3-D, 4- A
    (c) 1-A, 2-C, 3-D, 4-B (d) 1-D, 2-A, 3-C, 4-B

76. If there are 1000 cells per ml in the middle of the log phase, and the generation time of the cells is 30 min, how many cells will be there 2 hours later?
    (a) 2000 (b) 8000 (c) 10,000,000 (d) 16000

77. Which of the following is NOT involved in photorespiration?
    (a) Rubisco (b) Peroxisome (c) PEP carboxylase (d) Mitochondria

78. A population A, each woman bears exactly two female children born at maternal ages 20 and 25. Population B is the same, except that the children are born when the mother is 25 and 30. Assuming that these are the only differences between A and B, when the sizes of the two population are monitored over time, we will find that
    (a) Both are stationary (b) Both grow at the same rate
    (c) A grows faster than B (c) B grows faster than A

79. A characteristic property of stem cells is
    (a) Metastasis (b) Terminal differentiation
    (c) Self-renewal (d) Senescence

80. Which of the following is NOT a cell death mechanism?
    (a) Apoptosis (b) Endocytosis (c) Necrosis (d) Autophagy

Made in the USA
Monee, IL
04 November 2026

67385474R00215